MW00389687

Effective CRM Using Predictive Analytics

Effective CRM Using Predictive Analytics

Antonios Chorianopoulos

This edition first published 2016

Registered Office
John Wiley & Sons, Ltd, The Atrium, Southern Gate, Chichester, West Sussex, PO19 8SQ, United Kingdom

For details of our global editorial offices, for customer services and for information about how to apply for permission to reuse the copyright material in this book please see our website at www.wiley.com.

Library of Congress Cataloging-in-Publication Data

Chorianopoulos, Antonios.
Effective CRM using predictive analytics / Antonios Chorianopoulos.
pages cm
Includes bibliographical references and index.
ISBN 978-1-119-01155-2 (cloth)
1. Customer relations–Management–Data processing. 2. Data mining. I. Title.
HF5415.5.C4786 2015
658.8′12–dc23

2015020804

A catalogue record for this book is available from the British Library.

Cover image: Baitong333/iStockphoto

Set in 10/12pt Times by SPi Global, Pondicherry, India

1 2016

To my daughter Eugenia and my wife Virginia

Contents

Preface

This book is in a way the "sequel" of the first book that I wrote together with Konstantinos Tsiptsis. It follows the same principles, aiming to be an applied guide rather than a generic reference book on predictive analytics and data mining. There are many excellent, well-written books that succeed in presenting the theoretical background of the data mining algorithms. But the scope of this book is to enlighten the usage of these algorithms in marketing applications and to transfer domain expertise and knowledge. That's why it is packed with real-world case studies which are presented with the use of three powerful and popular software tools: IBM SPSS Modeler, RapidMiner, and Data Mining for Excel.

Here are a few words on the book's structure and some tips on "how to read the book." The book is organized in three main parts:

Part I, the Methodology. Chapters 2 and 3: I strongly believe that these sections are among the strong points of the book. Part I provides a methodological roadmap, covering both the technical and the business aspects for designing and carrying out optimized marketing actions using predictive analytics. The data mining process is presented in detail along with specific guidelines for the development of targeted acquisition, cross-/deep-/up-selling and retention campaigns, as well as effective customer segmentation schemes.

Part II, the Algorithms. Chapters 4 and 5: This part is dedicated in introducing the main concepts of some of the most popular and powerful data mining algorithms for classification and clustering. The data mining algorithms are explained in a simple and comprehensive language for business users with no technical expertise. The intention is to demystify the main concepts of the algorithms rather than "diving" deep in mathematical explanations and formulas so that data mining and marketing practitioners can confidently deploy them in their everyday business problems.

Part III, the Case Studies. Chapters 6, 7, and 8: And then it's "action time"! The third part of the book is the "hands-on" part. Three case studies from banking, retail, and telephony are presented in detail following the specific methodological steps explained in the previous chapters. The concept is to apply the methodological "blueprints" of Chapters 2 and 3 in real-world applications and to bridge the gap between analytics and their use in CRM. Given the level of detail and the accompanying material, the case studies can be considered as "application templates" for developing similar applications. The software tools are presented in that context.

In the book's companion website, you can access the material from each case study, including the datasets and the relevant code. This material is an inseparable part of the book, and I'd strongly suggest exploring and experimenting with it to gain full advantage of the book.

Those interested in segmentation and its marketing usage are strongly encouraged to look for the previous title: Konstantinos Tsiptsis and Antonios Chorianopoulos. *Data Mining Techniques in CRM: Inside Customer Segmentation*. Wiley, New York, 2009.

Finally, I would really like to thank all the readers of the first book for their warm acceptance, all those who read or reviewed the book, and all those who contacted us to share kind and encouraging words about how much they liked it. They truly inspired the creation of this new book. I really hope that this title meets their expectations.

Acknowledgments

Special thanks to Ioanna Koutrouvis and Vassilis Panagos at PREDICTA (http://www.predicta.gr) for their support.

1

An overview of data mining: The applications, the methodology, the algorithms, and the data

1.1 The applications

Customers are the most important asset of an organization. That's why an organization should plan and employ a clear strategy for customer handling. Customer relationship management (CRM) is the strategy for building, managing, and strengthening loyal and long-lasting customer relationships. CRM should be a customer-centric approach based on customer insight. Its scope should be the "personalized" handling of the customers as distinct entities through the identification and understanding of their differentiated needs, preferences, and behaviors.

CRM aims at two main objectives:

1. Customer retention through customer satisfaction
2. Customer development

Data mining can provide customer insight which is vital for these objectives and for establishing an effective CRM strategy. It can lead to personalized interactions with customers and hence increased satisfaction and profitable customer relationships through data analysis. It can offer individualized and optimized customer management throughout all the phases of the customer life cycle, from acquisition and establishment of a strong relationship to attrition prevention and win-back of lost customers. Marketers strive to get a greater market share and a greater share of their customers. In plain words, they are responsible for getting, developing, and keeping the customers. Data mining can help them in all these tasks, as shown in Figure 1.1.

Effective CRM using Predictive Analytics, First Edition. Antonios Chorianopoulos.

Companion website: www.wiley.com/go/chorianopoulos/effective_crm

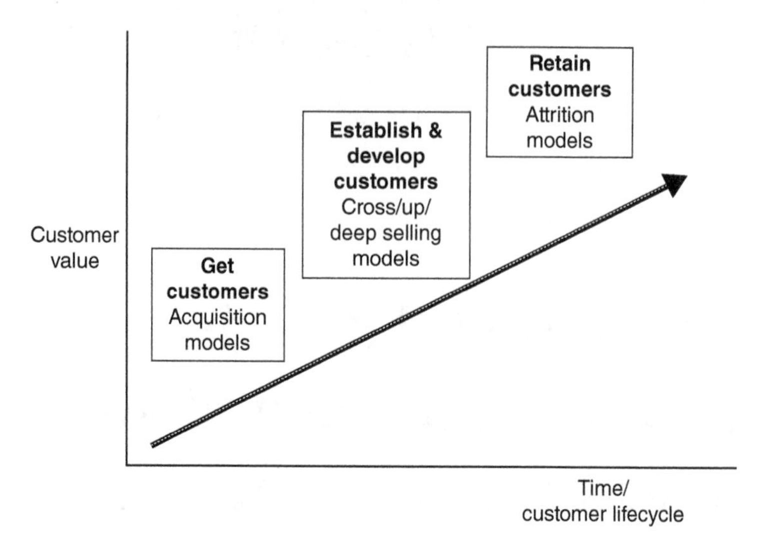

Figure 1.1 Data mining and customer life cycle management. Source: Tsiptsis and Chorianopoulos (2009). Reproduced with permission from Wiley

More specifically, the marketing activities that can be supported with the use of data mining include:

Customer segmentation

Segmentation is the process of dividing the customer base in distinct and homogeneous groups in order to develop differentiated marketing strategies according to their characteristics. There are many different segmentation types according to the specific criteria/attributes used for segmentation. In behavioral segmentation, customers are grouped based on behavioral and usage characteristics. Although behavioral segments can be created using business rules, this approach has inherent disadvantages. It can handle only a few segmentation fields, and its objectivity is questionable as it is based on the personal perceptions of a business expert. Data mining on the other hand can create data-driven behavioral segments. Clustering algorithms can analyze behavioral data, identify the natural groupings of customers, and suggest a grouping founded on observed data patterns. Provided it is properly built, it can uncover groups with distinct profiles and characteristics and lead to rich, actionable segmentation schemes with business meaning and value.

Data mining can also be used for the development of segmentation schemes based on the current or expected/estimated value of the customers. These segments are necessary in order to prioritize the customer handling and the marketing interventions according to the importance of each customer.

Direct marketing campaigns

Marketers carry out direct marketing campaigns to communicate a message to their customers through mail, Internet, e-mail, telemarketing (phone), and other direct channels in order to prevent churn (attrition) and drive customer acquisition and purchase of add-on products. More specifically, acquisition campaigns aim at drawing new and potentially valuable customers from the competition. Cross/deep/up-selling campaigns are rolled out to sell additional products, more of the same product, or alternative but more profitable products to the existing customers. Finally, retention campaigns aim at preventing valuable customers from terminating their relationship with the organization.

These campaigns, although potentially effective, when not refined can also lead to a huge waste of resources and to the annoyance of customers with unsolicited communication. Data mining and classification (propensity) models in particular can support the development of targeted marketing campaigns. They analyze the customer characteristics and recognize the profile of the target customers. New cases with similar profiles are then identified, assigned a high propensity score, and included in the target lists. Table 1.1 summarizes the use of data mining models in direct marketing campaigns.

When properly built, propensity models can identify the right customers to contact and lead to campaign lists with increased concentrations of target customers. They outperform random selections as well as predictions based on business rules and personal intuitions.

Table 1.1 Data mining models and direct marketing campaigns

Business objective	**Marketing campaign**	**Data mining models**
Getting customers	• Acquisition: finding new customers and expanding the customer base with new and potentially profitable customers	• Acquisition classification models can be used to recognize potentially profitable prospect customers by finding "clones" of existing valuable customers in lists of contacts
Developing customers	• Cross selling: promoting and selling additional products/services to existing customers • Up selling: offering and switching customers to premium products, other products more profitable than the ones that already have • Deep selling: increasing usage of the products/services that customers already have	• Cross/up/deep-selling classification models can reveal the existing customers with purchase potentials
Retaining customers	• Retention: prevention of voluntary churn, with priority given to presently or potentially valuable customers at risk	• Voluntary attrition (churn) models can identify early churn signals and discern the customers with increased likelihood of voluntary churn

Source: Tsiptsis and Chorianopoulos (2009).

Market basket and sequence analysis Data mining and association models in particular can be used to identify related products, typically purchased together. These models can be used for market basket analysis and for the revealing of bundles of products/services that can be sold together. Sequence models take into account the order of actions/purchases and can identify sequences of events.

1.2 The methodology

The modeling phase is just one phase in the implementation process of a data mining project. Steps of critical importance precede and follow the model building and have a significant effect in the success of the project. An outline of the basic phases in the development of a data mining project, according to the Cross Industry Standard Process for Data Mining (CRISP-DM) process model, is presented in Table 1.2.

Data mining projects are not simple. They may end in business failure if the engaged team is not guided by a clear methodological framework. The CRISP-DM process model charts the steps that should be followed for successful data mining implementations. These steps are:

Business understanding. The data mining project should start with the understanding of the business objective and the assessment of the current situation. The project's parameters should be considered, including resources and limitations. The business objective should be translated to a data mining goal. Success criteria should be defined and a project plan should be developed.

Data understanding. This phase involves considering the data requirements for properly addressing the defined goal and an investigation on the availability of the required data. This phase also includes an initial data collection and exploration with summary statistics and visualization tools to understand the data and identify potential problems of availability and quality.

Data preparation. The data to be used should be identified, selected, and prepared for inclusion in the data mining model. This phase involves the data acquisition, integration, and formatting according to the needs of the project. The consolidated data should then be cleaned and properly transformed according to the requirements of the algorithm to be applied. New fields such as sums, averages, ratios, flags, etc. should be derived from the original fields to enrich the customer information, better summarize the customer characteristics, and therefore enhance the performance of the models.

Modeling. The processed data are then used for model training. Analysts should select the appropriate modeling technique for the particular business objective. Before the training of the models and especially in the case of predictive modeling, the modeling dataset should be partitioned so that the model's performance is evaluated on a separate validation dataset. This phase involves the examination of alternative modeling algorithms and parameter settings and a comparison of their performance in order to find the one that yields the best results. Based on an initial evaluation of the model results, the model settings can be revised and fine-tuned.

Evaluation. The generated models are then formally evaluated not only in terms of technical measures but, more importantly, in the context of the business success criteria set in the business understanding phase. The project team should decide whether the

Table 1.2 The CRISP-DM phases

1. Business understanding	**2. Data understanding**	**3. Data preparation**
• Understanding of the business goal • Situation assessment • Translating the business goal to a data mining objective • Development of a project plan	• Considering data requirements • Initial data collection/ exploration and quality assessment	• Selection of required data • Data acquisition • Data integration and formatting (merge/joins, aggregations) • Data cleaning • Data transformations and enrichment (regrouping/ binning of existing fields, creation of derived attributes, and KPIs: ratios, flag fields, averages, sums, etc.)
4. Modeling	**5. Model evaluation**	**6. Deployment**
• Selection of the appropriate modeling technique • Especially in the case of predictive models, splitting of the dataset into training and testing subsets for evaluation purposes • Development and examination of alternative modeling algorithms and parameter settings • Fine-tuning of the model settings according to an initial assessment of the model's performance	• Evaluation of the models in the context of the business success criteria • Model approval	• Create a report of findings • Planning and development of the deployment procedure • Deployment of the data mining model • Distribution of the model results and integration in the organization's operational CRM system • Development of a maintenance–update plan • Review of the project • Planning of next steps

Source: Tsiptsis and Chorianopoulos (2009). Reproduced with permission from Wiley.

results of a given model properly address the initial business objectives. If so, this model is approved and prepared for deployment.

Deployment. The project's findings and conclusions are summarized in a report, but this is hardly the end of the project. Even the best model will turn out to be a business failure if its results are not deployed and integrated in the organization's everyday marketing operations. A procedure should be designed and developed that will enable the scoring of customers and the update of the results. The deployment procedure should also enable the distribution of the model results throughout the enterprise and their incorporation in the organization's data warehouse and operational CRM system. Finally, a maintenance plan should be designed and the whole process should be reviewed. Lessons learned should be taken into account and next steps should be planned.

The aforementioned phases present strong dependencies, and the outcomes of a phase may lead to revisiting and reviewing the results of preceding phases. The nature of the process is cyclical since the data mining itself is a never-ending journey and quest, demanding continuous reassessment and update of completed tasks in the context of a rapidly changing business environment.

This book contains two chapters dedicated in the methodological framework of classification and behavioral segmentation modeling. In these chapters, the recommended approach for these applications is elaborated and presented as a step-by-step guide.

1.3 The algorithms

Data mining models employ statistical or machine-learning algorithms to identify useful data patterns and understand and predict behaviors. They can be grouped in two main classes according to their goal:

1. **Supervised/predictive models**
   In supervised, also referred to as predictive, directed, or targeted, modeling, the goal is to predict an event or estimate the values of a continuous numeric attribute. In these models, there are input fields and an output or target field. Inputs are also called predictors because they are used by the algorithm for the identification of a prediction function for the output. We can think of predictors as the "X" part of the function and the target field as the "Y" part, the outcome.

   The algorithm associates the outcome with input data patterns. Pattern recognition is "supervised" by the target field. Relationships are established between the inputs and the output. An input–output "mapping function" is generated by the algorithm that associates predictors with the output and permits the prediction of the output values, given the values of the inputs.

2. **Unsupervised models**
   In unsupervised or undirected models, there is no output, just inputs. The pattern recognition is undirected; it is not guided by a specific target field. The goal of the algorithm is to uncover data patterns in the set of inputs and identify groups of similar cases, groups of correlated fields, frequent itemsets, or anomalous records.

1.3.1 Supervised models

Models learn from past cases. In order for predictive algorithms to associate input data patterns with specific outcomes, it is necessary to present them cases with known outcomes. This phase is called the training phase. During that phase, the predictive algorithm builds the function that connects the inputs with the target. Once the relationships are identified and the model is evaluated and proved of satisfactory predictive power, the scoring phase follows. New records, for which the outcome values are unknown, are presented to the model and scored accordingly.

Some predictive algorithms such as regression and Decision Trees are transparent, providing an explanation of their results. Besides prediction, these algorithms can also be used for insight and profiling. They can identify inputs with a significant effect on the target attribute, for example, drivers of customer satisfaction or attrition, and they can reveal the type and the magnitude of their effect.

According to their scope and the measurement level of the field to be predicted, supervised models are further categorized into:

1. **Classification or propensity models**
   Classification or propensity models predict categorical outcomes. Their goal is to classify new cases to predefined classes, in other words to predict an event. The classification algorithm estimates a propensity score for each new case. The propensity score denotes the likelihood of occurrence of the target event.

2. **Estimation (regression) models**
   Estimation models are similar to classification models with one big difference. They are used for predicting the value of a continuous output based on the observed values of the inputs.

3. **Feature selection**
   These models are used as a preparation step preceding the development of a predictive model. Feature selection algorithms assess the predictive importance of the inputs and identify the significant ones. Predictors with trivial predictive power are discarded from the subsequent modeling steps.

1.3.1.1 Classification models

Classification models predict categorical outcomes by using a set of inputs and a historical dataset with preclassified data. Generated models are then used to predict event occurrence and classify unseen records. Typical examples of target categorical fields include:

- Accepted a marketing offer: yes/no
- Defaulted: yes/no
- Churned: yes/no

In the heart of all classification models is the estimation of confidence scores. These scores denote the likelihood of the predicted outcome. They are estimates of the probability of occurrence of the respective event, typically ranging from 0 to 1. Confidence scores can be translated to propensity scores which signify the likelihood of a particular target class: the propensity of a customer to churn, to buy a specific add-on product, or to default on his loan. Propensity scores allow for the rank ordering of customers according to their likelihood. This feature enables marketers to target their lists and optimally tailor their campaign sizes according to their resources and marketing objectives. They can expand or narrow their target lists on the base of their particular objectives, always targeting the customers with the relatively higher probabilities.

Popular classification algorithms include:

- **Decision Trees**. Decision Trees apply recursive partitions to the initial population. For each split (partition), they automatically select the most significant predictor, the predictor that yields the best separation in respect to the target filed. Through successive partitions, their goal is to produce "pure" subsegments, with homogeneous behavior in terms of the output. They are perhaps the most popular classification technique. Part of their popularity is because they produce transparent results that are easily interpretable, offering insight in the event under study. The produced results can have two equivalent formats. In a rule format, results are represented in plain English, as ordinary rules:

IF (PREDICTOR VALUES) ***THEN*** (TARGET OUTCOME & CONFIDENCE SCORE)

For example:

IF (Gender = Male and Profession = White Colar and SMS_Usage > 60 messages per month) ***THEN*** Prediction = Buyer and Confidence = 0.95

In a tree format, rules are graphically represented as a tree in which the initial population (root node) is successively partitioned into terminal (leaf) nodes with similar behavior in respect to the target field.

Decision Tree algorithms are fast and scalable. Available algorithms include:

- C4.5/C5.0
- CHAID
- Classification and regression trees (CART)

- **Decision rules**. They are quite similar to Decision Trees and produce a list of rules which have the format of human understandable statements: IF (PREDICTOR VALUES) THEN (TARGET OUTCOME & CONFIDENCE SCORES). Their main difference from Decision Trees is that they may produce multiple rules for each record. Decision Trees generate exhaustive and mutually exclusive rules which cover all records. For each record, only one rule applies. On the contrary, decision rules may generate an overlapping set of rules. More than one rule, with different predictions, may hold true for each record. In that case, through an integrated voting procedure, rules are evaluated and compared or combined to determine the final prediction and confidence.

- **Logistic regression**. This is a powerful and well-established statistical algorithm that estimates the probabilities of the target classes. It is analogous to simple linear regression but for categorical outcomes. Logistic regression results have the form of continuous functions that estimate membership probabilities of the target classes:

$$\ln\left(\frac{p_j}{p_k}\right) = b_0 + \sum_i b_i X_i$$

where p_j = probability of the target class $\boldsymbol{j}$, p_k probability of the reference target class k, X_i the predictors, b_i the regression coefficients, and b_0 the intercept of the model. The regression coefficients represent the effect of predictors.

For example, in the case of a binary target denoting churn,

$$\ln\left(\frac{\text{churn probability}}{\text{no churn probability}}\right) = b_0 + b_1 \cdot \text{tenure} + b_2 \cdot \text{num of products} + \cdots$$

In order to yield optimal results, it may require special data preparation, including potential screening and transformation (optimal binning) of the predictors. It demands some statistical experience yet, provided it is built properly, it can produce stable and understandable results.

- **Neural networks**. Neural networks are powerful machine-learning algorithms that use complex, nonlinear mapping functions for estimation and classification. They consist of neurons organized in layers. The input layer contains the predictors or input neurons. The output layer includes the target field. These models estimate weights that connect predictors (input layer) to the output. Models with more complex topologies may also include intermediate, hidden layers, and neurons. The training procedure is an iterative process. Input records, with known outcome, are presented to the network, and model prediction is evaluated in respect to the observed results. Observed errors are used to adjust and optimize the initial weight estimates. They are considered as opaque or "black box" solutions since they do not provide an explanation of their predictions. They only provide a sensitivity analysis, which summarizes the predictive importance of the input fields. They require minimum statistical knowledge but, depending on the problem, may require long processing times for training.

- **Support Vector Machine (SVM)**. SVM is a classification algorithm that can model highly nonlinear complex data patterns and avoid overfitting, that is, the situation in which a model memorizes patterns only relevant to the specific cases analyzed. SVM works by mapping data to a high-dimensional feature space in which records become more easily separable (i.e., separated by linear functions) in respect to the target categories. Input training data are appropriately transformed through nonlinear kernel functions, and this transformation is followed by a search for simpler functions, that is, linear functions, which optimally separate cases. Analysts typically experiment with different kernel functions and compare the results. Overall, SVM is an effective yet demanding algorithm, in terms of processing time and resources. Additionally, it lacks transparency since the predictions are not explained, and only the importance of predictors is summarized.

- **Bayesian networks**. Bayesian networks are statistical models based on the Bayes theorem. They are probabilistic models as they estimate the probabilities of belonging to each target class. Bayesian belief networks, in particular, are graphical models which provide a visual representation of the attribute relationships, ensuring transparency and explanation of the model rationale.

1.3.1.2 Estimation (regression) models

Estimation models, also referred to as regression models, deal with continuous numeric outcomes. By using linear or nonlinear functions, they use the input fields to estimate the unknown values of a continuous target field.

Estimation algorithms can be used to predict attributes like the following:

- The expected balance of the savings accounts of the customers of a bank in the near future
- The estimated loss given default (LGD) incurred after a customer has defaulted
- The expected revenue from a customer within a specified time period

A dataset with historical data and known values of the continuous output is required for the model training. A mapping function is then identified that associates the available inputs to the output values. These models are also referred to as regression models, after the well-known and established statistical algorithm of *ordinary least squares regression (OLSR)*. The OLSR estimates the line that best fits the data and minimizes the observed errors, the so-called least squares line. It requires some statistical experience, and since it is sensitive to possible violations of its assumptions, it may require specific data examination and processing before building. The final model has the intuitive form of a linear function with coefficients denoting the effect of predictors to the outcome. Although transparent, it has inherent limitations that may affect its performance in complex situations of nonlinear relationships and interactions between predictors.

Nowadays, traditional regression is not the only available estimation algorithm. New techniques, with less stringent assumptions, which also capture nonlinear relationships, can also be employed to handle continuous outcomes. More specifically, *polynomial regression, neural networks, SVM, and regression trees such as CART* can also be employed for the prediction of continuous attributes.

1.3.1.3 Feature selection (field screening)

The feature selection (field screening) process is a preparation step for the development of classification and estimation (regression) models. The situation of having hundreds of candidate predictors is not an unusual case in complicated data mining tasks. Some of these fields though may not have an influence to the output that we want to predict.

The basic idea of feature selection is to use basic statistical measures to assess and quantify the relationship of the inputs to the output. More specifically, feature selection is used to:

- Assess all the available inputs and rank them according to their association with the outcome.
- Identify the key predictors, the most relevant features for classification or regression.
- Screen the predictors with marginal importance, reducing the set of inputs to those related to the target field.

Some predictive algorithms, including Decision Trees, integrate screening mechanisms that internally filter out the unrelated predictors. A preprocessing feature selection step is also available in Data Mining for Excel, and it can be invoked when building a predictive model. Feature selection can efficiently reduce data dimensionality, retaining only a subset of significant inputs so that the training time is reduced with no or insignificant loss of accuracy.

1.3.2 Unsupervised models

In unsupervised modeling, only input fields are involved. The scope is the identification of groupings and associations. Unsupervised models include:

1. **Cluster models**
   In cluster models, the groups are not known in advance. Instead, the algorithms analyze the input data patterns and identify the natural groupings of instances/cases. When new cases are scored by the generated cluster model, they are assigned into one of the revealed clusters.

2. **Association (affinity) and sequence models**
   Association and sequence models also belong to the class of unsupervised algorithms. Association models do not involve direct prediction of a single field. In fact, all fields have a double role, since they act as inputs and outputs at the same time. Association algorithms detect associations between discrete events, products, and attributes. Sequence algorithms detect associations over time.

3. **Dimensionality reduction models**
   Dimensionality reduction algorithms "group" fields into new compound measures and reduce the dimensions of data without sacrificing much of the information of the original fields.

1.3.2.1 Cluster models

Cluster models automatically detect the underlying groups of cases, the clusters. The clusters are not known in advance. They are revealed by analyzing the observed input data patterns. Clustering algorithms assess the similarity of the records/customers in respect to the clustering fields, and they assign them to the revealed clusters accordingly. Their goal is to detect groups with internal homogeneity and interclass heterogeneity.

Clustering algorithms are quite popular, and their use is widespread from data mining to market research. They can support the development of different segmentation schemes according to the clustering attributes used: behavioral, attitudinal, or demographical segmentation.

The major advantage of the clustering algorithms is that they can efficiently manage a large number of attributes and create data-driven segments. The revealed segments are not based on personal concepts, intuitions, and perceptions of the business people. They are induced by the observed data patterns, and provided they are properly built, they can lead to results with real business meaning and value. Clustering models can analyze complex input data patterns and suggest solutions that would not otherwise be apparent. They reveal customer typologies, enabling tailored marketing strategies.

Nowadays, various clustering algorithms are available which differ in their approach for assessing the similarity of the cases. According to the way they work and their outputs, the clustering algorithms can be categorized in two classes, the hard and the soft clustering algorithms. The hard clustering algorithms assess the distances (dissimilarities) of the instances. The revealed clusters do not overlap and each case is assigned to a single cluster.

Hard clustering algorithms include:

- **Agglomerative or hierarchical**. In a way, it is the "mother" of all clustering algorithms. It is called hierarchical or agglomerative since it starts by a solution where each record comprises a cluster and gradually groups records up to the point where all records fall into one supercluster. In each step, it calculates the distances between all pairs of records and groups the ones most similar. A table (agglomeration schedule) or a graph (dendrogram) summarizes the grouping steps and the respective distances. The analyst should then consult this information, identify the point where the algorithm starts to group disjoint cases, and then decide on the number of clusters to retain. This algorithm cannot effectively handle more than a few thousand cases. Thus, it cannot be directly applied in most business clustering tasks. A usual workaround is to a use it on a sample of the clustering population. However, with numerous other efficient algorithms that can easily handle even millions of records, clustering through sampling is not considered an ideal approach.

- **K-means**. K-means is an efficient and perhaps the fastest clustering algorithm that can handle both long (many records) and wide datasets (many data dimensions and input fields). In K-means, each cluster is represented by its centroid, the central point defined by the averages of the inputs. K-means is an iterative, distance-based clustering algorithm in which cases are assigned to the "nearest" cluster. Unlike hierarchical, it does not need to calculate distances between all pairs of records. The number of clusters to be formed is predetermined and specified by the user in advance. Thus, usually a number of different solutions should be tried and evaluated before approving the most appropriate. It best handles continuous clustering fields.

- **K-medoids**. K-medoids is a K-means variant which differs from K-means in the way clusters are represented during the model training phase. In K-means, each cluster is represented by the averages of inputs. In K-medoids, each cluster is represented by an actual, representative data point instead of using the hypothetical point defined by the cluster means. This makes this algorithm less sensitive to outliers.

- **TwoStep cluster**. A scalable and efficient clustering model, based on the BIRCH algorithm, included in IBM SPSS Modeler. As the name implies, it processes records in two steps. The first step of preclustering makes a single pass of the data, and records are assigned to a limited set of initial subclusters. In the second step, initial subclusters are further grouped, into the final segments.

- **Kohonen Network/Self-Organizing Map (SOM)**. Kohonen Networks are based on neural networks, and they typically produce a two-dimensional grid or map of the clusters, hence the name SOM. Kohonen Networks usually take longer time to train than K-means and TwoStep, but they provide a different and worth trying view on clustering.

The soft clustering techniques on the other end use probabilistic measures to assign the cases to clusters with a certain probabilities. The clusters can overlap and the instances can belong to more than one cluster with certain, estimated probabilities. The most popular probabilistic clustering algorithm is *Expectation Maximization (EM) clustering*.

1.3.2.2 Association (affinity) and sequence models

Association models analyze past co-occurrences of events and detect associations and frequent itemsets. They associate a particular outcome category with a set of conditions. They are typically used to identify purchase patterns and groups of products often purchased together. Association algorithms generate rules of the following general format:

IF (ANTECEDENTS) ***THEN*** CONSEQUENT

For example:

IF (product A and product C and product E and...) ***THEN*** product B

More specifically, a rule referring to supermarket purchases might be:

IF EGGS & MILK & FRESH FRUIT ***THEN*** VEGETABLES

This simple rule, derived by analyzing past shopping carts, identifies associated products that tend to be purchased together: when eggs, milk, and fresh fruit are bought, then there is an increased probability of also buying vegetables. This probability, referred to as the rule's confidence, denotes the rule's strength.

The left or the IF part of the rule consists of the *antecedents* or conditions: a situation that when holds true, the rule applies and the consequent shows increased occurrence rates. In other words, the antecedent part contains the product combinations that usually lead to some other product. The right part of the rule is the *consequent* or the conclusion: what tends to be true when the antecedents hold true. The rule complexity depends on the number of antecedents linked with the consequent.

These models aim at:

- Providing insight on product affinities. Understand which products are commonly purchased together. This, for instance, can provide valuable information for advertising, for effectively reorganizing shelves or catalogues and for developing special offers for bundles of products or services.
- Providing product suggestions. Association rules can act as a recommendation engine. They can analyze shopping carts and help in direct marketing activities by producing personalized product suggestions, according to the customer's recorded behavior.

This type of analysis is referred to as *market basket analysis* since it originated from point-of-sale data and the need of understanding consuming shopping patterns. Its application was extended though to also cover any other "basketlike" problem from various other industries. For example:

- In banking, it can be used for finding common product combinations owned by customers.
- In telecommunications, for revealing the services that usually go together.
- In web analytics, for finding web pages accessed in single visits.

Association models are unsupervised since they do not involve a single output field to be predicted. They analyze product affinity tables: multiple fields that denote product/service possession. These fields are at the same time considered as inputs and outputs. Thus, all products are predicted and act as predictors for the rest of the products.

Usually, all the extracted rules are described and evaluated in respect to three main measures:

- The support: it assesses the rule's coverage or "how many records constitute the rule." It denotes the percentage of records that match the antecedents.
- The confidence: it assesses the strength and the predictive ability of the rule. It indicates "how likely is the consequent, given the antecedents." It denotes the consequent percentage or probability, within the records that match the antecedents.
- The lift: it assesses the improvement of the predictive ability when using the derived rule compared to randomness. It is defined as the ratio of the rule confidence to the prior confidence of the consequent. The prior confidence is the overall percentage of the consequent within all the analyzed records.

The *Apriori* and the *FP-growth* algorithms are popular association algorithms.

Sequence algorithms analyze paths of events in order to detect common sequences. They are used to identify associations of events/purchases/attributes over time. They take into account the order of events and detect sequential associations that lead to specific outcomes. Sequence algorithms generate rules analogous to association algorithms with one difference: a sequence of antecedent events is strongly associated with the occurrence of a consequent. In other words, when certain things happen with a specific order, a specific event has increased probability to follow. Their general format is:

IF (ANTECEDENTS with a specific order) ***THEN*** CONSEQUENT

Or, for example, a rule referring to bank products might be:

IF SAVINGS ***& THEN*** CREDIT CARD ***& THEN*** SHORT TERM DEPOSIT ***THEN*** STOCKS

This rule states that bank customers who start their relationship with the bank as savings customers and subsequently acquire a credit card and a short-term deposit present increased likelihood to invest in stocks. The support and confidence measures are also applicable in sequence models.

The origin of sequence modeling lies in web mining and click stream analysis of web pages; it started as a way to analyze web log data in order to understand the navigation patterns in web sites and identify the browsing trails that end up in specific pages, for instance, purchase checkout pages. The use of these algorithms has been extended, and nowadays, they can be applied to all "sequence" business problems. They can be used as a mean for predicting the next expected "move" of the customers or the next phase in a customer's life cycle. In banking, they can be applied to identify a series of events or customer interactions that may be associated with discontinuing the use of a product; in telecommunications, to identify typical purchase paths that are highly associated with the purchase of a particular add-on service; and in manufacturing and quality control, to uncover signs in the production process that lead to defective products.

1.3.2.3 Dimensionality reduction models

As their name implies, dimensionality reduction models aim at effectively reducing the data dimensions and remove the redundant information. They identify the latent data dimensions and replace the initial set of inputs with a core set of compound measures which simplify subsequent modeling while retaining most of the information of the original attributes.

Factor, Principal Components Analysis (PCA), and Independent Component Analysis (ICA) are among the most popular data reduction algorithms. They are unsupervised, statistical algorithms which analyze and substitute a set of continuous inputs with representative compound measures of lower dimensionality.

Simplicity is the key benefit of data reduction techniques, since they drastically reduce the number of fields under study to a core set of composite measures. Some data mining techniques may run too slow or may fail to run if they have to handle a large number of inputs. Situations like these can be avoided by using the derived component scores instead of the original fields.

1.3.2.4 Record screening models

Record screening models are applied for anomaly or outlier detection. They try to identify records with odd data patterns that do not "conform" to the typical patterns of the "normal" cases.

Unsupervised record screening models can be used for:

- Data auditing, as a preparation step before applying subsequent data mining models
- Fraud discovery

Valuable information is not only hidden in general data patterns. Sometimes rare or unexpected data patterns can reveal situations that merit special attention or require immediate actions. For instance, in the insurance industry, unusual claim profiles may indicate fraudulent cases. Similarly, odd money transfer transactions may suggest money laundering. Credit card transactions that do not fit the general usage profile of the owner may also indicate signs of suspicious activity.

Record screening algorithms can provide valuable help in fraud discovery by identifying the "unexpected" data patterns and the "odd" cases. The unexpected cases are not always suspicious. They may just indicate an unusual yet acceptable behavior. For sure though, they require further investigation before being classified as suspicious or not.

Record screening models can also play another important role. They can be used as a data exploration tool before the development of another data mining model. Some models, especially those with a statistical origin, can be affected by the presence of abnormal cases which may lead to poor or biased solutions. It is always a good idea to identify these cases in advance and thoroughly examine them before deciding for their inclusion in subsequent analysis.

The examination of record distances as well as standard data mining techniques, such as clustering, can be applied for anomaly detection. Anomalous cases can often be found among cases distant from their "neighbors" or among cases that do not fit well in any of the emerged clusters or lie in sparsely populated clusters.

1.4 The data

The success of a data mining project strongly depends on the breadth and quality of the available data. That's why the data preparation phase is typically the most time consuming phase of the project. Data mining applications should not be considered as one-off projects but rather as ongoing processes, integrated in the organization's marketing strategy. Data mining has to be "operationalized." Derived results should be made available to marketers to guide them in their everyday marketing activities. They should also be loaded in the organization's frontline systems in order to enable "personalized" customer handling. This approach requires the setting up of well-organized data mining procedures, designed to serve specific business goals, instead of occasional attempts which just aim to cover sporadic needs.

In order to achieve this and become a "predictive enterprise," an organization should focus on the data to be mined. Since the goal is to turn data into actionable knowledge, a vital step in this "mining quest" is to build the appropriate data infrastructure. Ad hoc data extraction and queries which just provide answers to a particular business problem may soon end up into a huge mess of unstructured information. The proposed approach is to design and build a central mining datamart that will serve as the main data repository for the majority of the data mining applications.

1.4.1 The mining datamart

All relevant information should be taken into account in the datamart design. Useful information from all available data sources, including internal sources such as transactional, billing and operational systems, and external sources such as market surveys and third-party lists, should be collected and consolidated in the datamart framework. After all, this is the main idea of the datamart: to combine all important blocks of information in a central repository that can enable the organization to have a complete a view of each customer.

The mining data mart should:

- Integrate data from all relevant sources.
- Provide a complete view of the customer by including all attributes that characterize each customer and his/hers relationship with the organization.
- Contain preprocessed information, summarized at the minimum level of interest, for instance, at an account or at a customer level. To facilitate data preparation for mining purposes, preliminary aggregations and calculations should be integrated in the loading of the datamart.
- Be updated on a regular and frequent basis to contain the current view of the customer.
- Cover a sufficient time period (enough days or months, depending on the specific situation) so that the relevant data can reveal stable and nonvolatile behavioral patterns.
- Contain current and historical data so that the view of the customer can be examined in different points in time. This is necessary since in many data mining projects and in classification models in particular, analysts have to associate behaviors of a past observation period with events occurring in a subsequent outcome period.
- Cover the requirements of the majority of the upcoming mining tasks, without the need of additional implementations and interventions from the IT.

1.4.2 The required data per industry

Table 1.3 presents an indicative, minimum list of required information that should be loaded and available in the mining datamart of retail banking.

Table 1.4 lists the minimum blocks of information that should reside in the datamart of a mobile telephony operator (for residential customers).

Table 1.5 lists the minimum information blocks that should be loaded in the datamart of retailers.

1.4.3 The customer "signature": from the mining datamart to the enriched, marketing reference table

Most data mining applications require a one-dimensional, flat table, typically at a customer level. A recommended approach is to consolidate the mining datamart information, often spread in a set of database tables, into one table which should be designed to cover the key mining as well as marketing and reporting needs. This table, also referred to as the marketing

Table 1.3 The minimum required data for the mining datamart of retail banking

Product mix and product utilization: ownership and balances	
Product ownership and balances per product groups/subgroups For example: • Deposits/savings • Time deposits • Investments • Insurances • Corporate Loans • Small business loans (SBL) • Mortgages • Consumer loans • Cards and open loans	
Frequency (number) and volume (amount) of transactions	
Transactions by transaction type For example: • Deposits • Credit cards • Withdrawals • Payments • Transfers • Queries • Other	***Transactions by transaction channel*** For example: • Branch • ATM • Automatic Payment System • Internet • Phone • SMS • Standing order • Other
Product (account) openings/terminations	
For the specific case of credit cards, frequency and volume of purchases by type (one-off, installments, etc.), and merchant category	
Credit score and arrears history	
Information on the profitability of customers	
Customer status history (active, inactive, dormant, etc.) and core segment membership (retail, corporate, private banking, affluent, mass, etc.)	
Registration and sociodemographical information of customers	

reference table, should integrate all the important customer information, including demographics, usage, and revenue data, providing a customer "signature": a unified view of the customer. Through extensive data processing, the data retrieved from the datamart should be enriched with derived attributes and informative KPIs to summarize all the aspects of the customer's relationship with the organization.

The reference table typically includes data at a customer level, though the data model of the mart should support the creation of custom reference tables of different granularities. It should be updated at a regular basis to provide the most recent view of each customer, and it should be

Table 1.4 The minimum required data for the mining datamart of mobile telephony (residential customers)

Phone usage: number of calls/minutes of calls/traffic			
Usage by call direction and network type	**Usage by core service type**	**Usage by origination/ destination operator**	**Usage by call day/time**
• Incoming Outgoing • International Roaming	For example: • Voice • SMS • MMS • Internet	For example: • On-net • Mobile telephony competitor operator A, B, etc. • Fixed telephony operator A, B, etc.	For example: • Peak Off-peak • Work Nonwork
Customer communities: distinct telephone numbers with calls from or to			
Information by call direction (incoming/outgoing) and operator (on-net/mobile telephony competitor operator/fixed telephony operator)			
Top-up history (for prepaid customers), frequency, value, and recency of top-ups			
Information by top-up type			
Rate plan history (opening, closings, migrations)			
Billing, payment, and credit history (average number of days till payment, number of times in arrears, average time remaining in arrears, etc.)			
Financial information such as profit and cost for each customer (ARPU, MARPU)			
Status history (active, suspended, etc.) and core segmentation information (consumer postpaid/contractual, consumer prepaid customers)			
Registration and sociodemographical information of customers			

stored to track the customer view over time. The key idea is to comprise a good starting point for the fast creation of a modeling file for all the key future analytical applications.

Typical data management operations required to construct and maintain the marketing reference tables include:

Filtering of records. Filtering of records based on logical conditions to include only records of interest.

Joins. Consolidation of information of different data sources using key fields to construct a total view of each customer.

Aggregations/group by. Aggregation of records to the desired granularity. Replacement of the input records with summarized output records, based on statistical functions such as sums, means, count, minimum, and maximum.

Table 1.5 The minimum required data for the mining datamart of retailers

RFM attributes: recency, frequency, monetary overall, and per product category			
Time since the last transaction (purchase event) or since the most recent visit day ***Frequency (number) of transactions or number of days with purchases*** ***Value of total purchases***			
Spending: purchases amount and number of visits/transactions			
Relative spending by product hierarchy and private labels	***Relative spending by store/ department***	***Relative spending by date/time***	***Relative spending by channel of purchase***
For example: • Apparel/shoes/jewelry • Baby • Electronics • Computers • Food and wine • Health and beauty • Pharmacy • Sports and Outdoors • Books/Press • Music, movies, and video games • Toys • Home • Other	For example: • Store A, B, C, etc. For example: • Department A, B, C, etc.	• Weekday • Month • Timezone • Special occasion (e.g., sales or holiday season)	For example: • Store • Phone • Internet • Other
Usage of vouchers and reward points			
Payments by type (e.g., cash, credit card)			
Registration data collected during the loyalty scheme application process			

Restructure/pivoting. Rearrangement of the original table with the creation of multiple separate fields. Multiple records, for instance, denoting credit card purchases by merchant type, can be consolidated into a single record per customer, summarized with distinct purchase fields per merchant type.

Derive. Construction of new, more informative fields which better summarize the customer behavior, using operators and functions on the existing fields.

The derived measures in the marketing reference table should include:

Sums/averages. Behavioral fields are summed or averaged to summarize usage patterns over a specific time period. For instance, a set of fields denoting the monthly average number of transactions by transaction channel (ATM, Internet, phone, etc.) or by transaction type (deposit, withdrawal, etc.) can reveal the transactional profile of each customer. Sums and particularly (monthly) averages take into account the whole time period examined and not only the most recent past, ensuring the capture of stable, non-volatile behavioral patterns.

Ratios. Ratios (proportions) can be used to denote the relative preference/importance of usage behaviors. As an example, let's consider the relative ratio (the percentage) of the total volume of transactions per transaction channel. These percentages reveal the channel preferences while also adjust for the total volume of transactions of each customer. Other examples of this type of measures include the relative usage of each call service type (voice, SMS, MMS, etc.) in telecommunications or the relative ratio of assets versus lending balances in banking. Apart from ratios, plain numeric differences can also be used to compare usage patterns.

Flags. Flag fields are binary (dichotomous) fields that directly and explicitly record the occurrence of an event of interest or the existence of an important usage pattern. For instance, a flag field might show whether a bank customer has made any transactions through an alternative channel (phone, Internet, SMS, etc.) in the period examined or whether a telecommunications customer has used a service of particular interest.

Deltas. Deltas capture changes in the customer's behavior over time. They are especially useful for supervised (predictive) modeling since a change in behavior in the observation period may signify the occurrence of an event such as churn, default, etc. in the outcome period. Deltas can include changes per month, per quarter, or per year of behavior. As an example, let's consider the monthly average amount of transactions. The ratio of this average over the last 3 months to the total average over the whole period examined is a delta measure which shows whether the transactional volume has changed during the most recent period.

1.5 Summary

In this chapter, we've presented a brief overview of the main concepts of data mining. We've outlined how can data mining help an organization to better address the CRM objectives and achieve "individualized" and more effective customer management through customer insight. We've introduced the main types of data mining models and algorithms and a process model, a methodological framework for designing and implementing successful data mining projects. We've also presented the importance of the mining datamart and provided indicative lists of required information per industry.

Chapters 2 and 3 are dedicated in the detailed explanation of the methodological steps for classification and segmentation modeling. After clarifying the roadmap, we describe the tools, the algorithms. Chapters 4 and 5 explain in plain words some of the most popular and powerful data mining algorithms for classification and clustering. The second part of the book is the "hands-on" part. The knowledge of the previous chapters is applied in real-world applications of various industries using three different data mining tools. A worth noting lesson that I've learned after concluding my journey in this book is that it is not the tool that matters the most but the roadmap. Once you have an unambiguous business and data mining goal and a clear methodological approach, then you'll most likely reach your target, yielding comparable results, regardless of the data mining software used.

Part I

The Methodology

2

Classification modeling methodology

2.1 An overview of the methodology for classification modeling

In this chapter, we present the methodological steps for classification modeling. Classification modeling is a form of supervised modeling in which the analytical task is to classify new instances in known classes. Historical data are used for model training. Classified cases are presented to the model which analyzes the data patterns and the associations of the input attributes (predictors) with the observed outcome. When the model is applied to new, unseen (unlabeled) cases, it assigns them to a class based on the predictors' values and the identified data patterns associated with each class. Along with the class prediction, the model also estimates prediction confidence which denotes the likelihood of prediction. Many marketing applications can be "translated" to classification problems and be tackled with classification modeling, including optimization of targeted marketing campaigns for acquiring new customers, cross-/up-/deep-selling, and attrition prevention. Since the scope of this book is to be used as a guide for real analytical applications in marketing, in this chapter, we try to go beyond a generic presentation of the classification methodology. Therefore, we dedicate a large part of this chapter in explaining how classification modeling can be applied to support and optimize specific marketing applications in all major industries.

Effective CRM using Predictive Analytics, First Edition. Antonios Chorianopoulos.

Companion website: www.wiley.com/go/chorianopoulos/effective_crm

The following are the five main steps of the classification modeling procedure, in accordance with the main CRISP-DM phases presented in Section 1.2:

1. Understand and design
2. Explore, prepare, and enrich the data
3. Build model(s)
4. Evaluate model(s)
5. Deploy to new cases

The first task is to understand the marketing needs and objectives and design the appropriate mining process. Then, the data should be understood, transformed, and enriched with informative predictors which could boost the predictive accuracy of the model. After data are appropriately prepared, the classification model is build and evaluated before being used for scoring new, unlabeled instances. The close collaboration of analysts and marketers in all these phases is required for the success of the project since the analytical skills and capabilities should be guided by in-depth domain knowledge. The classification methodology steps are presented in a more refined way in Table 2.1, and they are thoroughly explained in the next paragraphs.

2.2 Business understanding and design of the process

In the first phase of the project, the business need should be understood and translated into a tangible data mining goal. More specifically, the tasks typically carried out in this phase include the following.

2.2.1 Definition of the business objective

I. Business understanding and design of the process	***I.1. Understanding of the business situation: definition of the business objective***

Before designing the data mining approach, the business situation should be understood, and the marketing goal should be defined. In this preliminary step of the process, the project team of marketers and data miners, through a series of meetings and in-depth discussions, should share their business and technical knowledge and achieve thorough understanding of the current business situation, including solutions in place, available data resources, and possible limitations.

In this phase, the business objective should also be defined. In many situations, the analysts may face very ambitious and/or vague business objectives set by the marketers of the organization who may regard data mining as the silver bullet to magically eliminate all their problems at once. Therefore, in order to avoid overexpectations, it is important for the analysts to clearly define the business questions to be addressed by the subsequent mining models, translate the marketing problem to data mining tasks, and communicate right from the start what can be achieved by data mining.

Table 2.1 The classification methodology

I. *Business understanding and design of the process*	II. *Data understanding, preparation, and enrichment*	III. *Classification modeling*	IV. *Model evaluation*	V. *Model deployment*
I.1. Understanding of the business situation: definition of the business objective	**II.1. Investigation of data sources**	**III.1. Trying different models and parameter settings**	**IV.1. Thorough evaluation of the model accuracy**	**V.1. Scoring customers to roll the marketing campaign**
I.2. Definition of the mining approach and of the data model	**II.2. Selecting the data sources to be used**	**III.2. Combing models: ensemble/Boosting/ Bagging/Random forests**	*IV.1.1. Confusion matrix and accuracy measures*	V.1.1. Building propensity segments
I.3. Design of the modeling process	**II.3. Data integration and aggregation**	**III.3. Proceed to a first rough evaluation of the models**	*IV.1.2. Gains/Response/ Lift charts*	**V.2. Designing a deployment procedure and disseminating the results**
I.3.1. Definition of the modeling population	**II.4. Data exploration, validation, and cleaning**		*IV.1.3. ROC curves*	
I.3.2. Determining the modeling level (analysis level)	**II.5. Data transformations and enrichment**		*IV.1.4. Profit/ROI charts*	
I.3.3. Definition of the target event and population	**II.6. Applying a validation technique**		**IV.2. Evaluating a deployed model: test/ control groups**	
I.3.4. Deciding on time frames	**II.7. Dealing with imbalanced outcomes**			

Apart from setting expectations and defining the business and the corresponding data mining objective, in this phase, the project team should also be appointed with a clear task assignment and a plan for coworking. Finally, the expected data mining deliverables should be made clear from the start to everyone involved as well as the method for measuring the success of the project and the success criteria.

2.2.2 Definition of the mining approach and of the data model

I. Business understanding and design of the process	***I.2. Definition of the mining approach and of the data model***

The marketing application and the business target call for a certain mining approach and a corresponding data model. However, there are different mining approaches applicable for the same marketing application, and the project team has to decide on the one to be used according to the business situation, the priorities, and the available resources.

Although the appropriate mining approaches for targeted marketing campaigns (customer acquisition, cross-/up-/deep-selling, and customer churn) are presented in detail and with examples by industry later in this chapter, we briefly introduce the main concepts here to outline the significance of this step which defines the entire project.

Consider, for example, an organization that wants to promote a specific product to existing customers. It plans to carry out a cross-selling campaign to stimulate customers to uptake this product and decides to rely on data mining models to target the campaign. The thing is that there are numerous different classification mining approaches for the specific marketing objective, all with their pros and cons. For instance, an approach could be to run a test campaign on a sample of customers. Then, "go after" actual responders, analyze their defining characteristics, and try to identify customers with the same profile among the entire customer base. This "pilot campaign" approach, although highly effective since it simulates the exact planned customer interaction, is time consuming and requires resources for the test campaign. An alternative mining approach could be to consider as the target population those customers who purchased the product of interest in the near past, the so-called "product uptake" approach. Instead, due to the difficulties of the aforementioned approaches, the organization may decide to simply build a profile of the customers who currently own the product of interest and then identify their "clones" among those customers who don't have the product. This "product possession" approach seems attractive due to its simplicity. However, its effectiveness is limited since the respective models are analyzing data patterns and behaviors that may be the result and not the "cause" of the event of interest.

Each of the aforementioned mining approaches requires different data sources and data setup. They define from the start the entire procedure and the data model of the project, hence deciding which one to follow is critical.

The mining approaches for classification in support of targeted marketing campaigns can be grouped according to the need of historical data. Although more complicated, the ones using an observation and a different outcome period tend to stand out in terms of validity and predictive ability compared to the simple profiling approaches. Another group of effective, yet demanding in terms of resources, mining setups includes the ones that are based on pilot campaigns or other interactions with the customers, for instance, a communication for information request.

2.2.3 Design of the modeling process

I. Business understanding and design of the process	***I.3. Design of the modeling process***

After selecting the appropriate mining approach, the next step is the design of the modeling process. Having discussed and understood the specific business situation and goal, a number of key decisions should be made concerning:

1. The modeling population
2. The modeling/analysis level
3. The target event and population
4. The attributes to be used as predictors
5. The modeling time frames

All these substeps are presented in detail in the following paragraphs.

2.2.3.1 Defining the modeling population

I.3 Design of the modeling process	***I.3.1. Definition of the modeling population***

The modeling population is comprised of the instances (customers) to be used for model training. Since the goal is to build a classification model for a targeted marketing campaign, the modeling population also corresponds to the customers who'll be reached by the marketing action. The decision on who to include should be made with the final goal in mind. Thus, the question that mainly determines the modeling population is "how results will be used." For instance, which customers do we want to contact to promote a product, or which customers do we want to prevent from leaving? Do we want to reach all customers or only customers of a specific group? For a voluntary churn prevention campaign, for example, we should decide if we want the model and the campaign to cover the entire customer base or just the group of valuable customers.

Usually, not-typical customers such as staff members of VIP customers should be identified and excluded from both the training and the scoring processes.

Quite often, the modeling population is based on the organization's core segments. In retail banking, for example, customers are typically grouped in segments like affluent, mass, and inactive according to their balances, product ownership, and total number of transactions. These segments contain customers of different characteristics and are usually managed by different marketing teams or even departments. For operational as well as analytical reasons, a cross-segment classification model and marketing campaign would be unusual and probably ineffective. On the other hand, a separate approach and the development of distinct models for each segment may enhance the classification accuracy and the campaign's effect.

Sometimes, it is worthwhile to go beyond the core business segments and build separate models for subsegments identified by cluster analysis. The cluster model in this approach is a preparatory step which can identify customer groups with different characteristics. The training of separate models for each cluster is an extra workload but might also significantly improve the performance of subsequent classification models due to the similar profiles of the analyzed instances.

In churn modeling, the "customer" status is taken into account (active, inactive, dormant) in the definition of the modeling population. A churn model is usually trained on cases that were active at the end of the observation period. These cases comprise the training population, and their behavior in the outcome period, after the observation period, is examined to define the target event.

In cross-selling, the modeling population is typically defined based on the ownership of products/services. According to the "product uptake" mining approach presented in Section 2.9.1.2, the cross-selling modeling population typically includes all customers not in possession of a product in the observation period who purchased or not the product in the outcome period that followed.

2.2.3.2 Determining the modeling (analysis) level

I.3. Design of the modeling process	*I.3.2. Determining the modeling level*

A decision tightly connected with the selection of the modeling population is the one concerning the level of the model. Apart from deciding whom to analyze, the project team should also agree on the appropriate analysis level which in turn defines the aggregation level and granularity of the modeling file—in plain words, what each record of the modeling file summarizes: customers, accounts, contracts, telephone lines, etc.

Once again, this choice depends on the type of the subsequent marketing campaign. Typically, a customer has a composite relationship with an organization, owning, for instance, more than one credit cards or telephone lines or insurance contracts. The modeling level depends on the specific campaign's objective. For instance, if the plan is to target customers taking into account their entire relationship with the bank instead of cardholders of individual cards, then a customer-level approach should be followed.

Obviously, the selected modeling level and population delineate the form of the final modeling file and the relevant data preparation procedures that have to be carried out. For a customer-level campaign, all collected data, even if initially retrieved at lower levels, should be aggregated at the desired level for model training and scoring. Original information should be transformed at the granularity level necessary for answering the business question and addressing the specific business objective.

2.2.3.3 Definition of the target event and population

I.3. Design of the modeling process	*I.3.3. Definition of the target event and population*

In classification modeling, the task is to assign new cases to known classes by using a model which had been trained on preclassified cases. The goal is to predict an event with known outcomes. Hence, in the design of the modeling process, it is necessary to define the target event and its outcomes in terms of available data—in other words, to reach a decision about the target, also known as the output or label or class attribute of the subsequent model.

As explained in Chapter 2, the target attribute in classification modeling is a categorical/symbolic field recording class membership.

Although still at an early stage of the process, it is necessary to outline the target event to be predicted. Obviously, this selection should be made in close cooperation with the marketers of the organization that might already have their definitions for relative processes.

The target population includes the instances belonging to the target event outcome, the target class. These instances are typically the customers we are trying to identify since they belong in the class of special interest. Most of the times, the model's target attribute is a flag (dichotomous or binary) field denoting membership (Yes/No or True/False) into the target population. In a churn model, for example, the target population is consisted of those that have voluntary churned within the event outcome period, as opposed to those who stayed loyal. In a cross-selling model, those that purchased a product or responded in a promotional campaign comprise the target population.

The definition of the target event should take into account the available data. The analysts involved should also ensure that the designed data model will allow the construction of the target field.

Although a disconnection event might be an obvious and well-defined target for most churn applications, in some cases, it is not applicable. In the case of prepaid customers in mobile telephony, for example, there isn't a recorded disconnection event to be modeled. The separation between active and churned customers is not evident. In such cases, a target event could be defined in terms of specific customer behavior. This handling requires careful data exploration and cooperation between the data miners and the marketers. For instance, prepaid customers with no incoming or outgoing phone usage within a certain time period could be considered as churners. In a similar manner, certain behaviors or changes in behavior, for instance, substantial decrease in usage or a long period of inactivity, could be identified as signals of specific events and then be used for the definition of the respective target.

Moreover, the same approach could also be followed when analysts want to act proactively. For instance, even when a churn/disconnection event could be directly identified through a customer's action, a proactive approach would be to analyze model behaviors before the typical attrition by identifying early signals of defection and not waiting for the official termination of the relationship with the customer.

When using behavioral information to build the target event, it is crucial to analyze relevant data and come up with a stable definition. For instance, preliminary analysis and descriptive statistics might reveal the inactivity time period which signifies permanent churn and trivial chances of "revival." A simple graph will most probably show that the percentage of returning customers becomes insignificant after a specific number of inactivity months. This information can reveal the appropriate time window for defining a stable, behavioral churn event.

2.2.3.4 Deciding on time frames

I.3. Design of the modeling process	*I.3.4. Deciding on time frames*

Most classification models make use of historical, past data. Customer profiles are examined in the observation period before the occurrence of the target event, and input data patterns are associated with the event which follows, whether purchase of an add-on product or churn.

This approach requires the examination of the customer view in different points in time. Therefore, the analyzed data should cover different time frames. The time frame refers to the time period summarized by the relevant attributes. The general strategy is to have at least two distinct time frames. The first time frame, the observation window (also known as the historical period), is a past snapshot of the customer, used for constructing predictors which summarize the customer profile before the event occurrence. The second time frame, referred as event outcome period, is used for recording the event outcome and building the target attribute. Typically, an additional time period, in between the observation and the event outcome period, is taken into account, corresponding to the time needed to gather the data, prepare the model, score new cases, and roll out the targeted marketing campaign. Apparently, this approach requires that predictors should come from a time frame preceding the one used for examining the target event.

Let's consider, for example, a typical classification model trained on historical data. The data setup is presented in Figure 2.1.

The respective time frames are:

1. Observation (historical) period: Used for building the customer view in a past time period, before the occurrence of the target event. It refers to the distant past, and corresponds to the time frame used for building the model inputs (predictors).

2. Latency period: It refers to the gap between the observation and the event outcome period, reserved for taking into account the time needed to collect all necessary information, score new cases, predict class assignment, and execute the relevant campaign. A latency

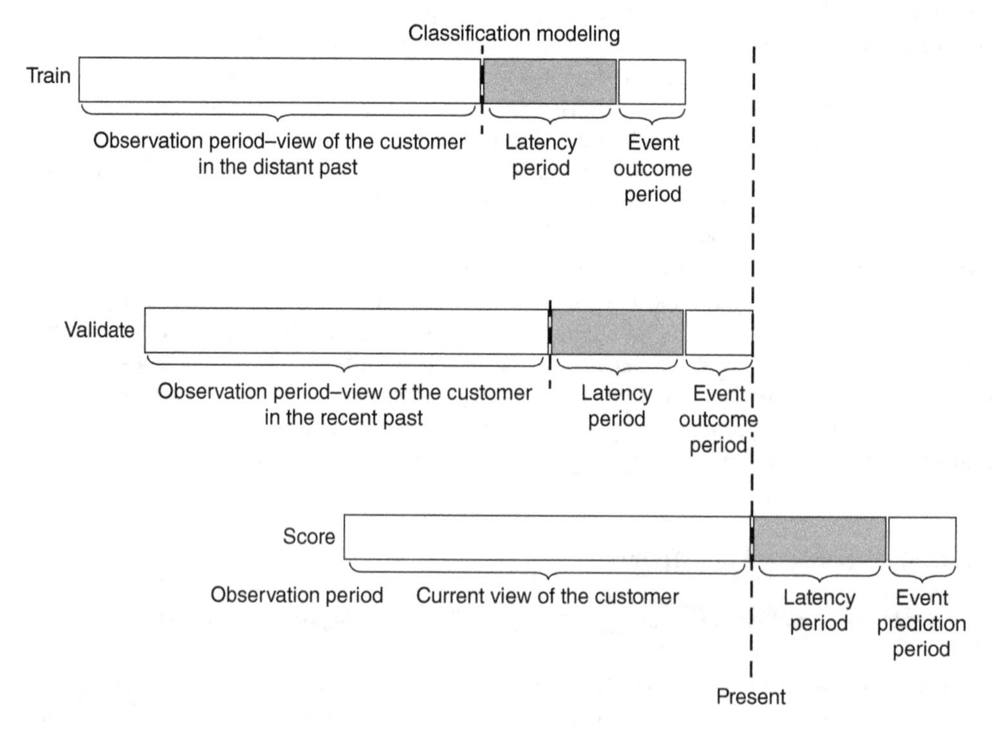

Figure 2.1 The data setup and time frames in a classification model trained on historical data. Source: Tsiptsis and Chorianopoulos (2009). Reproduced with permission from Wiley

period also ensures that the model is not trained to identify "immediate" event occurrence, for instance, immediate churners. Even if we manage to identify those customers, chances are that by the time they are contacted, they could already be gone or it will be too late to change their minds. The goal of the model should be long term: the recognition of early churn signals and the identification of customers with increased likelihood to churn in the near but not in the immediate future, since for them there is a chance of retention.

3. Event outcome period: Used for recording the target event and the class assignment. It follows the observation and the latency period, and it is only used for defining the target (output) attribute of the classification model. Predictors should not extend in this time period.

The model is trained by associating input data patterns of the observation period with the event outcomes recorded in the event outcome period.

Typically, in the validation phase, the model's classification accuracy is evaluated in disjoint time periods. Yet again, the observation period, used for constructing predictors, should precede the event outcome period. In the deployment phase, the generated model scores new cases according to their present view. Now, the observation period covers the period right before the present, and the event outcome period corresponds to the future. The event outcome is unknown and its future value is predicted.

A more specific example of a voluntary churn model setup is illustrated in Figure 2.2.

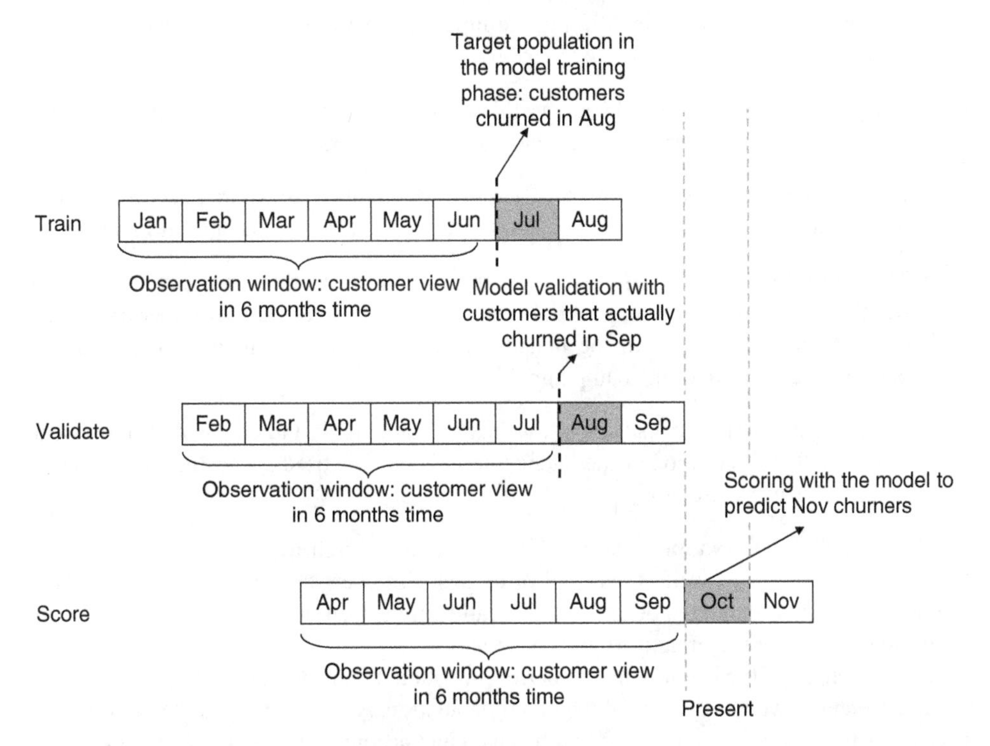

Figure 2.2 The data setup and time frames in a churn model. Source: Tsiptsis and Chorianopoulos (2009). Reproduced with permission from Wiley

In this example, the goal of a mobile telephony network operator is to set up a model for the early identification of potential voluntary churners. This model will be the base for a respective targeted retention campaign and predicts voluntary attrition 1 month ahead. We assume that the model is built in October:

- The model is trained on a 6-month observation window covering January to June. The inputs summarize all the aspects of the customer relationship with the organization in that period providing an integrated customer view also referred to as the customer signature. Yet, they come strictly from the historical period, and they do not overlap with the target event which is defined in the outcome period.
- The model is trained on customers that were active at the end of the observation window (end of the 6-month historical period). These customers comprise the modeling population.
- 1 month has been reserved in order to allow for scoring and campaign preparation. This month is shown as a grayed box in the respective illustrations and it corresponds to the latency period. No inputs were used from the latency period (July).
- The target event is attrition, recorded as application for disconnection in the event outcome period, that is, within August.
- The target population is consisted of those that have voluntary churned in August.
- The model is trained by identifying the input data patterns associated with voluntary churn.
- The generated model is validated on a disjoint data set of a different time period, before being deployed for classifying currently active customers.
- In the deployment phase, customers active in September, are scored according to the model. Remember that the model is built in October which is the current month.
- The profile of customers is built using information from the 6 most recent months, from April to September. The model predicts who will leave in November. If October churners become known in the meantime, they are filtered-out from the target list before the execution of the campaign.
- The presented time frames used in this example are purely indicative. A different time frame for the observation or the latency period could be used according to the specific task and business situation.

The length of the observation window should be long enough to capture and represent a stable customer profile as well as seasonal variations. However, it should not be extended that much to also capture outdated behaviors. Typically a time span of 12–24 months, depending on the specificities of the organization, is adequate.

For the length of the event outcome period, we should consider a possible periodicity of the target event as well as the size of the target population. Apparently, for a target event with seasonal patterns, this seasonality should be taken into account when selecting the event outcome period. In terms of the size of the target population, ideally, we'd prefer to have at least

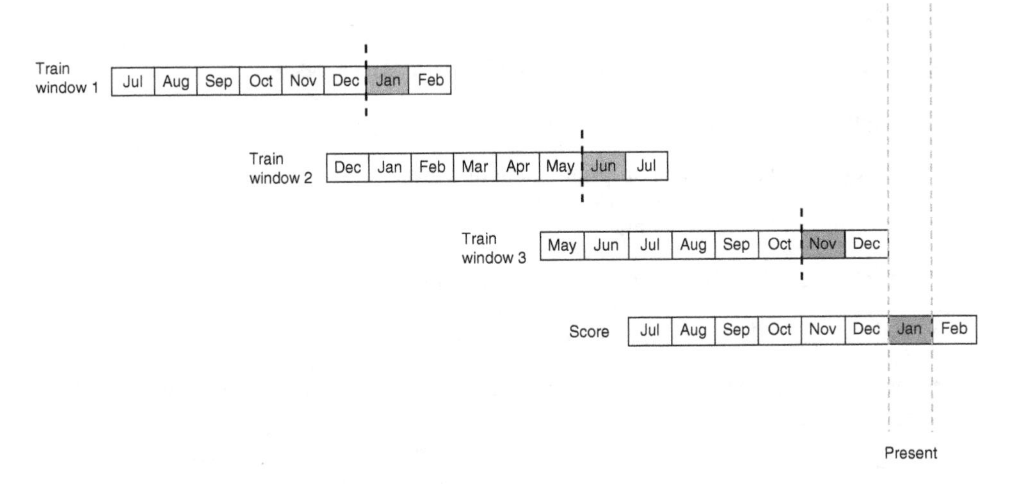

Figure 2.3 Using multiple time frames in model training

some hundreds of customers for model training. This might not be the case if the target event is rare. Although there are methods for boosting the number of target population cases and they are discussed later in this chapter, sometimes we can capture more positive cases by simply extending the time span of the event outcome period obviously at a cost of using and analyzing more data.

An advisable approach for training a classification model is to incorporate multiple time windows. With this approach, we can capture a more extended view of customer behaviors. Furthermore, we can also capture possible seasonal effects, avoiding the pitfall of using data from a particular season to train the model.

In the setup illustrated in Figure 2.3, 18 months of history are analyzed with three overlapping observation windows. In each setup, 6 months of historical data are used to predict churn 1 month ahead, and a single month is reserved as the latency period.

In the first time window, input data patterns recorded in past year's July to December are used to predict February churners. The second time window is shifted 5 months ahead. This time, the customer profiles are based on the period from December to May, and the target population includes July's churners. Finally, December's churners are predicted from historical data covering May to October. With this approach, available data are partitioned to build the three data setups allowing a broader capturing of behaviors. The patterns captured by the model do not reflect a single period. Moreover, this approach increases the number of cases in the model set and the frequency of the target population, a desirable result, especially in the case of limited training instances.

2.3 Data understanding, preparation, and enrichment

The second phase of the classification project covers the demanding and time-consuming phase of data preparation which, is critical for the success of the application. By data preparation, we refer to all the tasks necessary for transforming raw data to attributes ready to be included as inputs or output in the subsequent classification model.

These tasks include:

- Evaluating exiting sources of data for inclusion in the process
- Merging data from different sources
- Aggregating data to the appropriate modeling level
- Exploring and validating data
- Enriching existing attributes with informative indicators
- Reducing data dimensionality by retaining only informative predictors
- Incorporating a data validation procedure to avoid an overoptimistic evaluation of the accuracy of the classification model
- Tackling class imbalances

In classification modeling, all pieces of information outlining customer profiles are used to predict the target attribute. The final modeling file has the form of a one-dimensional flat table with a set of attributes, as inputs and a single target output. But preparing the modeling file might prove a heavy burden, especially when starting from raw data. This is where the mining datamart, described in Section 1.4.1, comes in. Having in place a mining datamart facilitates the data preparation procedure. However, each classification project has its particularities, calling for special data handling.

2.3.1 Investigation of data sources

II. Data understanding, preparation, and enrichment	***II.1. Investigation of data sources***

The prior action of data preparation is the investigation of available data sources in order to select the ones relevant to the project's scope. All data sources, even those not covered in the mining datamart, should be considered as candidates for inclusion. Data understanding should be followed by an initial, rough exploration of data in terms of availability, quality, and usefulness for the specific project.

2.3.2 Selecting the data sources to be used

II. Data understanding, preparation, and enrichment	***II.2. Selecting the data sources to be used***

A good classification model is trained on attributes providing a complete view of the customer. The retrieved data should adequately and thoroughly summarize all relevant aspects of the relationship of the customer with the organization. In brief, all essential resources should be lined up from all sources and systems. All information that could contribute in enriching what we know about the customers and enhance the predictive ability of the model, whether residing in the mining datamart, stored in the organization's data warehouse, collected from market research surveys, or logged from web visits, should be retrieved and prepared for analysis.

2.3.3 Data integration and aggregation

II. Data understanding, preparation, and enrichment	***II.3. Data integration and aggregation***

Required data, usually residing at different locations or even systems, should be brought together and be integrated to form the modeling file. For instance, for the needs of the cross-selling model of the retail case study presented in Chapter 7, two types of data were used. Transactional data were used to summarize customer purchase behaviors as well as customer demographics. Purchase data were collected at point of sales (POS), and they were used to summarize customer spending habits. Customer information on the other hand was recorded upon registration of a customer at the loyalty program of the retailer. When combined, the aforementioned data sources provided an integrated view of the customer. Since information from both sources is critical for training the model, the first data preparation task was their consolidation to a single modeling file.

Additionally, input data should be transformed at the selected modeling level. This typically requires the aggregation of the original inputs into the appropriate granularity. By aggregation, multiple detailed data records are replaced with summarized records. Returning to the retail cross-selling example of Chapter 7, the business objective was to identify customers with increased probability to uptake an add-on product. The final modeling file should be at the customer level. However, purchase data were at a transactional level. Each original POS data record corresponded to a purchased item with information regarding price of the item, day/time, and location of purchase. Therefore, transactional records had to be grouped by customer to designate customer purchase patterns. Besides merging and aggregating data, the data preparation phase also involves the creation of informative KPIs from the original attributes.

2.3.4 Data exploration, validation, and cleaning

II. Data understanding, preparation, and enrichment	***II.4. Data exploration, validation, and cleaning***

After all relevant data are consolidated into the modeling file, it is time for exploring and inspecting the retrieved fields. Data are explored with reports and graphs of basic statistical information.

Continuous (range) attributes are typically explored with:

Measures of central tendency, such as:

- The mean: The standard arithmetic average of values.
- The median: The middle data value which separates the data into two sets of equal size after data have been rank ordered. The median is a measure robust to extreme values.

Measures of variability (dispersion of values), such as:

- The minimum and maximum data values.
- The range: The difference of the maximum minus the minimum value.

- The variance: The variance is a measure of the variability of the attribute. It summarizes the dispersion of the field values around the mean. It is calculated by summing the squared deviations from the mean and dividing them to the total number of records minus 1.
- The standard deviation which is the square root of the variance. It expresses variability in the units of the attribute.

The distribution of continuous attributes is graphically investigated with histograms. In histograms, continuous attributes are binned into bands of equal width, and bars depict the frequency of each band. By studying histograms, analysts can explore the shape of a distribution and spot potential outliers. The skewness measure can also provide an indication about the shape of the distribution. Skewness is a measure of the symmetry of a distribution. Positive skewness indicates large values and a longer tail on the right, while negative skewness indicates the reverse pattern.

The distribution of categorical attributes is typically explored with tables and charts (pie charts, bar charts) of *frequencies*.

Figure 2.4 presents the output of the main data exploration tool of IBM SPSS Modeler, the Data Audit, which provides a first, quick exploration of the data.

Initial data exploration should be supplemented with an assessment of data quality. A health check on data should be performed looking for:

- **Fields with too many nulls (missing values)**. Analysts should be very cautious with missing values since some algorithms exclude incomplete records (records with null values in any of the input fields) from model training as well as scoring.

  Fields with too many missing values can be omitted from model training. Furthermore, missing values can be imputed with simple statistics like the mean or median for a continuous attribute or the mode, the modal category, for a categorical attribute. A more sophisticated approach is to impute nulls with estimations generated by respective supervised models trained on the rest of the attributes.

- **Categorical fields with too many categories**. An extreme example of such a field is the customer ID field which has a unique value for each customer and hence has

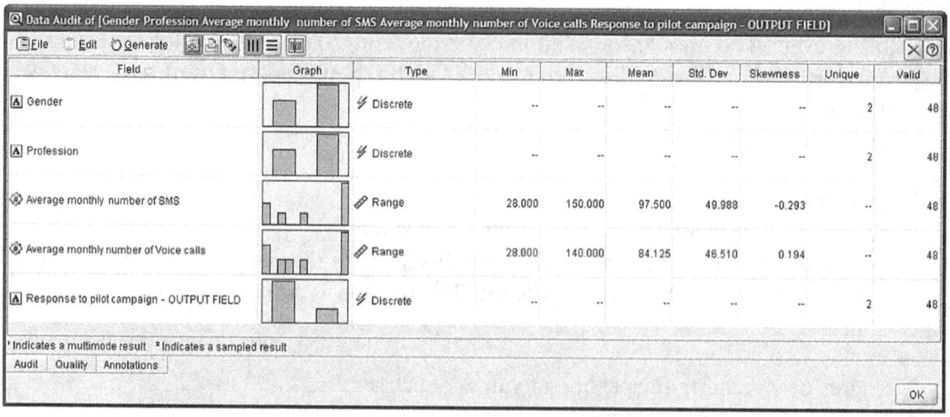

Data Audit of [Gender Profession Average monthly number of SMS Average monthly number of Voice calls Response to pilot campaign - OUTPUT FIELD]

Field	Graph	Type	Min	Max	Mean	Std. Dev	Skewness	Unique	Valid
Gender		Discrete	--	--	--	--	--	2	48
Profession		Discrete	--	--	--	--	--	2	48
Average monthly number of SMS		Range	28.000	150.000	97.500	49.988	-0.293	--	48
Average monthly number of Voice calls		Range	28.000	140.000	84.125	46.510	0.194	--	48
Response to pilot campaign - OUTPUT FIELD		Discrete	--	--	--	--	--	2	48

Indicates a multimode result   Indicates a sampled result

Audit | Quality | Annotations   OK

Figure 2.4 The Data Audit node of IBM SPSS Modeler for data exploration

nothing to contribute in pattern finding. In general, categorical fields with a large number of categories are hard to manage and add complexity to the model. Moreover, if used as predictors, their relationship with the target event would be hard to interpret and communicate. Therefore, the grouping of their categories is advisable.

Categories can be regrouped with conceptual criteria by merging categories that are considered similar. An even better approach is the supervised or "optimal" grouping of the categories. With this approach, categories which are similar in respect to the target attribute are merged. This handling of categorical predictors is inherent in some classification techniques such as Decision Trees which may decide to merge the categories of a splitting categorical attribute before branching.

- **Categorical fields with unfixed categories which frequently change**. Categorical predictors with categories that change over time can produce unstable models. The customer view and the predictor attributes used for scoring should correspond to the attributes used for model training. If in the meantime the categories of a predictor are replaced with new ones, a new training of the model is required to associate the new categories with the event outcomes.

  Such predictors can be replaced with continuous ones denoting the historical relationship of the categories with the outcome. For instance, in a churn model developed for a mobile telephony operator, analysts can use the churn rate per rate plan, calculated over a specific period of time, instead of using the unstable rate plan information itself. Likewise, for the needs of a cross-selling model, the rate plan predictor can be substituted by the cross-selling index of each rate plan, denoting the average number of add-on services for each rate plan.

- **Categorical fields with all cases in a single category**. These "constant" fields present no variation. Since both members and nonmembers of the target population share the same value, such an attribute has no discriminating power and should be omitted from model training.

- **Outlier values**. Data should be scanned for odd outlier values. Outliers need special handling. They may represent unexpected, but interesting, explainable, and acceptable patterns. In fact, in many applications such as in fraud detection, they may indicate a data pattern associated with the target population. However, in many cases, outlier values may just be inexplicable, noisy data, observed only in the specific modeling dataset. A model built to identify general patterns may be misguided by such data. In such cases, a more active approach is needed in terms of outliers.

  Outlier values can be coerced with the minimum or the maximum of the "normal" values. For example, if an outlier is defined as three standard deviations above the overall mean, then all values above this threshold are replaced with the highest value within this range. Other ways of handling outliers include discretization and standardization. With discretization or binning, the values of a continuous attribute are grouped into bands. A standardized attribute with the z-score method is derived by subtracting the overall average from each value and dividing with the standard deviation. We must stress however that specific classification algorithms such as the Decision Trees are robust to outliers as well as skewed distributions since they internally discretize the continuous predictors or identify an optimal split point before branching.

The data validation process should also include exploring the data for the identification of logical inconsistencies which might indicate a problematic input or even errors in the data integration and aggregation.

2.3.5 Data transformations and enrichment

II. Data understanding, preparation, and enrichment	***II.5. Data transformations and enrichment***

Classification modeling requires preclassified instances and hence the presence of an output, also known as label or target attribute. Assuming that all relevant information is lined up, the target attribute should be constructed to denote class membership in accordance with the target event definitions discussed in Section 2.2.3.3 and the modeling time frames discussed in Section 2.2.3.4. For example, by appropriately transforming information concerning disconnection applications such as application type and date data, a binary target attribute can be derived to designate customers who churned within the event outcome period. Similarly, by working on data about product openings/purchases, a cross-selling target attribute can be derived which discriminates recent purchasers.

Besides building a target attribute, analysts should finalize the customer view by appropriately transforming and enriching predictors. One thing should be outlined before discussing predictors. Predictors should by no means be confounded with the target attribute. Predicting a target event using attributes directly related with it is pointless. For instance, using the usage of a product to predict its ownership might lead to a model with astonishingly but erroneously high classification accuracy since we use a variant of the target attribute as a predictor. This pitfall is common in the case of models with distinct periods for observation and outcome. In such cases, it is crucial not to build predictors which extend in the event outcome period. This might also erroneously boost the classification accuracy of the model; however, the best predictors won't be available for deployment. Take, for example, a churn model in which predictors also extend in the event outcome period. These predictors may mistakenly appear significant; however, when trying to predict ahead in the future and score unseen cases, they'll be unavailable and impossible to reconstruct.

A transformation usually applied for increasing the predictive power of the inputs is "optimal" discretization or binning. "Optimal" *binning* is applied to continuous inputs and may attain significantly higher predictive accuracy when using algorithms such as logistic regression which calls for a linear relationship among the predictor and the probability (logit) of the target outcome.

But above all, the data preparation phase is about *data enrichment*. By data enrichment, we refer to the derivation of new, informative KPIs which intelligently summarize the customer behavior. It is the art of constructing new attributes which convey significant information and portray the customer profile. Common data transformations applied for data enrichment include calculations of sums and averages to summarize continuous attributes over time, ratios to denote relative preference/usage, flags to denote ownership/usage, and deltas to capture trends of behaviors over time. Deltas might prove especially useful predictors when using historical data since changes in behavior might signify a subsequent action such as a purchase or a decision to terminate the relationship with the organization. They denote the relative change in the most recent compared to the overall or the most distant observation period, and they indicate if the usage is increasing or decreasing over time.

Domain expertise is required for efficient data enrichment. To understand the logic and the significance of this step, let's consider the example of a retailer who plans to use purchase data recorded at POS to identify prospects for its new house and furniture department. In fact, this is the case study presented in Chapter 7. The input data records all purchase details including purchase date, store, payment method, item bought, and amount spent. After properly transforming the data, the retailer constructed a series of KPIs for each customer indicating:

- Relative spending per product group
- Preferred stored
- Preferred payment method
- Preferred time zone
- Preferred day of the week
- Purchases at special occasions such as in sales periods
- Frequency of visits
- Monthly average spending amount
- Recency of visits, indicating time since last visit
- Basket size defined as the average amount spent per visit

The above fields summarize all major aspects of purchase habits, they define the customer signature, and they are appropriate for intelligent reporting as well as modeling.

In most classification applications, analysts end up with tens or even hundreds of candidate predictors. However, some or even most of them turn out to be unrelated to the target event and of trivial predictive efficiency. Specific classification algorithms, such as Decision Trees, integrate internal screening mechanisms to select and use only those attributes which are relevant to the target event. Other algorithms such as neural networks, Bayes networks, or Support Vector Machines lack this feature. Feeding these complex algorithms with wide datasets and a large number of potential inputs consumes unnecessary resources and may hinder or even prevent the model training. In such situations, a field screening preprocessing step, also known as *feature selection*, is recommended. It includes the application of a feature selection algorithm to assess all the available inputs, find those with marginal predictive power, and exclude them from model training.

The dimensionality of data and the number of inputs can also be reduced with the application of a *data reduction* algorithm. The principal component analysis (PCA), for instance, identifies sets of continuous fields and extracts components which can be used as inputs in subsequent classification. Apart from simplification, PCA also offers an additional benefit. The extracted components are uncorrelated, a great advantage when dealing with statistical models such as logistic regression which are sensitive to multicollinearity (the case of correlated predictors).

Finally, here is a note about *the naming of the attributes*. In most cases when using time series fields, the role of each attribute may vary according to the data setup and the used time frame. For the application illustrated in Figure 2.2, for example, data from June refer to the most recent month of the observation period during model training. In deployment however which takes place in October, June corresponds to the fourth month of the observation period. Instead of naming all June attributes with a month-name indicator (prefix or suffix), it is recommended to follow a naming typology which indicates the time sequence of the month. Such a naming typology is consistent, indicates the role of the attribute in the model, and can be easily shifted in time. For example, using names such as balance_m1, balance_m2, etc. is preferred instead of using names such as balance_Jan, balance_Feb, etc.

2.3.6 Applying a validation technique

II. Data understanding, preparation, and enrichment	***II.6. Applying a validation technique***

The model is trained on instances with known outcome. Before deployed on new, unseen cases, its accuracy is evaluated with a series of measures. However, the evaluation of the classification model is optimistic when based on the dataset used for training. We want to be sure that the model will correctly classify unseen cases and the performance on the training set is not a good indicator for future performance. We have to test the model on cases that played no role in the creation of the model. An independent testing file for validation purposes is required. Therefore, before training the model, a validation technique should be applied to ensure an unbiased, subsequent model evaluation.

In this chapter, we present three widely used validation techniques: the Split (Holdout) validation, the Cross- or n-fold validation, and the Bootstrap validation method.

2.3.6.1 Split or Holdout validation

A Split (Holdout) validation method works by partitioning the modeling file into two distinct, exhaustive, and mutually exclusive parts with random sampling. The training part usually contains the majority of training instances, and it is used for model training. The testing part is reserved for model evaluation. Its instances do not take part in model training. A common practice is to use a split ratio of about 70–75% for the training file and hence allocate approximately 70–75% of the cases in the training dataset and hold out about 25–30% of instances for evaluation.

Apparently, analysts should mainly focus on the examination of performance metrics in the testing dataset. A model underperforming in the testing dataset should be reexamined since this is a typical sign of overfitting and of memorizing the specific training data. Models with this behavior do not provide generalizable results. They provide solutions that only work for the particular data on which they were trained.

To better illustrate the logic of Split validation, let's have a look at Table 2.2 which presents a simple modeling file for 20 customers. A pilot cross-selling campaign has been carried out, and an indicative list of inputs is used to classify customers to responders and nonresponders. The modeling file is partitioned in training and testing files through random

Table 2.2 A modeling dataset partitioned into training and testing samples

	Input fields				Output field	Split validation
Customer ID	Gender	Occupation	Average monthly number of SMS	Average monthly number of voice calls	Response to pilot campaign	Training/ testing file
1	Female	Blue collar	28	134	No	**Testing**
2	Male	Blue collar	45	54	No	**Training**
3	Female	Blue collar	57	65	No	**Training**
4	Male	White collar	134	156	Yes	**Testing**
5	Female	White collar	87	87	No	**Training**
6	Male	Blue collar	143	28	Yes	**Training**
7	Female	White collar	150	140	Yes	**Training**
8	Male	Blue collar	56	67	No	**Testing**
9	Female	Blue collar	67	32	No	**Training**
10	Male	Blue collar	75	78	No	**Training**
11	Female	White collar	87	145	Yes	**Training**
12	Male	Blue collar	32	45	No	**Testing**
13	Male	Blue collar	80	90	No	**Training**
14	Female	Blue collar	120	130	Yes	**Training**
15	Female	White collar	40	70	No	**Testing**
16	Male	Blue collar	120	126	Yes	**Training**
17	Female	White collar	130	160	Yes	**Testing**
18	Male	Blue collar	15	62	No	**Training**
19	Female	White collar	77	45	No	**Training**
20	Male	Blue collar	71	51	No	**Training**

sampling. The rightmost column of the table denotes allocation to the two partitions. Roughly 70% of the training instances are assigned to the training partition, leaving about 30% of the instances (six customers) for model evaluation.

The distribution of the target attribute should be analogous into both the training and the testing dataset. Normally, a partitioning based on simple random sampling will not distort the overall distribution of the target attribute. However, for even more accurate representation of all classes in the partitions, a Split validation with proportionate stratified sampling can be applied. With this approach, random samples are drawn independently from each class. The same sample ratios are used for each class, ensuring that a proportionate number of cases from all outcomes is allocated in the both the training and testing data files.

Some analysts split the modeling file into three distinct parts. Alongside the training and the testing files, an additional partition, the validation dataset is drawn. The training file is used for model training. The parameters of the generated model are refined on the validation file. Finally, the fine-tuned model is evaluated on the testing dataset.

@Tech tips for applying the methodology

Split validation in IBM SPSS Modeler

In IBM SPSS Modeler, the Partition node, shown in Figure 2.5, is used for Split validation. Apart from defining the split ratio for the training and the testing files, users can set a random seed to ensure the same partitions on every run.

The Partition node (Figure 2.6) should be placed before the model training node in the Modeler process (stream) so that only training instances are used for modeling, leaving aside the testing instances for evaluation.

Split validation in RapidMiner

In RapidMiner, Split validation is performed with the Split Validation operator shown in Figure 2.7. Users can set the split ratio which defines the relative size of the training set. Additionally, they can set a specific random seed as well as employ a stratified random sampling to ensure an even representation of all classes in the partitions.

The Split Validation operator encompasses two subprocesses (Figure 2.8). The left subprocess corresponds to the training sample and covers the model training phase. Obviously, this subprocess expects a model training operator. The right subprocess corresponds to the testing dataset and covers all the evaluation steps.

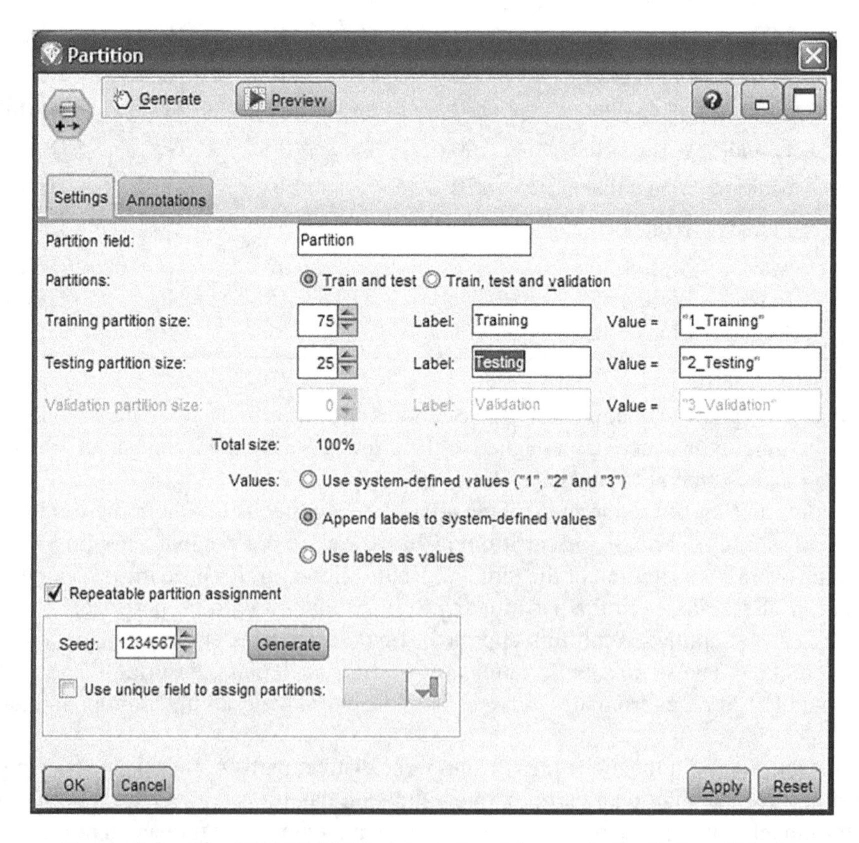

Figure 2.5 The IBM SPSS Modeler's Partition node for Split validation

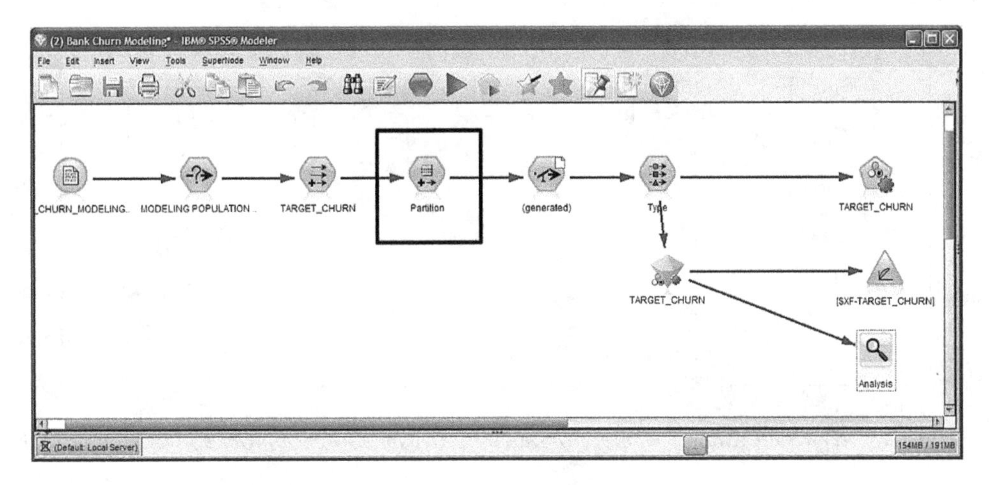

Figure 2.6 A Modeler stream with Split validation

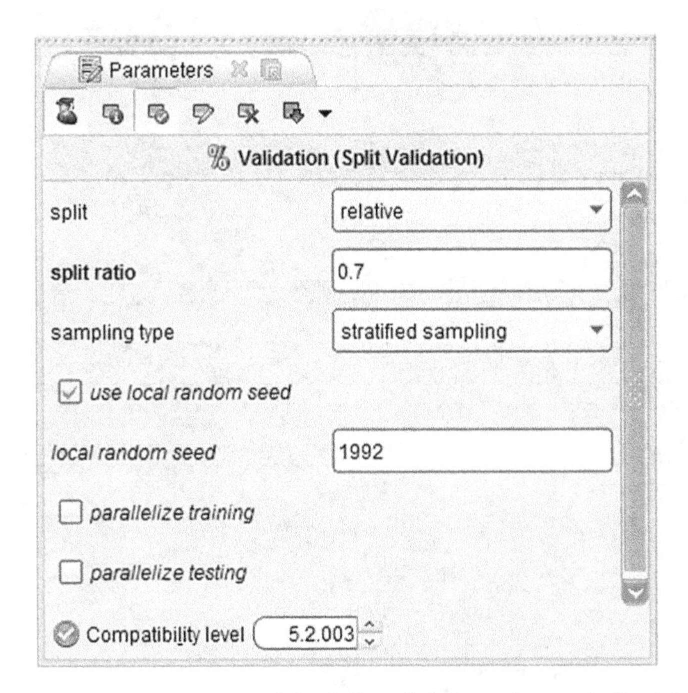

Figure 2.7 The parameters of the Split Validation operator in RapidMiner

Split validation in Data Mining for Excel

In the Data Mining for Excel, the Classify Wizard incorporates a Split validation step in which users can set the proportion of testing cases as shown in Figure 2.9. Along with the model, the training and the test data are stored in the mining structure. The saving of the training dataset allows the addition of new models in the mining structure (through the Add Model to Structure wizard). The test dataset is saved for subsequent validation on the holdout sample through the "test data from model" option in the evaluation wizards.

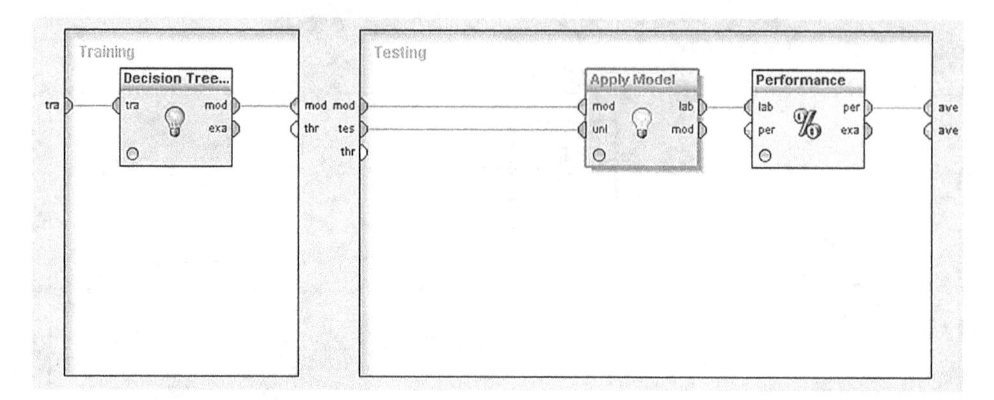

Figure 2.8 The Split Validation operator and the corresponding subprocesses in RapidMiner

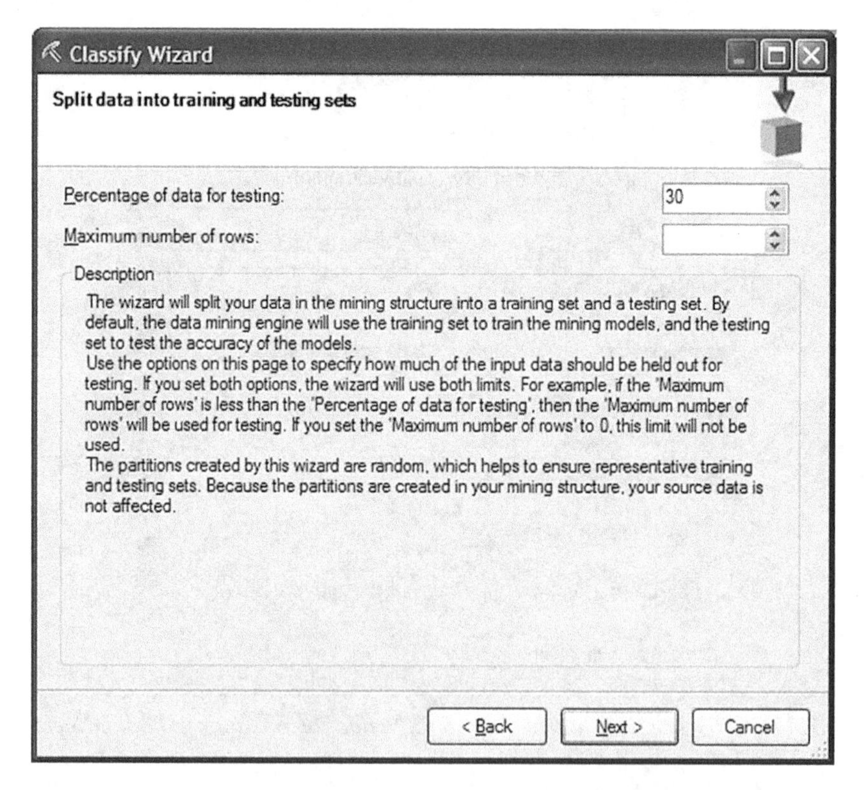

Figure 2.9 The Split step in the Classify Wizard of the Data Mining for Excel

Once model evaluation is done, the model training may be repeated, this time using all available data, including the testing data. This approach, followed by RapidMiner, generates a classifier which has "seen" more training cases and hence has increased possibilities to behave better on unseen cases.

2.3.6.2 Cross or n-fold validation

Although in most data mining applications we normally have enough training instances, there may be situations with data limitations. In such cases, we'd like to use all instances for model training without reserving a part for evaluation. This is what Cross- or n-fold validation does.

Initially, the cases are split into n distinct random subsamples or folds approximately of equal size. The process includes n iterations. In the first iteration, one of the folds is set apart, and the model is trained on the remaining training instances. The generated model is evaluated on the holdout fold. This process is repeated n times. In each iteration, a different fold is hold out for evaluation, and the model is trained on the other $(n-1)$ folds. In the end, all folds have been used $(n-1)$ times for model training and once for model evaluation.

Finally, the number of correct classifications and misclassified cases is counted across all the testing folds, and they are divided by the total number of cases to calculate the accuracy (percentage of correct classifications) and the error rate (percentage of misclassifications) of the full model. In general, the individual evaluation measures calculated on each fold are combined, typically averaged, to assess the performance of the full model.

Cross validation is the preferred validation in cases with data limitations. But it also has another advantage. The model evaluation is not based on a single random subsample as in Split validation. It is repeated n times on n different folds, providing a more reliable and unbiased evaluation of the model.

The standard approach is to use a 10-fold Cross validation. Table 2.3 presents a modeling file of 20 instances on which a fourfold Cross validation has been applied. Each fold contains five customers. In the first iteration, customers 1, 11, 14, and 18 of Fold 1 are hold out for assessing the performance of the classification model trained on Folds 2, 3, and 4. The procedure is repeated for three more times, each time with a different fold holdout for validation.

@Tech tips for applying the methodology

Cross validation in RapidMiner

Cross validation is applied through the X-Validation operator in RapidMiner (Figure 2.10).

Users can specify the number of folds (validations). They can also use stratified cross validation to ensure representative folds in terms of the class distribution.

Similarly to Split validation, the cross validation "nests" two subprocesses, one for model training and one for model testing as shown in Figure 2.11. N models are trained in the left subprocess and evaluated in the right subprocess after being applied to each testing fold separately.

Cross validation in Data Mining for Excel

In the Data Mining for Excel, Cross validation is implemented with the Cross Validation wizard. After a model is trained by the Classify Wizard, it can be selected and loaded to the Cross validation wizard for evaluation. Users can specify the number of folds for partitioning as shown in Figure 2.12. A set of evaluation measures is returned for each fold separately as well as averaged metrics for the full model.

Table 2.3 A modeling dataset with fourfold for Cross Validation

	Input fields				Output field	Cross Validation
Customer ID	Gender	Occupation	Average monthly number of SMS	Average monthly number of voice calls	Response to pilot campaign	Folds
1	Female	Blue collar	28	134	No	**Fold-1**
2	Male	Blue collar	45	54	No	**Fold-2**
3	Female	Blue collar	57	65	No	**Fold-4**
4	Male	White collar	134	156	Yes	**Fold-3**
5	Female	White collar	87	87	No	**Fold-4**
6	Male	Blue collar	143	28	Yes	**Fold-2**
7	Female	White collar	150	140	Yes	**Fold-3**
8	Male	Blue collar	56	67	No	**Fold-4**
9	Female	Blue collar	67	32	No	**Fold-4**
10	Male	Blue collar	75	78	No	**Fold-2**
11	Female	White collar	87	145	Yes	**Fold-1**
12	Male	Blue collar	32	45	No	**Fold-2**
13	Male	Blue collar	80	90	No	**Fold-2**
14	Female	Blue collar	120	130	Yes	**Fold-1**
15	Female	White collar	40	70	No	**Fold-3**
16	Male	Blue collar	120	126	Yes	**Fold-4**
17	Female	White collar	130	160	Yes	**Fold-3**
18	Male	Blue collar	15	62	No	**Fold-1**
19	Female	White collar	77	45	No	**Fold-3**
20	Male	Blue collar	71	51	No	**Fold-4**

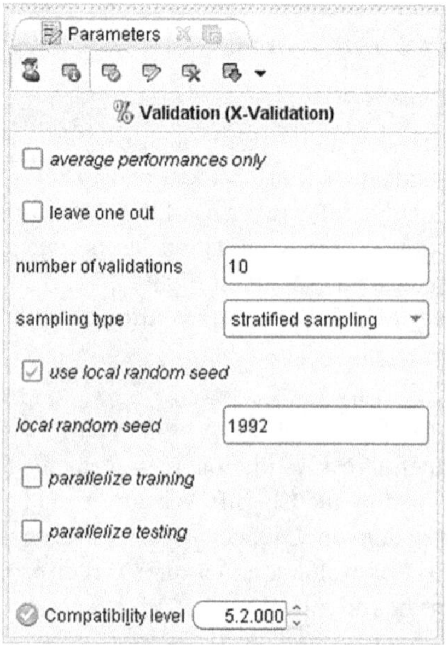

Parameters
Validation (X-Validation)
average performances only
leave one out
number of validations 10
sampling type stratified sampling
use local random seed
local random seed 1992
parallelize training
parallelize testing
Compatibility level 5.2.000

Figure 2.10 The RapidMiner Cross Validation operator

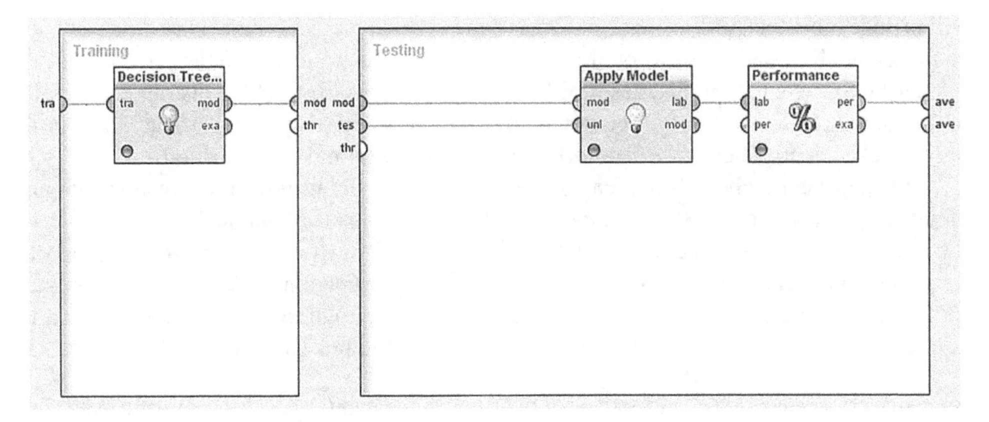

Figure 2.11 The subprocesses created by the RapidMiner Cross-Validation operator

Cross - Validation

Specify Cross-Validation Parameters

Fold Count: 10

Maximum Rows: 0

Target Attribute: CAMPAIGN_RESPONSE

Target State: T

Target Threshold:

Description

This task performs cross-validation on the selected models from the mining structure 'XSELL Structure.'
The task generates a report which describes how well the models predict the attribute CAMPAIGN_RESPONSE. For certain models and attributes you can also choose a specific state to be tested for this attribute. In addition, you can choose a threshold probability value at which the model should consider a prediction to be correct.

If the Maximum Rows property is set to 0, cross-validation uses all the rows in the data set. Therefore, the time needed to perform cross-validation might increase significantly.

< Back Finish Cancel

Figure 2.12 The Cross validation wizard in the Data Mining for Excel

2.3.6.3 Bootstrap validation

Bootstrap validation uses sampling with replacement; hence, a training case can be included more than once in the resulting training file.

More specifically, the initial modeling file of size n is sampled n times with replacement to give a training file of size n. Due to the sampling with replacement, the bootstrap sample will include more than one instance of some of the original n cases, while other cases will not be picked at all. In fact, it turns out (and it can be shown if we calculate the probabilities of selection) that with the method described here, referred to as the 0.632 bootstrap, on average 63.2% of the original cases, will be included in the bootstrap sample and around 36.8% will not be picked at all. These cases comprise the testing file for model evaluation.

Although the number of total cases of the resulting training file (the bootstrap sample) remains equal to the original size n, the fact is that it contains less "unique" cases compared, for instance, with a 10-fold cross validation training file. Using only the bootstrap testing file for evaluation would lead to pessimistic measures of classification accuracy. To compensate for that, the estimated error rate (misclassification percentage) of the full model is also taking into account the training cases of the bootstrap sample, and it is calculated as follows:

$$\text{Bootstrap error rate} = 0.368^{*}\left(\text{Error rate of bootstrap sample}\right) + 0.632\left(\text{Error rate of testing file}\right)$$

The overall bootstrap procedure can be repeated for k times and the results can be averaged over iterations. The Bootstrap validation method works well with very small datasets.

2.3.7 Dealing with imbalanced and rare outcomes

II. Data understanding, preparation, and enrichment	***II.7. Dealing with imbalanced outcomes***

Most classification algorithms work better when analyzing data with a balanced outcome distribution. However, in real-world applications, the distributions of the output classes differ substantially, and the target class is rare and underrepresented. Ideally, we'd prefer a percentage of about 20–50% for the target class and nonetheless above 10%, but this is rarely the case. Two widely used methods to tackle this issue are the balancing of the distribution of the target classes and the application of case weights to the training file. The "adjustment" of class distribution helps the classifier to discover hot spots consisting of very high concentrations of the target class.

Apart from the distribution, there may be an issue with the target class frequency. Ideally, we'd like sufficient instances of the target class, at least a few hundreds; if this is not the case, a simple extension of the event outcome period might prove sufficient to improve the situation. For example, by increasing the span of the event outcome period to 6 instead of 3 months, we can capture more actual churners for our target class.

2.3.7.1 Balancing

Balancing is the application of disproportionate stratified sampling on the training file to adjust class imbalances. Balancing changes the distribution of the training file. The frequent classes, typically denoting nonmembers of the target population, are undersampled. A random percentage of these instances is included in the balanced training file by applying a sample ratio, also referred to as balance factor, less than 1.0. Therefore, the frequency of the common cases is reduced. In the end, the training file will include all target class cases and a fraction of the other cases. Additionally or alternatively, the density of the rare class can be "inflated" by oversampling with a sample ratio higher than 1.0. Target class cases are resampled and may be included in the training file with more than one instance.

Either way, the rare class is "boosted in the balanced training file and its frequency is increased. Therefore, models should be evaluated on unbalanced data. Additionally, when scoring with a model trained on balanced data, we should be cautious with the generated scores. When a model is applied to new cases, along with the class prediction, it estimates its confidence and hence the propensity, the likelihood of the target outcome. If balanced data are used for training, the resulting propensities do not correspond to the actual probabilities of belonging to the target class. However, propensities are comparable and can be used to rank customers according to their likelihood of belonging to the target class.

The calculated probabilities for a binary target field can be adjusted for oversampling using the formula below:

$$\hat{p}_i = \frac{\hat{p}*_i}{\hat{p}*_i} + (1 - \hat{p}*_i)\left[\frac{(1-\pi_i)\rho_i}{\pi_i(1-\rho_i)}\right]$$

where $\hat{p}_i$ is the adjusted probability estimate of class i, $\hat{p}*_i$ the unadjusted probability estimate, π_i the actual population proportion of class i in the original training dataset, and ρ_i the observed proportion of class i in the balanced sample. The adjusted probabilities can be used as estimates of the actual class probabilities.

IBM SPSS Modeler offers an interesting solution to this issue. A generated model estimates adjusted propensities. These propensities are based on the unbalanced testing file (partition), and they are not affected by the balance.

The recommended balancing approach is to reduce the common outcomes through undersampling instead of boosting the rare ones with oversampling. The scope is to attain a target class percentage of about 25–50%.

Table 2.4 presents a class-imbalanced dataset for cross-selling. Three out of 30 customers have responded to the pilot campaign and were classified in the target class. Since the target class percentage is a poor 10%, the file was adjusted with balancing.

The training file after balancing is listed in Table 2.5. A balance factor of 1.0 was used for the target class, and all responders were retained. On the other hand, nonresponders were undersampled with a sample ratio of 0.26, and a random sample of seven nonresponders was selected for model training. After balancing, the proportion of the target class has been raised to 30% (3 out of 10).

@Tech tips for applying the methodology

Balancing in IBM SPSS Modeler

The Balance node, shown in Figure 2.13, is used in IBM SPSS Modeler for adjusting the distribution with over-/undersampling.

It should be placed in the stream process before a model training node and after a validation node as shown in Figure 2.14.

Users can set the balance factor for each class, which is the corresponding sample ratio. Since the node incorporates random sampling, it is advisable to temporarily save the balanced file through a Cache node. Users should also request to apply the balancing only on the training partition, leaving the testing partition unchanged for validation. Balancing is also available through the Sample node by using stratified sampling within each class.

Table 2.4 A class-imbalanced modeling file for cross-selling

	Input fields				Output field
Customer ID	Gender	Occupation	Average monthly number of SMS	Average monthly number of voice calls	Response to pilot campaign
1	Male	White collar	28	140	No
2	Male	Blue collar	32	54	No
3	Female	Blue collar	57	30	No
4	**Male**	**White collar**	**143**	**140**	**Yes**
5	Female	White collar	87	81	No
6	Male	Blue collar	143	28	No
7	Female	White collar	150	140	No
8	Male	Blue collar	15	60	No
9	Female	Blue collar	85	32	No
10	Male	Blue collar	75	32	No
11	Female	White collar	42	140	No
12	Male	Blue collar	32	62	No
13	Female	Blue collar	80	20	No
14	**Female**	**White collar**	**120**	**130**	**Yes**
15	Female	White collar	40	70	No
16	Male	Blue collar	120	30	No
17	Female	White collar	130	95	No
18	Male	Blue collar	15	62	No
19	Female	White collar	78	45	No
20	Male	Blue collar	71	51	No
21	Male	Blue collar	20	15	No
22	Male	White collar	62	52	No
23	Male	Blue collar	72	52	No
24	Female	Blue collar	70	50	No
25	**Female**	**Blue collar**	**90**	**110**	**Yes**
26	Female	White collar	40	30	No
27	Male	Blue collar	30	20	No
28	Female	Blue collar	80	40	No
29	Male	Blue collar	75	68	No
30	Female	White collar	63	43	No

Balancing in RapidMiner

Balancing, with the application of stratified sampling, is available in RapidMiner though the Sample operator. As shown in Figure 2.15, users can set the sample fraction for each class of the target attribute as well as a random seed.

Balancing in RapidMiner should be placed in the training subprocess of the Cross or the Split Validation operator to cover only the training partition(s), leaving the testing partition(s) unaffected (Figure 2.16). In n-fold Cross validation, the n-folds are sampled as usual. In each

Table 2.5 The balanced modeling file

	Input fields				Output field
Customer ID	Gender	Occupation	Average monthly number of SMS	Average monthly number of voice calls	Response to pilot campaign
3	Female	Blue collar	57	30	No
4	**Male**	**White collar**	**143**	**140**	**Yes**
8	Male	Blue collar	15	60	No
13	Female	Blue collar	80	20	No
14	**Female**	**White collar**	**120**	**130**	**Yes**
19	Female	White collar	78	45	No
23	Male	Blue collar	72	52	No
25	**Female**	**Blue collar**	**90**	**110**	**Yes**
28	Female	Blue collar	80	40	No
29	Male	Blue collar	75	68	No

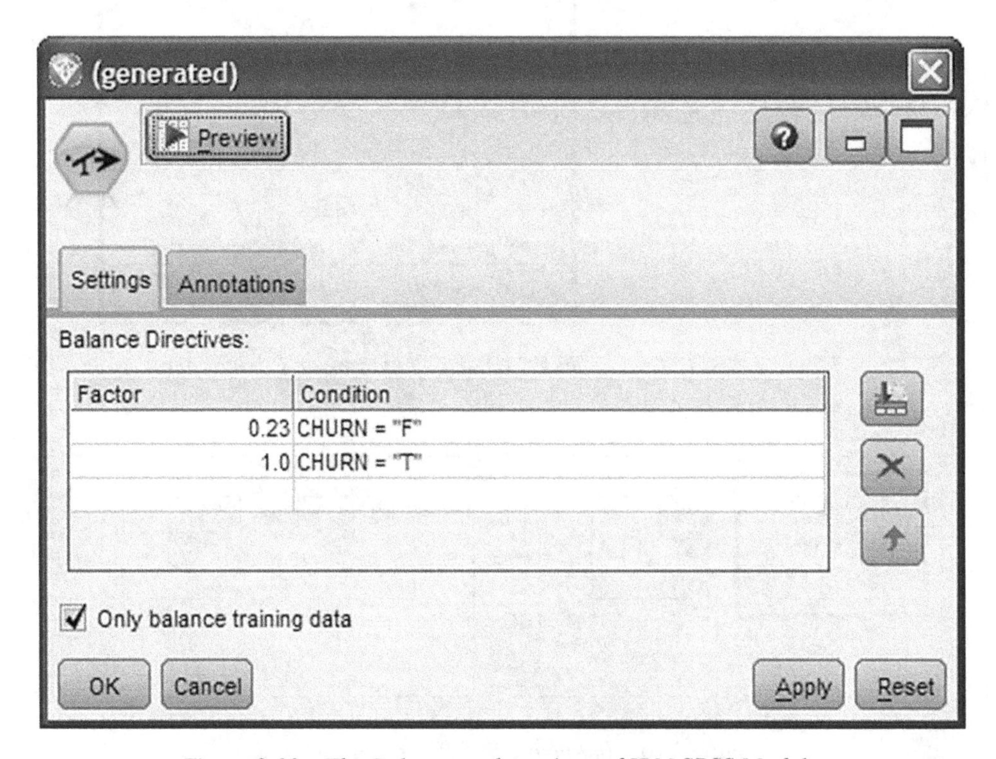

Figure 2.13 The Balance node options of IBM SPSS Modeler

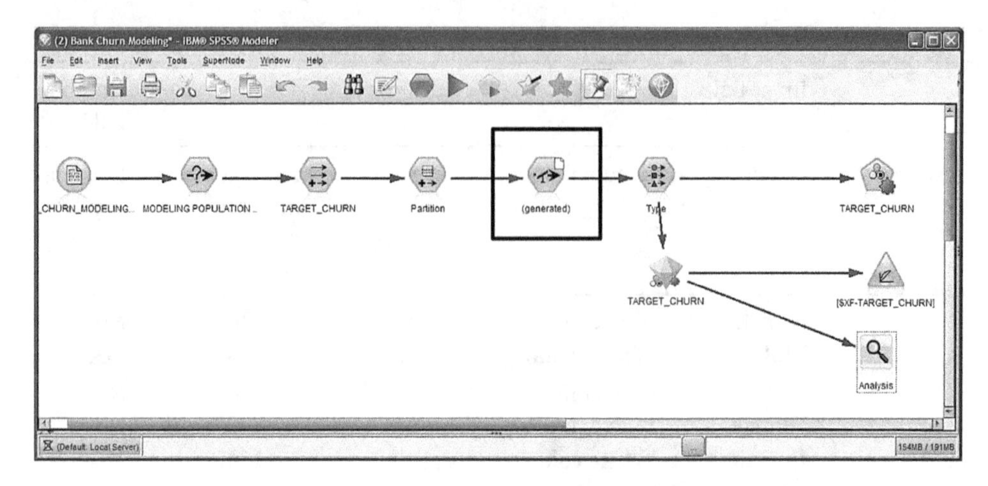

Figure 2.14 The Balance node in a Modeler stream

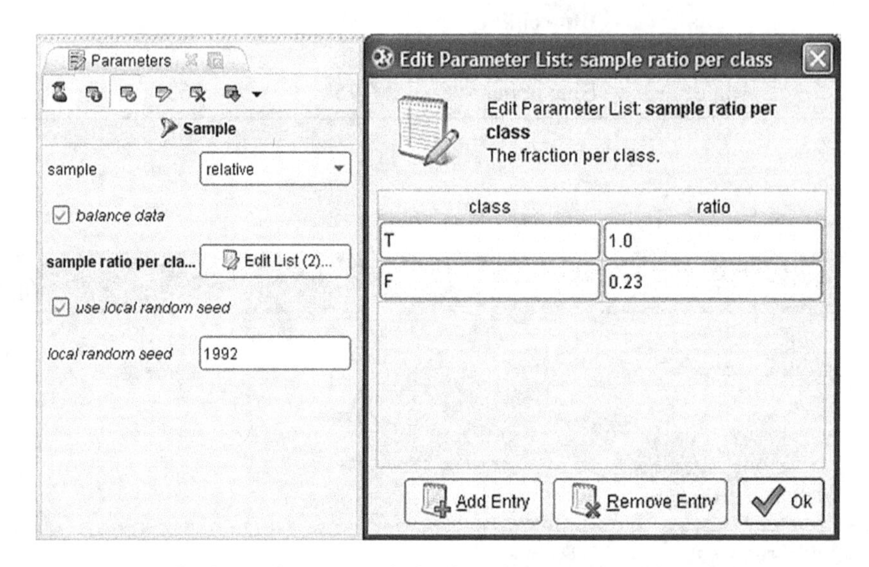

Figure 2.15 Balancing with the Sample operator of RapidMiner

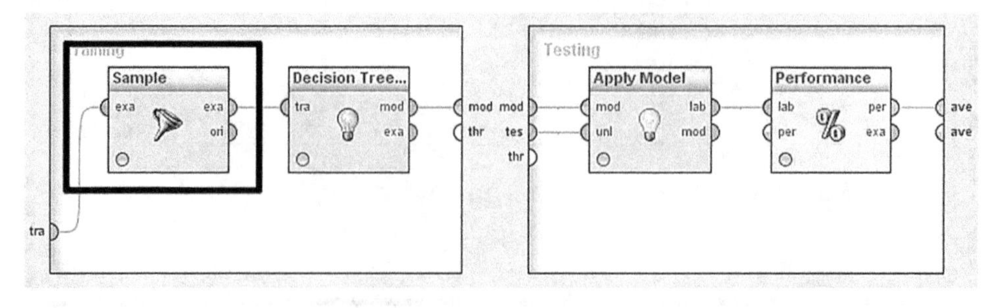

Figure 2.16 Balancing in a RapidMiner process

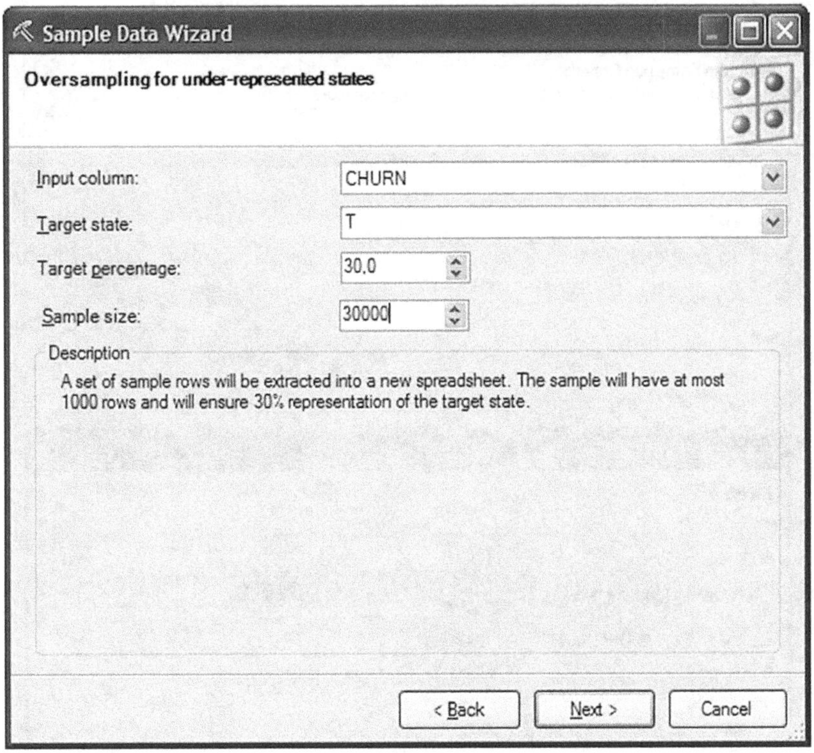

Figure 2.17 Balancing in Data Mining for Excel

iteration, the testing fold is used unbalanced for model evaluation, while the training cases are balanced for model training.

Balancing in Data Mining for Excel

In the Data Mining for Excel, balancing can be applied with the Sample Data wizard as shown in Figure 2.17.

Users specify the source data, the target attribute and class ("state"), and the target percentage, which is the desired proportion of the rare class in the result balanced dataset. Then, it's just a matter of undersampling the frequent category.

For balancing only the training dataset, initially the modeling dataset is partitioned with the Sample wizard into training and testing parts of designated size as shown in Figure 2.18. Then, balancing is applied only on the training dataset to adjust the proportion of the under-represented class, leaving the testing dataset unbalanced for evaluation.

2.3.7.2 Applying class weights

An alternative approach to deal with class imbalances is with the application of class weights. The contribution of each case to the model training is defined by its weighting factor. This method has the advantage of using all of the available training data, provided of course that

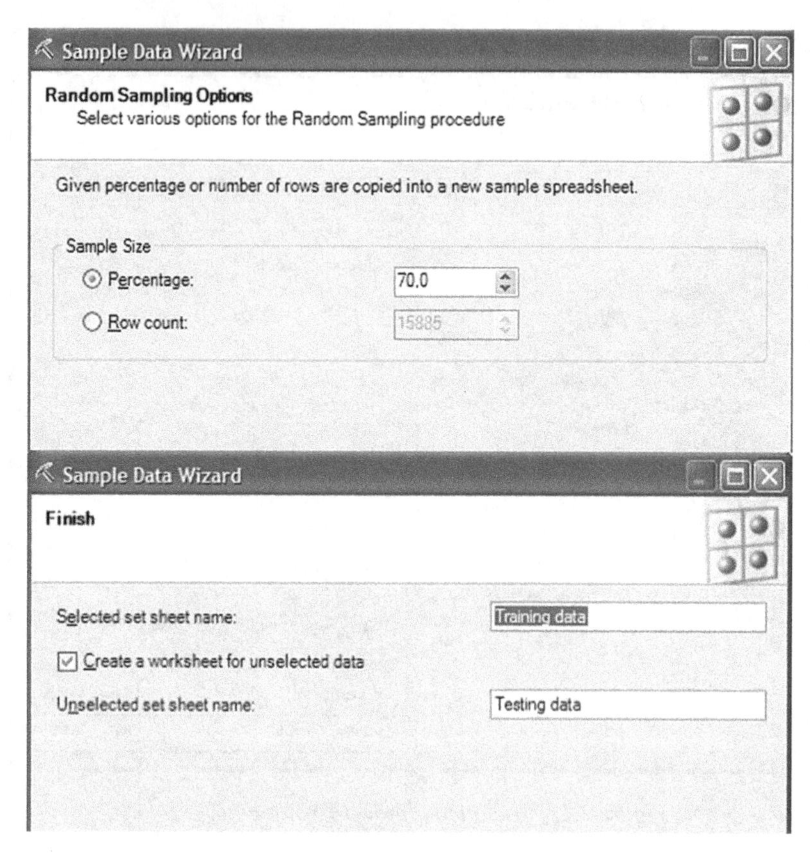

Figure 2.18 Using the Sample wizard in Data Mining for Excel for Split validation and for balancing only the training file

the modeling algorithm supports case weights. In the end, estimated propensities should be adjusted to correspond to the actual probabilities of the target class.

A recommended approach is to assign a weight of 1.0 to the rare class and a weight lower than 1.0 to the frequent class to give extra importance to the underrepresented cases. This weighting is shown in Table 2.6. A weighting factor of 1.0 is assigned to responders of the pilot cross-selling campaign and a weight of 0.26 to the 27 nonresponders, adjusting the distribution of the target class to 30–70%.

Class weights should be ignored in the model evaluation procedure, and the model should be tested on unweighted data.

@Tech tips for applying the methodology

Class weighting in IBM SPSS Modeler

C&RT, CHAID, and C5.0 classifiers support class weights in IBM SPSS Modeler. The weight field can be specified in the Fields tab of the respective modeling nodes. Analysis weight can also be defined in the Fields tab of the Auto Classifier node as shown in Figure 2.19. In this case, case weights are used only for model building and are not considered when evaluating or scoring models.

Table 2.6 A class-imbalanced modeling file with weights

	Input fields				Output field	
Customer ID	Gender	Occupation	Average monthly number of SMS	Average monthly number of Voice calls	Response to pilot campaign	Weight
1	Male	White collar	28	140	No	0.26
2	Male	Blue collar	32	54	No	0.26
3	Female	Blue collar	57	30	No	0.26
4	**Male**	**White collar**	**143**	**140**	**Yes**	**1.0**
5	Female	White collar	87	81	No	0.26
6	Male	Blue collar	143	28	No	0.26
7	Female	White collar	150	140	No	0.26
8	Male	Blue collar	15	60	No	0.26
9	Female	Blue collar	85	32	No	0.26
10	Male	Blue collar	75	32	No	0.26
11	Female	White collar	42	140	No	0.26
12	Male	Blue collar	32	62	No	0.26
13	Female	Blue collar	80	20	No	0.26
14	**Female**	**White collar**	**120**	**130**	**Yes**	**1.0**
15	Female	White collar	40	70	No	0.26
16	Male	Blue collar	120	30	No	0.26
17	Female	White collar	130	95	No	0.26
18	Male	Blue collar	15	62	No	0.26
19	Female	White collar	78	45	No	0.26
20	Male	Blue collar	71	51	No	0.26
21	Male	Blue collar	20	15	No	0.26
22	Male	White collar	62	52	No	0.26
23	Male	Blue collar	72	52	No	0.26
24	Female	Blue collar	70	50	No	0.26
25	**Female**	**Blue collar**	**90**	**110**	**Yes**	**1.0**
26	Female	White collar	40	30	No	0.26
27	Male	Blue collar	30	20	No	0.26
28	Female	Blue collar	80	40	No	0.26
29	Male	Blue collar	75	68	No	0.26
30	Female	White collar	63	43	No	0.26

Class weighting in RapidMiner

In RapidMiner, the Generate Weight (Stratification) operator, shown in Figure 2.20, can be used before a learner operator for assigning class weights. Infrequent classes will get higher weights than dominant ones. Weights are distributed over the training instances so that per label weights sum up equally to the specified total weight.

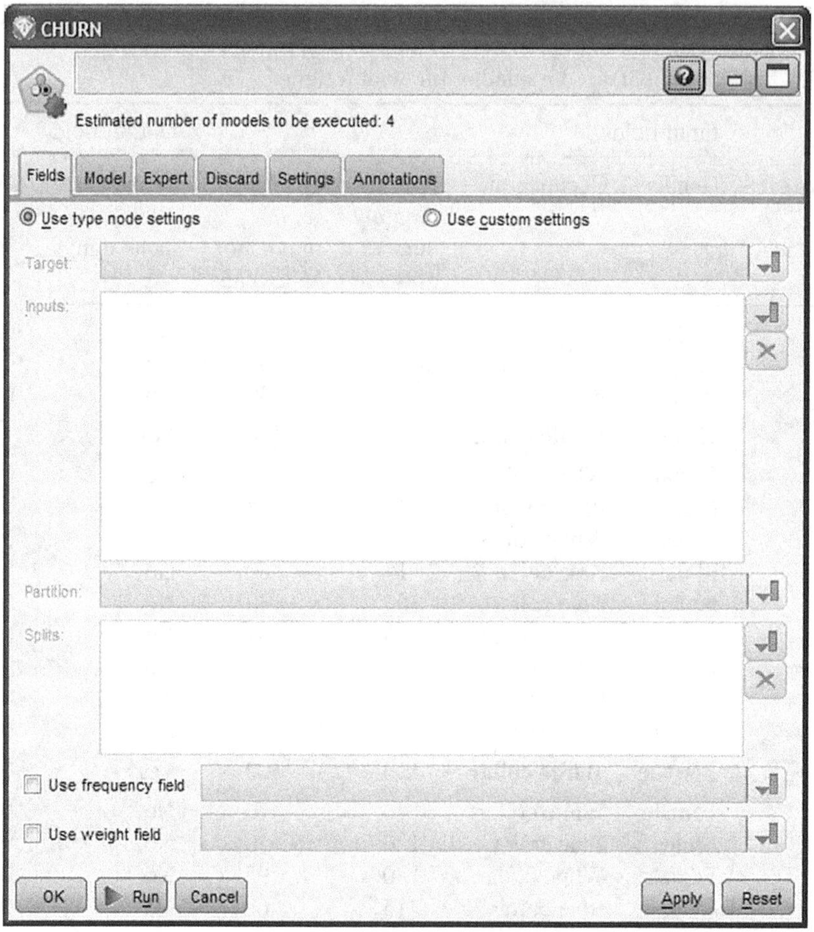

CHURN
Estimated number of models to be executed: 4
Fields
Model
Expert
Discard
Settings
Annotations
Use type node settings
Use custom settings
Target:
Inputs:
Partition:
Splits:
Use frequency field
Use weight field
OK
Run
Cancel
Apply
Reset

Figure 2.19 Applying class weight in IBM SPSS Modeler Decision Trees

Parameters
Generate Weight (Stratification)
total weight
1.0

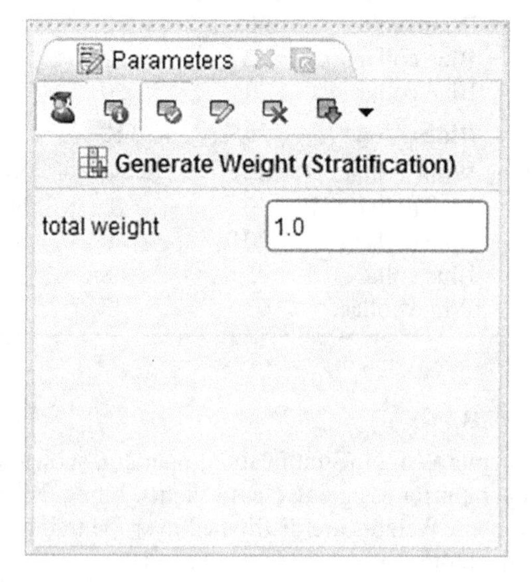

Figure 2.20 The Generate Weight (Stratification) used in RapidMiner for class weighting

Specifying a total weight of 6, twice the number of responders, for the file of Table 2.6, is equivalent to designating a weighting factor of about 1.0 for the target class and a weighting factor of about 0.11 for the frequent class.

Alternatively, users can construct a case weight attribute, for instance, through the Generate Attributes operator; assign the desired weights to each class; and then set a weight role to the derived attribute through a Set Role operator. In the case of holdout validation, the case weighting should be applied in the training subprocess of the Split Validation operator to cover only the training partition, leaving the testing partition unaffected.

2.4 Classification modeling

Modeling is a trial-and-error phase which involves experimentation. A number of different models should be trained and examined. Additionally, in a procedure called meta-modeling, multiple models can be combined to improve the classification accuracy. The initial examination of the derived models is followed by a formal and thorough evaluation of their predictive accuracy.

2.4.1 Trying different models and parameter settings

III. Classification modeling	***III.1. Trying different models and parameter settings***

Each attribute has a specific role in the training of the classifier. A set of selected attributes should be set as inputs for predicting the target attribute. Note that the target attribute is also referred to as the label or the output attribute. Then, at last, it is time for model training. Multiple classifiers with different parameter settings should be trained and examined before selecting the one for deployment.

The role of each attribute in IBM SPSS Modeler is specified with a Type node which should precede the model training node. Similarly, a Set Role operator can be used in RapidMiner to assign roles to attributes. In Data Mining for Excel, the attribute roles are specified in one of the steps of the Classify Wizard.

@Tech tips for applying the methodology

Classification modeling in IBM SPSS Modeler

The role of each attribute in IBM SPSS Modeler is specified with a Type node which should precede the model training node. The Auto Classifier node, shown in Figures 2.21 and 2.22, allows the simultaneous training and initial evaluation of multiple learners of different types. It also enables the estimation of combined predictions with an ensemble procedure which will be presented shortly.

Classification modeling in Data Mining for Excel

In Data Mining for Excel, the attribute roles are specified in an initial step of the Classify Wizard. In a subsequent step, users are asked to select the appropriate algorithm as shown in Figure 2.23.

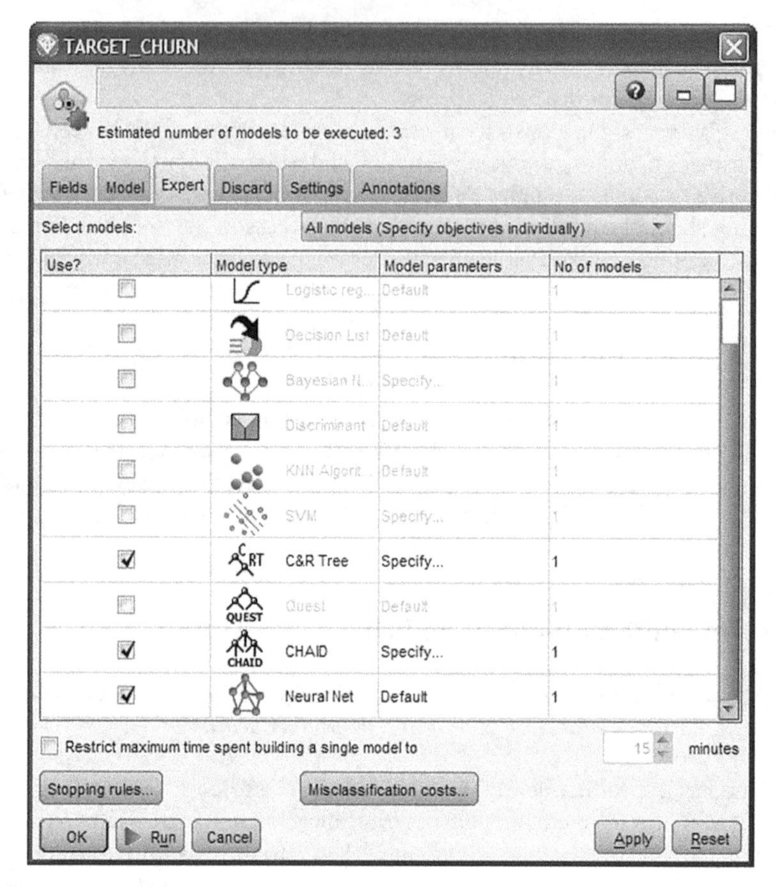

Figure 2.21 The Auto Classifier node of IBM SPSS Modeler for simultaneous training of multiple learners

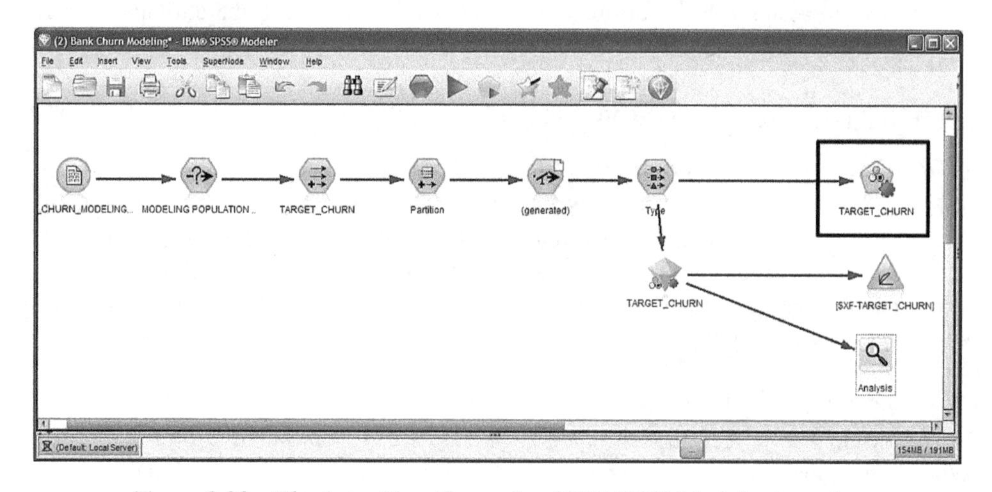

Figure 2.22 The Auto Classifier node of IBM SPSS Modeler in action

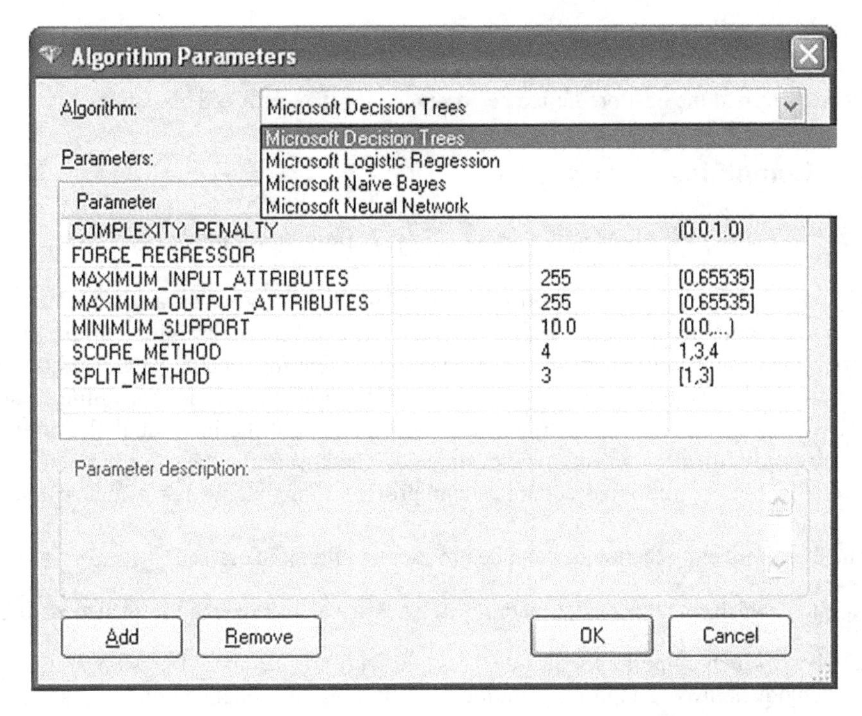

Figure 2.23 Selecting the classification algorithm and setting its parameters in Excel

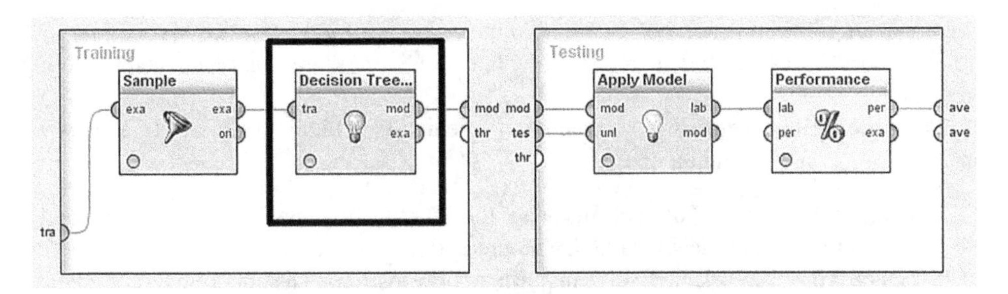

Figure 2.24 Training a classifier with RapidMiner

Classification modeling in RapidMiner

In RapidMiner, a Set Role operator can be used for assigning roles to the model attributes. The tool offers numerous operators/algorithms for the training of an individual model (Figure 2.24), as well as the Compare ROCs operator which enables the simultaneous training of multiple models and their comparison with receiver operating characteristic (ROC) curves (to be discussed later in this chapter). Additionally, the Vote operator builds a set of models and combines their predictions with a majority voting procedure (to be discussed later in this chapter as well).

Besides trying different types of models, analysts should also typically experiment with different parameter values for each model. Apart from the traditional manual setting and testing of parameters, RapidMiner offers an operator named Optimize Parameters (Grid) for optimization of parameters. This operator, through iterative training, finds and employs the optimal values for a set of parameters using a grid or list search. The best parameter values are those that yield the best predictive performance for the examined classifier.

2.4.2 Combining models

III. Classification modeling	***III.2. Combining models***

Individual models can be combined to improve the predictive accuracy. This approach, referred to as meta-modeling or ensemble modeling, involves the training of multiple base classifiers and the development of a composite classifier for scoring new cases. Imagine the different models as experts who vote for a decision. Instead of consulting a single expert, we might select to take into account the decisions of the individual classifiers and pool their predictions for scoring new customers. The ensemble classifier can often attain substantially better predictive accuracy and present more stable performance than the original models.

The meta-modeling techniques can be grouped in two main classes:

1. Those which combine models of different types, but all based on the original modeling file.

2. Those which combine models of the same type, for instance, Decision Trees. These techniques involve multiple iterations and training of models on different samples of the modeling file. Bagging, Boosting and Random Forests are meta-modeling techniques of this type.

The application of the first type of ensemble modeling is straightforward. Multiple models are developed as usual, and then the individual predictions are combined with a voting procedure.

There are various methods of combining the predictions of individual models. Below, we list some of the most common ones:

1. **(Majority) Voting**. For each instance, the number of times each class is predicted across the base models is tallied. The prediction of the ensemble model for each instance is the class selected by the majority of base learners. The prediction's confidence, that is, the likelihood of prediction, is calculated as the percentage of base models which returned that prediction. For example, if 2 out of 3 models predict *no* for a given instance, then the final prediction for this instance is *no* with 66.7% (2/3) confidence. This voting approach is incorporated by RapidMiner's Vote operator.

2. **Highest confidence**. This method selects as the prediction for each instance the class which presents the single highest confidence across all base models.

3. **Average propensity** (for two-class problems). The class with the highest value when each class' propensities (likelihoods) are averaged across base models is selected as the prediction.

4. **Confidence-/propensity weighted voting**. With simple voting, if two out of three models predict *yes*, then *yes* wins by a vote of two to one. In the case of confidence-weighted voting, the votes are weighted based on the confidence value of each prediction. Thus,

if one model predicts *no* with a higher confidence than the two *yes* predictions summed, then *no* wins. In other words, the *yes* votes, the confidences of models predicting *yes*, are summed and compared with the sum of votes (confidences) of models predicting *no*. The predicted class for each instance is the class with the highest total votes. The sum of weights (confidences) divided by the total number of base learners is the prediction's confidence.

@Tech tips for applying the methodology

Ensemble modeling in IBM SPSS Modeler

The Ensemble node of IBM SPSS Modeler, as well as the Auto Classifier node presented in Section 2.4.1, can be used for combining classifiers of different types. The individual models are combined and the ensemble model's predictions and confidences/propensities are generated.

Ensemble modeling in RapidMiner

RapidMiner offers the Vote operator for combining base learners into a compound classifier.

2.4.2.1 Bagging

With Bagging (Bootstrap Aggregation), multiple models of the same type are trained on different replicates of the original training file. A set of n bags, n bootstrap samples, typically of the same size as the original dataset, are drawn with replacement. Due to the resampling applied to the original training data, some training instances may be omitted, while others replicated. This technique involves n iterations. In each iteration, a different model is built on each bag. The set of n models generated are combined and vote to score new cases.

Bagging can improve the classification accuracy compared to individual models (although not as dramatically as Boosting which is presented immediately after). Additionally, since the ensemble model is based on more than one datasets, which unfortunately are not independent, it can be more robust to noisy data and hence more reliable and stable.

@Tech tips for applying the methodology

Bagging in IBM SPSS Modeler

In IBM SPSS Modeler, the Bagging technique can be requested in the Build options of the Decision Trees and Neural Network model nodes. Additionally, users can select a specific voting procedure for scoring (among majority voting, highest confidence, average propensity) in the Ensembles options.

Bagging in RapidMiner

In RapidMiner, Bagging is available through the dedicated Bagging operator as shown in Figure 2.25 which encloses an inner model training operator. Users can set the number of iterations and hence the number of individual models as well as the sample ratio for sampling.

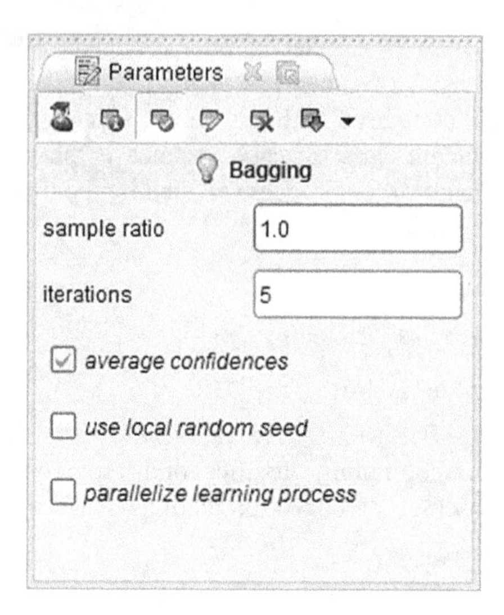

Figure 2.25 The Bagging operator in RapidMiner

2.4.2.2 Boosting

Boosting involves the training of a sequence of models of the same type which complement one another. In the first iteration, a classifier is built the usual way. The second model though depends on the first one as it focuses on the instances which the first model failed to classify correctly. This procedure is continued for *n* iterations with each model aiming at the "hard cases" of the previous models.

Adaptive Boosting (AdaBoost), a popular Boosting technique, works as follows: Each record is assigned a weight denoting its "hardness" to be classified. In the first iteration, all records are assigned equal weights. After the first model run, these weights are adjusted so that misclassified instances yield higher weights and correctly classified ones lower weights. The subsequent base model is trained on the weighted instances. Assuming the base learner supports case weights, the contribution of each case to the model training is defined by its weight. That's why each base model starts from where the previous models stopped.

Even in the case of a type of classifier which does not support case weights, the AdaBoost technique can be employed by using samples with replacement for each iteration. On the contrary to Bagging though, the inclusion probability of each instance should be proportional to its estimated weight.

A weighted voting procedure is finally applied for scoring with the boosted model. The weight of the vote of each base model depends on its error rate. Hence, more accurate classifiers have stronger influence on the prediction. The Boosting can substantially improve predictive accuracy; however, it is demanding in terms of training time and prone to overfitting.

@Tech tips for applying the methodology

Boosting in IBM SPSS Modeler

In IBM SPSS Modeler, the Boosting technique can be requested in the Build options of the Decision Trees and Neural Network model nodes as shown in Figure 2.26.

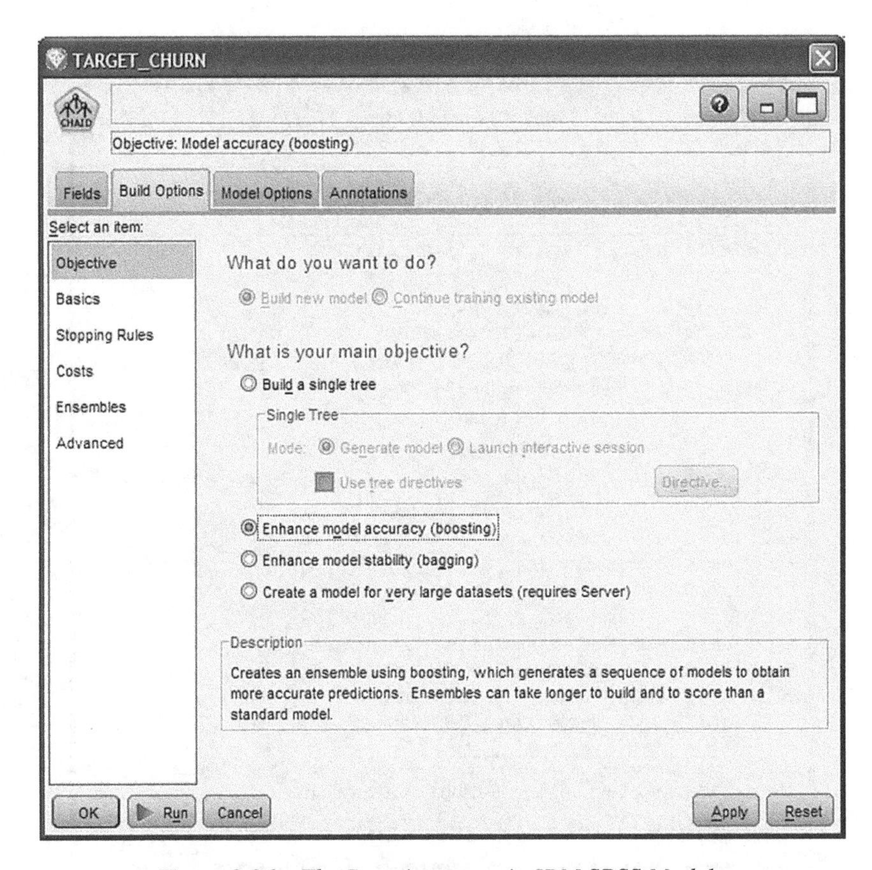

Figure 2.26 The Boosting menu in IBM SPSS Modeler

Figure 2.27 presents the 10 base Decision Tree models, in rule format, which were generated by Boosting with 10 iterations. The ensemble of the 10 base models can then be used for scoring.

Boosting in RapidMiner

In RapidMiner, the AdaBoost operator, which encompasses a model training operator, is available for Boosting.

2.4.2.3 Random Forests

Random Forests use Bagging and bootstrap samples to generate *n* different Decision Trees in n iterations. The Decision Tree models differ not only on their training instances but also on the used predictors. A random subset of predictors is considered for partition at each node. The predictor for the best split is chosen, according to the tree's attribute selection method, on the random list of attributes. A voting procedure is applied for scoring with the ensemble classifier.

Random Forests are faster than Bagging and Boosting and can improve the classification accuracy, especially if the different tree models are diverse and not highly correlated. This can be achieved by keeping the subset ratio relatively low.

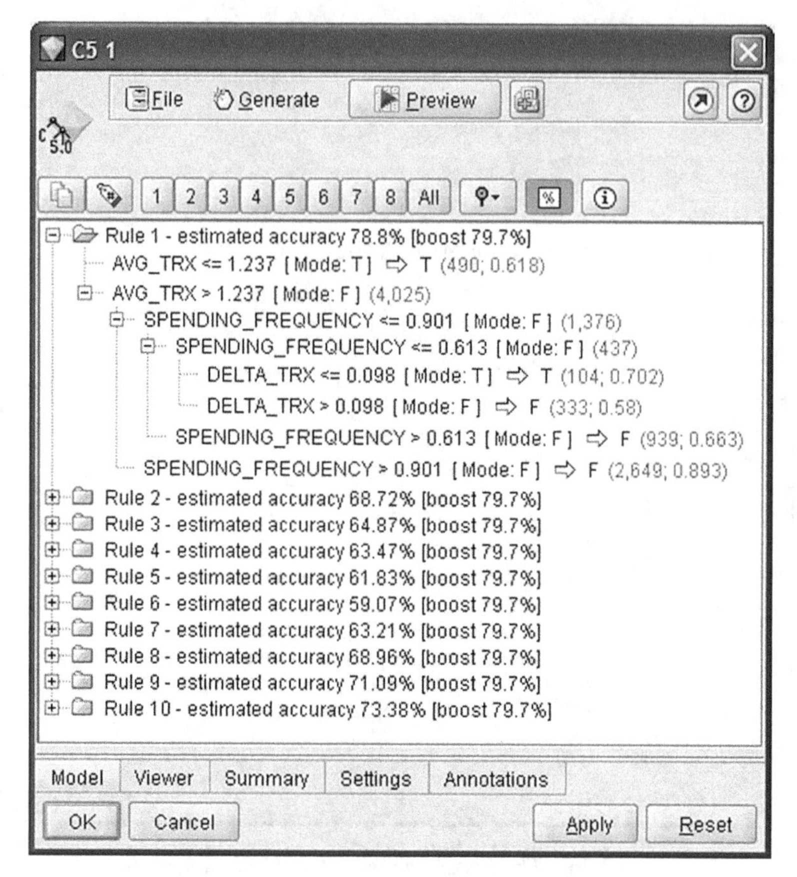

Figure 2.27 The base Decision Tree models generated by Boosting with 10 iterations

@Tech tips for applying the methodology

Random Forests in RapidMiner

A screenshot of the Random Forest operator of RapidMiner is shown in Figure 2.28. Users can specify the number of Decision Trees to be trained as well as the subset ratio of predictors to be considered for the splits. If a specific subset ratio value is not set, a value of $\log(m)+1$ is used where m is the number of the training instances.

2.5 Model evaluation

Before deploying the classifier on unseen cases, its predictive accuracy should be thoroughly evaluated. In general, the assessment of the predictive ability of the model is based on the comparison between the actual and the predicted event outcome. The model evaluation is carried out in two phases: before and after the rollout of the respective campaign.

The precampaign model validation typically includes a series of metrics and evaluation graphs which estimate the model's future predictive accuracy in unseen cases, for example,

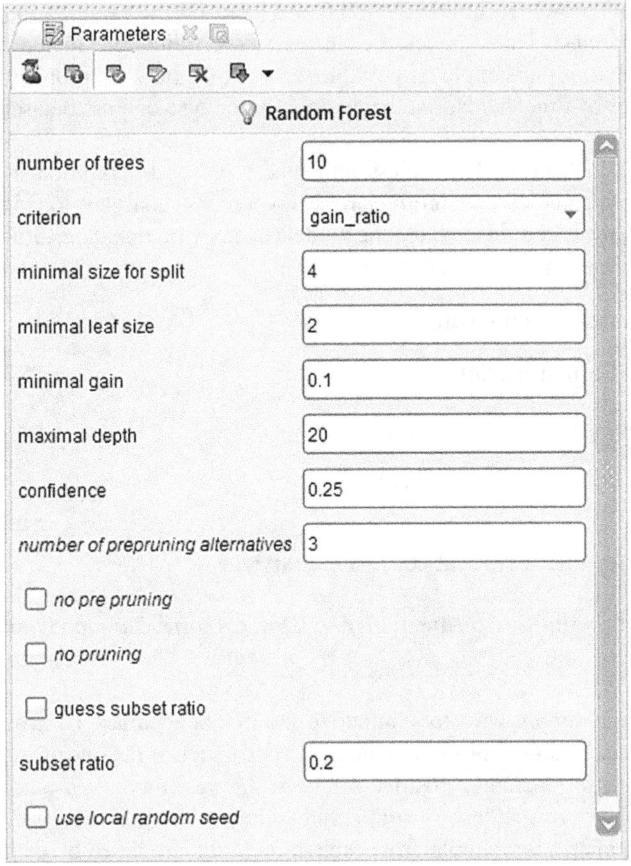

Figure 2.28 Random Forests in RapidMiner

in the holdout testing dataset. After the direct marketing campaign is executed, the deployed model's actual performance and the campaign design itself are evaluated with test–control groups. In Section 2.5.1, we focus on the precampaign evaluation.

2.5.1 Thorough evaluation of the model accuracy

IV. Model evaluation	***IV.1.* Thorough evaluation of the model accuracy**

In the precampaign evaluation, preclassified cases are scored and two new fields are derived: the predicted class and the prediction's confidence which denotes the likelihood of the prediction. In practice, all models make mistakes. There are always errors and misclassifications. The comparison of the predicted with the actual values is the first step for evaluating the model's performance.

The model should not be validated on the training instances since this would lead to optimistic assessment of its accuracy. Therefore, analysts should always employ a validation technique as those presented in Section 2.3.6 and focus on the examination of performance in the testing dataset(s). This is called the out-of-sample validation since it is based on cases not present in the training sample. If possible, an out-of-time validation should also be performed. In the out-of-time validation, the model is tested on a disjoint dataset from a different time period.

As discussed in Section 2.3.7, class imbalances are quite common in actual business problems. Balancing and case weighting are two common techniques to tackle this situation. In such cases, analysts should evaluate the model accuracy on unbalanced/unweighted cases.

Common evaluation techniques include:

- Confusion matrix and accuracy measures
- Gains/Lift/Response chart
- ROC curves
- Return on investment (ROI) curves

2.5.1.1 Accuracy measures and confusion matrices

IV.1. Thorough evaluation of the model accuracy	***IV.1.1. Confusion matrix and accuracy measures***

One of the most common ways to summarize the model accuracy is through a Confusion (also known as misclassification or coincidence) matrix such as the one presented in Table 2.7 for a two-class target attribute. Positive refers to the target event instances, for example, churned customers or responders to a direct marketing campaign.

A Confusion matrix is a simple cross-tabulation of the predicted by the actual classes. It is the base for the calculation of various accuracy measures including the accuracy and the error or misclassification rate.

The accuracy denotes the percentage of instances correctly classified and is calculated as

$$\text{Accuracy} = \text{TP} + \frac{\text{TN}}{P + N}$$

Table 2.7 Confusion matrix

		Predicted values		
		Positive	**Negative**	**Total**
Actual values	**Positive**	TP = *true positive record count (correct prediction)*	FN = *false negative record count (misclassification)*	*P = total positive records*
	Negative	FP = *false positive record count (misclassification)*	TN = *true negative record count (correct prediction)*	*N = total negative records*
	Total	*P′ = total records predicted as positive*	*N′ = total records predicted as negative*	$P + N = P' + N'$

In other words, it sums the table percentages of the Confusion matrix across the diagonal. In binary classification problems, a 50% estimated probability threshold is used to classify an instance as positive or negative. The error rate is the off-diagonal percentage—the proportion of records misclassified—and is calculated as

$$\text{Error rate} = 1 - \text{Accuracy} = \frac{\text{FP} + \text{FN}}{P + N}$$

Since some mistakes are more costly than others, accuracy percentages are also estimated for each category of the target field. The percentage of actual positives (target instances) correctly captured by the model defines sensitivity:

$$\text{Sensitivity} = \frac{\text{TP}}{P} = \text{True positive rate}$$

The percentage of negative instances correctly classified is the specificity:

$$\text{Specificity} = \frac{\text{TN}}{N} = \text{True negative rate}$$

while

$$1 - \text{Specificity} = \frac{\text{FP}}{N} = \text{False positive rate}$$

The false positive rate denotes the proportion of negative (nontarget) instances misclassified as positive (target). Remember the false positive rate when we discuss the ROC curve later in this section.

The aforementioned measures are very useful especially in the case of class-imbalanced problems where the target event is rare. In such cases, a model may present an acceptable accuracy rate while failing to identify the target instances. Imagine, for example, an organization with an overall churn rate of about 2%. A naïve model classifying all customers as nonchurners would yield an astonishingly high accuracy of 98%. Obviously, this model has no practical value as it fails to capture any churner as denoted by its 0 sensitivity.

Other accuracy measures include Precision, Recall, and the *F*-measure which stems from their combination.

The Precision measure represents the percentage of the predicted positives which were actual positives and can be thought of as a measure of exactness:

$$\text{Precision} = \frac{\text{TP}}{\text{TP} + \text{FP}}$$

The Recall measure is the same as sensitivity.

The *F*-measure is the harmonic mean of Precision (a measure of exactness) and Recall (a measure of completeness). It ranges from 0 to 1 with larger values corresponding to better models. It is defined as

$$F = 2 \times \text{Precision} \times \frac{\text{Recall}}{\text{Precision} + \text{Recall}}$$

@Tech tips for applying the methodology

Accuracy measures and Confusion matrices in IBM SPSS Modeler

In IBM SPSS Modeler, the Confusion matrix is built by selecting the Coincidence matrices option of the Analysis node (Figure 2.29).

Since the accuracy measures require scored instances with known outcomes, the Analysis node should be placed downstream a generated classification model to compare actual with predicted outcomes (Figure 2.30). In case a Split validation is in place, the Analysis node presents separate results for the training and the test partition. Obviously, analysts should mainly study the Confusion matrix of the testing dataset.

Accuracy measures and Confusion matrices in RapidMiner

RapidMiner's Performance (Binomial Classification) operator (Figure 2.31) estimates a series of accuracy measures in the case of a two-class target.

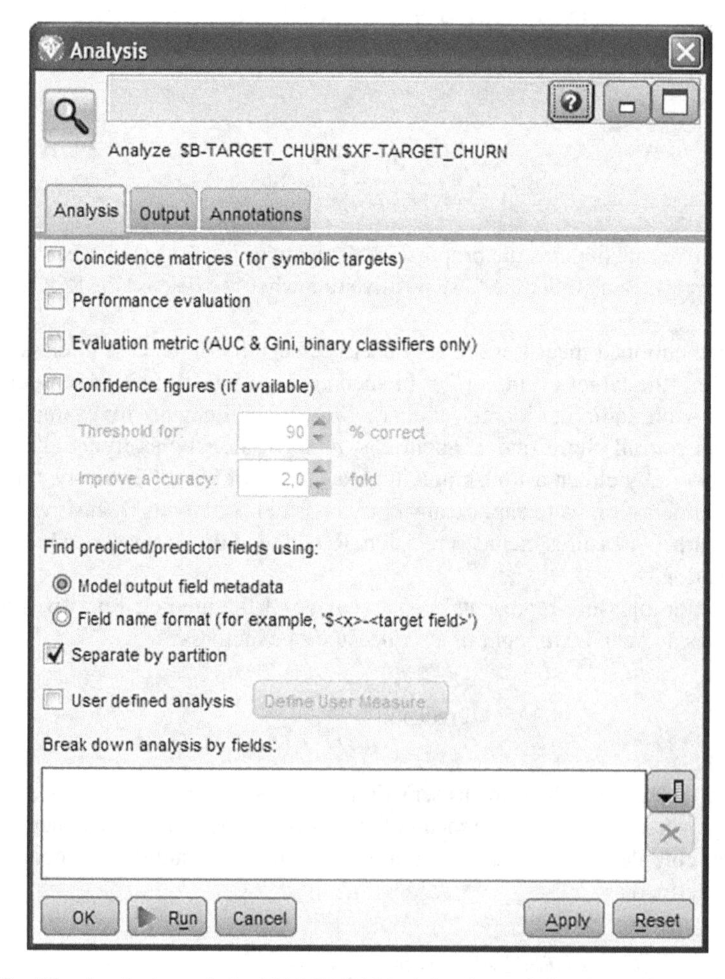

Figure 2.29 The Analysis node in IBM SPSS Modeler for validation with Confusion matrix

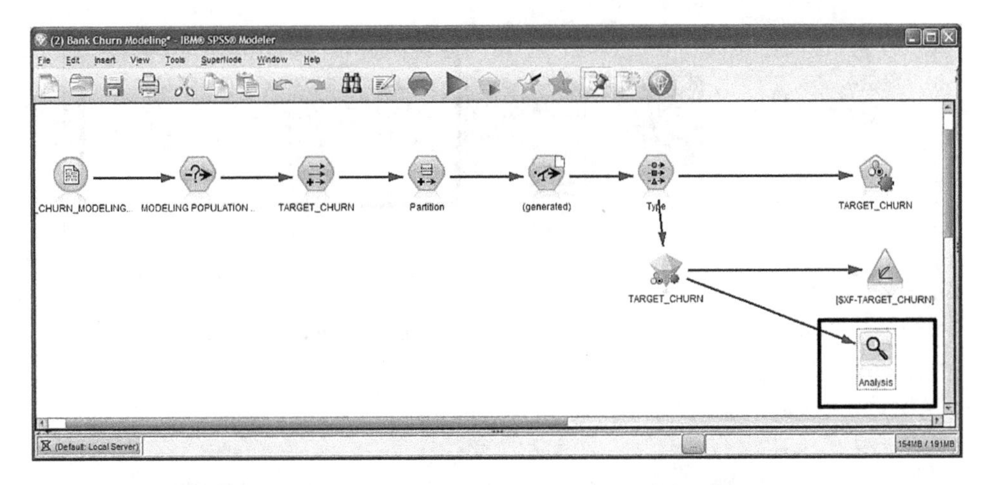

Figure 2.30 The Analysis node in a Modeler stream

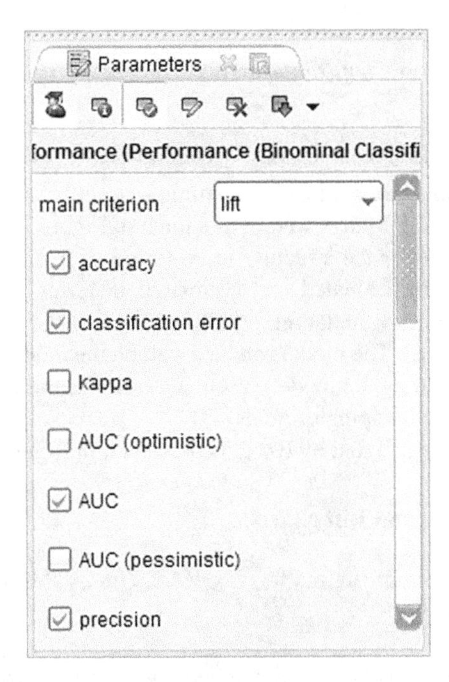

Figure 2.31 RapidMiner's Performance operator

The Performance operator should follow an Apply Model operator and a generated model. The Apply Model operator scores the instances and estimates the predicted class which is then analyzed along with the actual class by the Performance operator. In case of a validation operator, the Performance operator should be placed in the validation subprocess to evaluate the classifier on the testing dataset, as shown in Figure 2.32.

The Confusion matrix along with the accuracy, the Precision, and the Recall metrics generated by the Performance operator is shown in Figure 2.33.

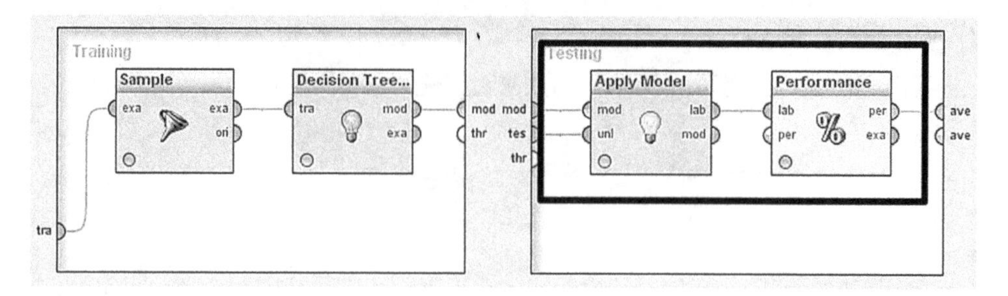

Figure 2.32 The Performance operator in a RapidMiner process

Table View Plot View

accuracy: 86.30%

	true F	true T	class precision
pred. F	5718	309	94.87%
pred. T	624	157	20.10%
class recall	90.16%	33.69%	

Figure 2.33 RapidMiner's Confusion matrix as generated by the Performance operator

Accuracy measures and Confusion matrices in Data Mining for Excel

In Data Mining for Excel, after a model and mining structure has been created, it can be tested using the Classification Matrix wizard as shown in Figure 2.34.

Users specify the model or the structure to be tested. If a structure is selected, then all models of the structure are evaluated and compared in terms of classification accuracy. Additionally, they must specify the target field, the format of the results (counts and/or percentages), and the test dataset. The model can be tested on the model test dataset (saved in the mining structure when building the model) or on any other external data (including database tables, after the appropriate mapping of fields).

The Confusion matrix generated by Excel is presented in Figure 2.35.

2.5.1.2 Gains, Response, and Lift charts

IV.1. Thorough evaluation of the model accuracy ***IV.1.2. Gains, Response, and Lift charts***

Many times, in the real world of classification modeling, we end up with models with moderate accuracy, especially in cases with a rare target event. Does this mean that the model is not adequate? Remember that in order to estimate the accuracy and the error rate of a model in a binary classification problem, we typically use a default propensity threshold of 50% for classifying an instance as positive. In problems with a rare target class, an increased FN rate and therefore an increased error rate may simply mean that the estimated propensities are often below 50% for actual positives. Does this mean that the model is not useful? To answer that, we must examine whether the positive instances yield higher estimated propensities compared to the negative ones. In other words, we must examine if the model propensities rank well and if they discriminate the actual positives from the actual negatives. The Gains, Response, and

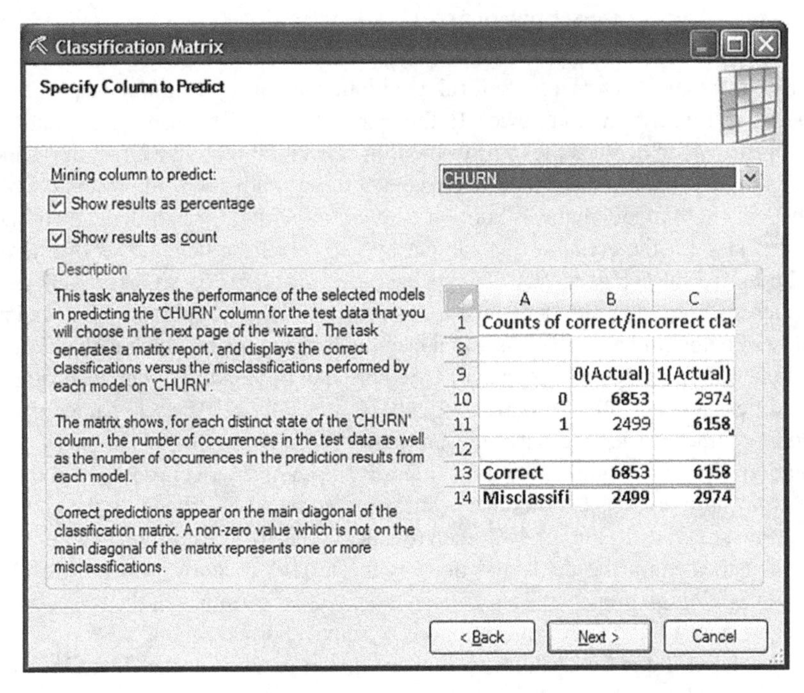

Figure 2.34 The Classification Matrix in Data Mining for Excel

Counts of correct/incorrect classification for model 'CC_Churn_BDE'		
Predicted Column 'CHURN'		
Columns correspond to actual values		
Rows correspond to predicted values		
Model name:	**CC_Churn_BDE**	**CC_Churn_BDE**
Total correct:	**93.11 %**	**6338**
Total misclassified:	**6.89 %**	**469**
Results as Percentages for Model 'CC_Churn_BDE'		
	F(Actual)	T(Actual)
F	100.00 %	100.00 %
T	0.00 %	0.00 %
Correct	100.00 %	0.00 %
Misclassified	0.00 %	100.00 %
Results as Counts for Model 'CC_Churn_BDE'		
	F(Actual)	T(Actual)
F	6338	469
T	0	0
Correct	6338	0
Misclassified	0	469

Figure 2.35 The Confusion matrix in Data Mining for Excel

Lift/Index tables and charts are helpful evaluation tools that summarize the model's performance and discrimination power. In this paragraph, we'll present these charts and the measure of Lift which denotes the improvement in concentration of the target class due to the model. To illustrate the basic concepts and usage of these charts, we will present the results of a hypothetical churn model that was built on a binary target field which flagged churners.

The first step for the creation of such charts is to select the target class of interest, also referred to as the hit category. For Gains/Response/Lift charts as well for ROI and Profit charts which will presented immediately after, we assume that the classifier can estimate the probability of belonging at each class and hence the hit propensity which is the likelihood of the target class. Records/customers can then be sorted in descending order according to their hit propensities and binned into groups of equal size, typically of 1% each, named percentiles. The model accuracy is then evaluated within these percentiles.

In our hypothetical example, the target class is the category of churners, and the hit propensity is the churn propensity—in other words, the estimated likelihood by the churn model. Customers are split into 10 tiles of 10% each (deciles). The 10% of customers with the highest churn propensities comprise tile 1, and those with the lowest churn propensities, tile 10. In general, we expect high propensities to correspond to actual members of the target population. Therefore, we hope to find large concentrations of actual churners among the top model tiles.

The cumulative Table 2.8 evaluates our churn model in terms of the Gain, Response, and Lift measures.

But what exactly do these performance measures represent and how they are used for model evaluation? A brief explanation is as follows:

- **Response %**: "How likely is the target class within the examined quantiles?" It denotes the likelihood of the target outcome, the percentage of the actual churners (positives) within the tiles. It is a measure of *exactness* and is the same as the *Precision* measure discussed in Section 2.5.1.1.

  In our example, 10.7% of the customers of the top 10% model tile were actual churners, yielding a Response % of the same value. Since the overall churn rate was about 2.9%, we expect that a random list would also have an analogous churn rate. However, the estimated churn rate for the top model tile was 3.71 times (or 371.4%) higher. This is called the Lift. Analysts have achieved about four times better results than

Table 2.8 The gains, response, and lift table

Model tiles	Cumulative % of customers	Propensity threshold (minimum value)	Gain (%)	Response (%)	Lift (%)
1	10	0.150	37.1	10.7	371.4
2	20	0.100	56.9	8.2	284.5
3	30	0.070	69.6	6.7	232.1
4	40	0.065	79.6	5.7	199.0
5	50	0.061	87.0	5.0	174.1
6	60	0.052	91.6	4.4	152.7
7	70	0.043	94.6	3.9	135.2
8	80	0.039	96.4	3.5	120.6
9	90	0.031	98.2	3.1	109.2
10	100	0.010	100.0	2.9	100.0

randomness in the examined model tile. As we move from the top to the bottom tiles, the model exactness decreases. The concentration of the actual churners is expected to decrease. Indeed, the first two tiles, which jointly account for the top 20% of the customers with the highest estimated churn scores, have a smaller percentage of actual churners (8.2%). This percentage is still 2.8 times higher than randomness.

- **Gain %**: "Which percentage of the target class falls in the tiles?" Gain % is defined as the percentage of the total target population that belongs in the tiles. It is the same as *Sensitivity* and *Recall* presented in Section 2.5.1.1; hence, it is a measure of *completeness*. It denotes the true positive rate, the percentage of true positives (actual churners) included in the tile.

  In our example, the top 10% model tile contains 37.1% of all actual churners, yielding a Gain % of the same value. A random list containing 10% of the customers would normally capture about 10% of all observed churners. However, the top model tile contains more than a third (37.1%) of all observed churners. Once again, we come upon the Lift concept. The top 10% model tile identifies about four times more target customers than a random list of the same size.

- **Lift**: "How much better is the classifier compared to randomness?" The Lift or Index assesses the factor of improvement in response rate due to the model. It is defined as the *ratio of the model Response % (or equivalently Gain %) to the Response % of a random model*. In other words, it compares the model quantiles with a random list of the same size. Therefore, it represents how much a trained classifier exceeds the baseline model of random selection.

The Gain, Response, and Lift evaluation measures can also be depicted in corresponding charts such as the ones shown in the following. The two added reference lines correspond to the top 5% and the top 10% tiles. The diagonal line in the Gains chart represents the baseline model of random guessing.

According to the cumulative Gains chart shown in Figure 2.36, when scoring an unseen customer list, data miners should expect to capture about 40% of all potential churners if they target the customers of the top 10% model tile. Narrowing the list to the top 5% percentile decreases the percentage of potential churners to be reached to approximately 25%. As we move to the right of the *X*-axis, the expected number of total churners (true positives) to be identified increases. But this comes at a cost of increased error rate and false positives. On the contrary, left parts of the *X*-axis lead to smaller but more targeted campaigns.

What we hope to see in a real model evaluation is a Gains curve steeply and smoothly rising above the diagonal along the top tiles before gradually easing off after a point.

Analysts can study Gains charts and compare the accuracy of models. A model closer to the diagonal line of random guessing is less accurate. A Gains chart typically also includes an Optimal or Best line which corresponds to the ideal model that classifies all records correctly.

Although we'd like a Gains curve to be close to the Optimal line, extreme proximity and absolute accuracy might indicate a problem with the model training such as using a predictor directly related with the target attribute.

By studying the Gains charts, analysts assess the model's discriminating power. They also gain valuable insight about its future predictive accuracy on new records. These charts can be used for deciding the optimal size of the respective campaign by choosing the top propensity-based tiles to target. Hence, they may choose to conduct a small campaign, limited to the top tiles, in order to address only customers with very high propensities and minimize

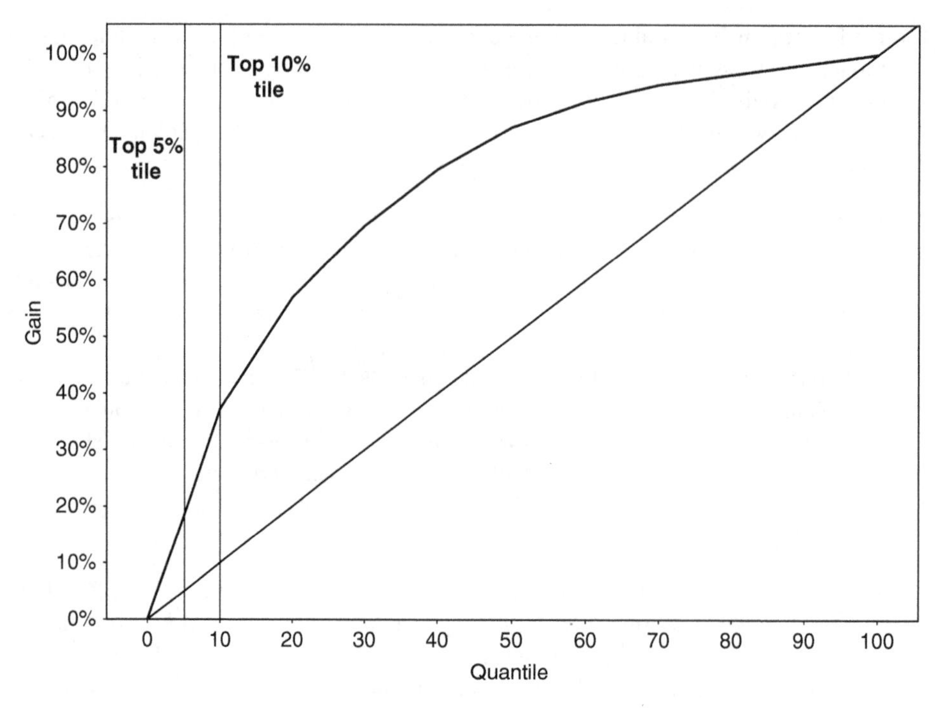

Figure 2.36 Gains chart

the false positive cases. Alternatively, especially if the cost of the campaign is small compared to the potential revenue, they may choose to expand their list by including more tiles and more customers with relatively lower propensities.

The maximum benefit point

An approach often referred to bibliography as a rule of thumb for selecting the optimal size for a targeted marketing list is to examine the Gains chart and select all top tiles up to the point where the distance between the Gains curve and the diagonal reference line becomes maximum. This is referred to as the maximum benefit point, and it is the point where the difference between the Gains curve and the diagonal reference line gets its maximum value. This point also corresponds to the point (propensity threshold) with the maximum Kolmogorov–Smirnov (KS) statistic. The KS test examines whether the distribution of propensities is different between the two target outcomes. It measures the maximum vertical separation between the two cumulative distributions.

The reasoning behind this approach is that from that point on, the model classifies worse than randomness. This approach usually yields large targeting lists. In practice, analysts and marketers should take into consideration the particular business situation, objectives, and resources and possibly consider as a classification threshold the point of lift maximization. If possible, they should also incorporate in the Gains chart cost (per offer) and revenue (per acceptance) information and select the cut point that best serves their specific business needs and maximizes the expected ROI and Profit.

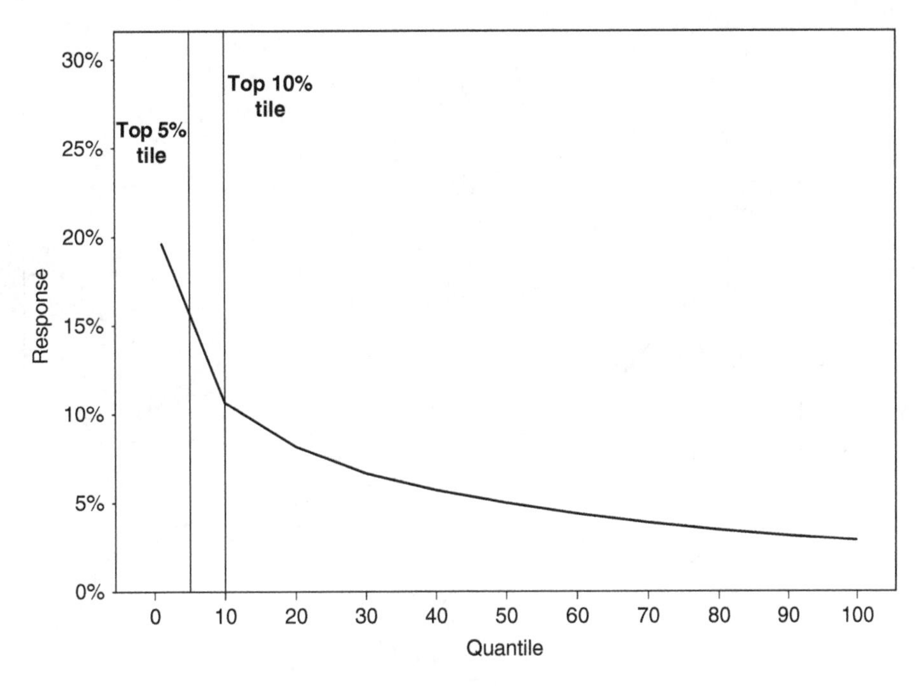

Figure 2.37 Response chart

Figure 2.37 presents the cumulative Response chart for our hypothetical example.

It illustrates the estimated churn likelihood along the model tiles. As we move to the left of the *X*-axis and toward the top tiles, we have increased churn probabilities. These tiles would result more targeted lists and smaller error rates. Expanding the list to the right part of the *X*-axis, toward the bottom model tiles, would increase the expected false positive error rate by including in the targeting list more customers with low likelihood to churn.

The cumulative Lift or Index chart (Figure 2.38) directly compares the model predictive performance with the baseline model of random selection. The concentration of churners is estimated to be four times higher than randomness among the top 10% customers and about six times higher among the top 5% customers.

@Tech tips for applying the methodology

Gains charts in IBM SPSS Modeler

In IBM SPSS Modeler, the Evaluation node, following a generated model node, can be used for the creation of Gains, Response, and Lift charts as shown in Figure 2.39. Users can choose to include the diagonal line of random guessing as well as the optimal line of perfect accuracy to be helped in their visual evaluation. The default option takes into account the Split validation, if present, and plots the charts for both the training and the testing dataset. Obviously, users should mainly examine the plots of the testing dataset. The Evaluation node also permits the investigation of Profit and ROI charts which will be presented shortly.

A sample Modeler Gains chart which compares a series of classifiers is shown in Figure 2.40.

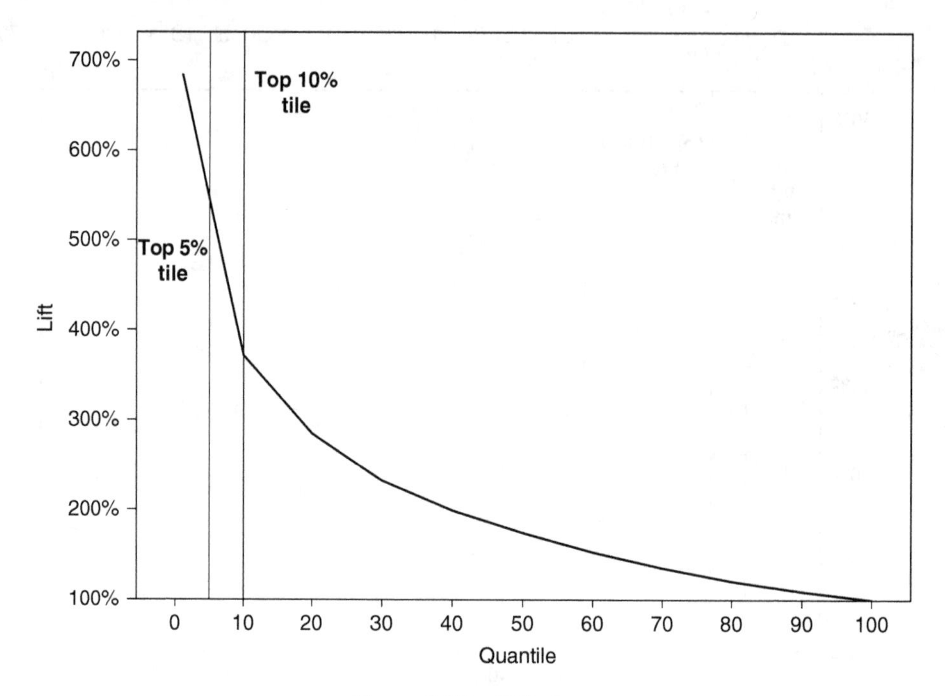

Figure 2.38 Lift chart

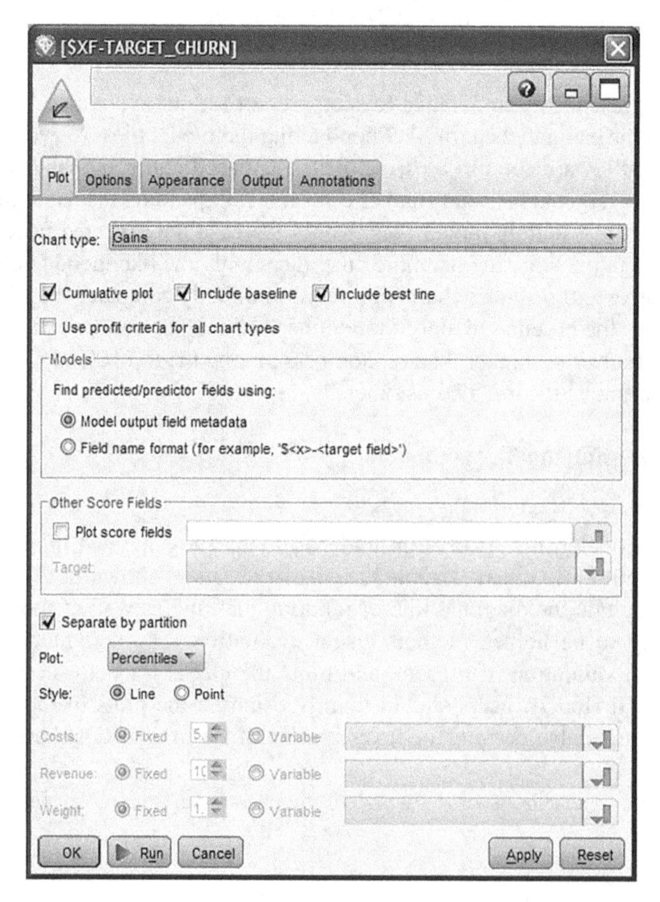

Figure 2.39 The Evaluation Modeler node for Gains, Lift, and Response charts

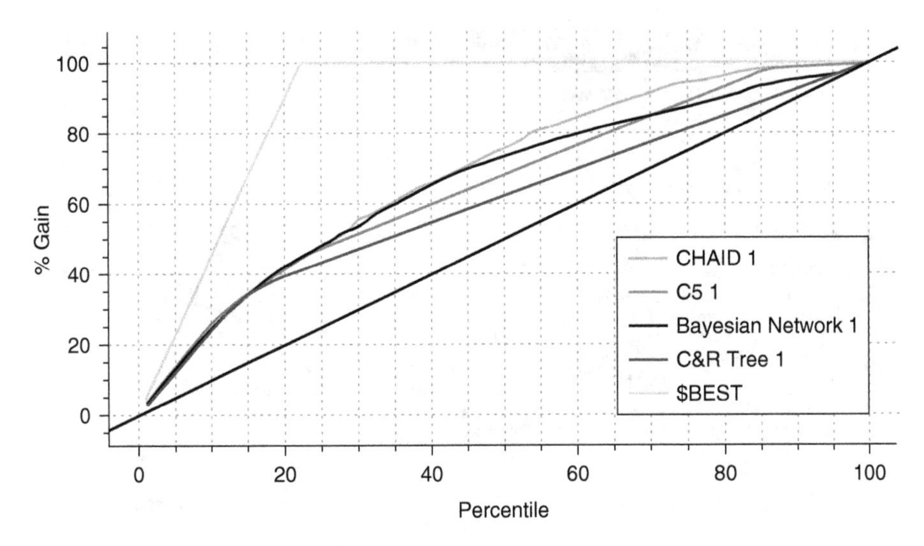

Figure 2.40 A Modeler Gains chart for the comparison of a series of models

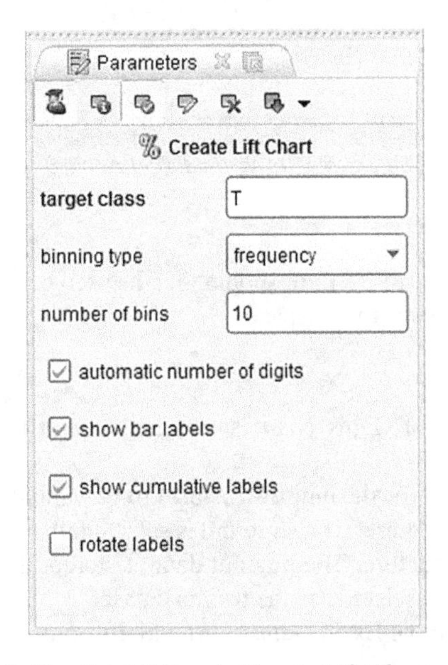

Figure 2.41 RapidMiner's Create Lift Chart operator

Gains charts in RapidMiner

RapidMiner's Create Lift Chart operator (Figure 2.41) creates a Lift Pareto chart which is a combined plot of a cumulative Gains and a noncumulative Response chart. It expects as inputs a model operator and a testing dataset. Users must specify the target class.

In case a validation technique and operator is present, the Lift chart operator should be placed within the validation subprocess in order to evaluate the classifier on the testing dataset. In such a case, a Remember operator should be used inside the validation subprocess to

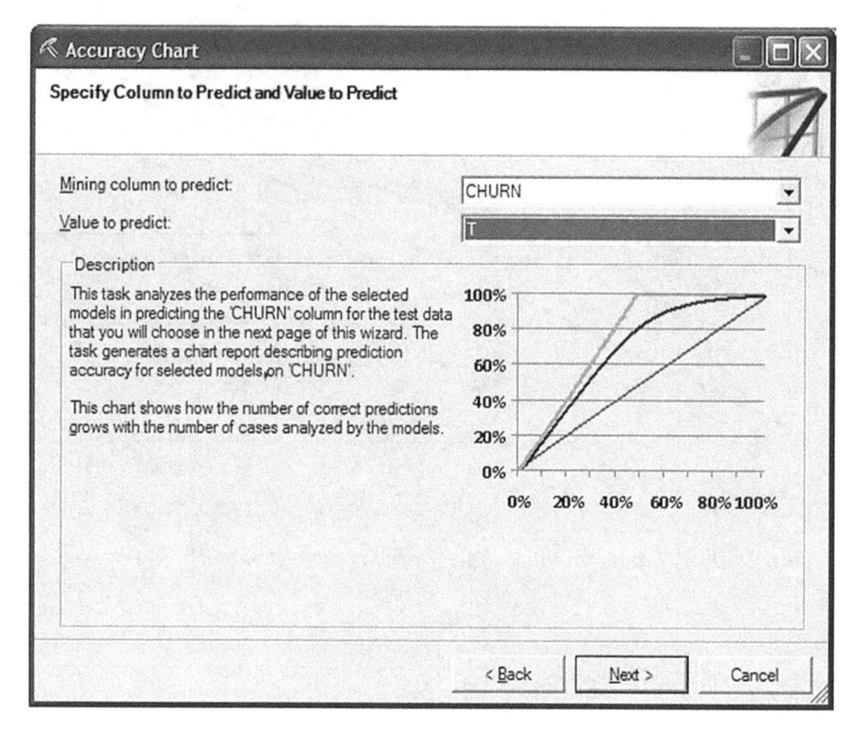

Figure 2.42 The Data Mining for Excel Accuracy Chart wizard

store the Lift chart, and a Recall operator should then be used outside the Validation operator to deliver the chart to the Results workspace.

Gains charts in Data Mining for Excel

In Data Mining for Excel, a Gains chart is created through the Accuracy Chart wizard (Figure 2.42).

Users specify the model (or the entire structure) to be validated, the target attribute, the target class, and the testing dataset. If a structure is selected, then Gains charts all plotted for all stored models of the structure. The holdout dataset stored along with the model or any other external dataset can be selected as the testing dataset.

Figure 2.43 shows a sample Excel Gains chart with its optimal line and random guessing diagonal.

2.5.1.3 ROC curve

IV.1. Thorough evaluation of the model accuracy	***IV.1.3. ROC curve***

The ROC curve also visualizes the performance and the discriminating power of the classifier. It plots the model's true positive rate, the sensitivity, in the vertical axis. Its horizontal axis corresponds to the false positive rate (1—specificity), the proportion of negative (nontarget)

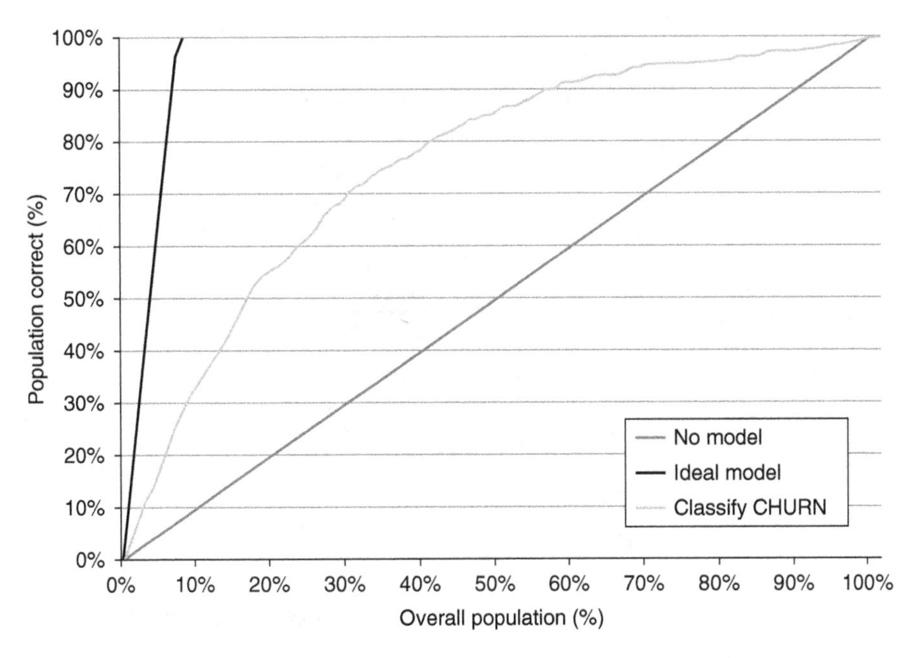

Figure 2.43 A sample Data Mining for Excel Gains chart

instances misclassified as positive. Therefore, the ROC curve depicts the trade-off between capturing more positives but with an increased cost of false positives.

Just as the Gains chart, the ROC curve is based on the rank ordering of the test instances in decreasing order according to their propensities. The Gains chart and the ROC curve also have the same vertical axis, the true positive percentage. Their difference is at the horizontal axis. In Gains charts, it plots the percentage of the total test population, while in ROC curves, the proportion of false positives. That's why the ROC curve shape does not depend on the overall distribution of the target category, and hence, it is unaffected by oversampling. However, since in many real-world applications the density of the target category is small (for instance, below 1%), there is little difference between the proportion of the total test population and the proportion of total negatives, and hence, the ROC curve and the Gains chart have similar shapes.

If the model is adequate, its ROC curve will rise sharply near the vertical axis before easing off. In the case of a trivial model, the curve will approximate a diagonal line from the lower left to the upper right corner of the graph. A measure of the accuracy of the model is the area under the curve (AUC) which is equivalent to the c-statistic. It ranges between 0 and 1.0. The closer the AUC is to 1.0, the better the model. A model with AUC close to 0.5 is not better than random guessing. An ideal model will have a value of 1.0, while values above 0.7 can be considered adequate.

The AUC measure is also equivalent to the Gini index. The Gini index is calculated as the area between the ROC curve and the diagonal line of the random model divided by the area between the optimal curve and the diagonal. It ranges between 0 and 1.0 with values above 0.4 denoting acceptable efficiency. In fact:

$$\text{Gini index} = 2^{*}\left(\text{AUC} - 0.5\right).$$

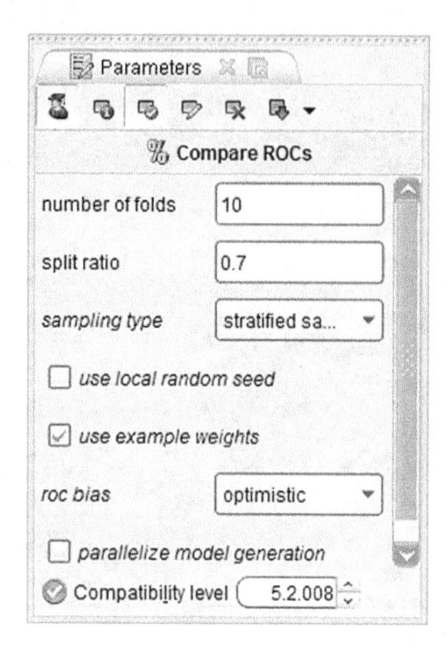

Figure 2.44 The Compare ROCs operator of RapidMiner

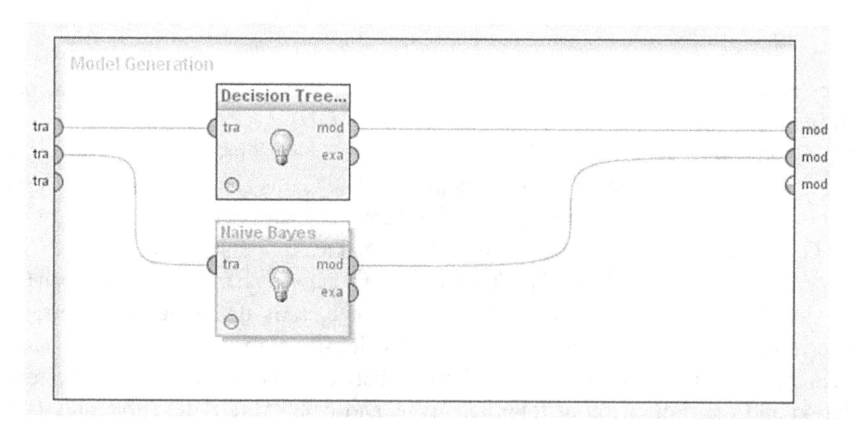

Figure 2.45 Comparing a series of classifiers with the Compare ROCs operator of RapidMiner

@Tech tips for applying the methodology

ROC curve in IBM SPSS Modeler

In IBM SPSS Modeler, a ROC chart can be produced through the Evaluation node. Additionally, the AUC and the Gini measures are included in the output of the Analysis node.

ROC Curve in RapidMiner

RapidMiner includes the Compare ROCs operator for visual comparison of a series of classifiers (Figure 2.44). It is a nested operator with a subprocess. A series of models is trained within the subprocess (Figure 2.45), and their ROC curves are jointly plotted on the same

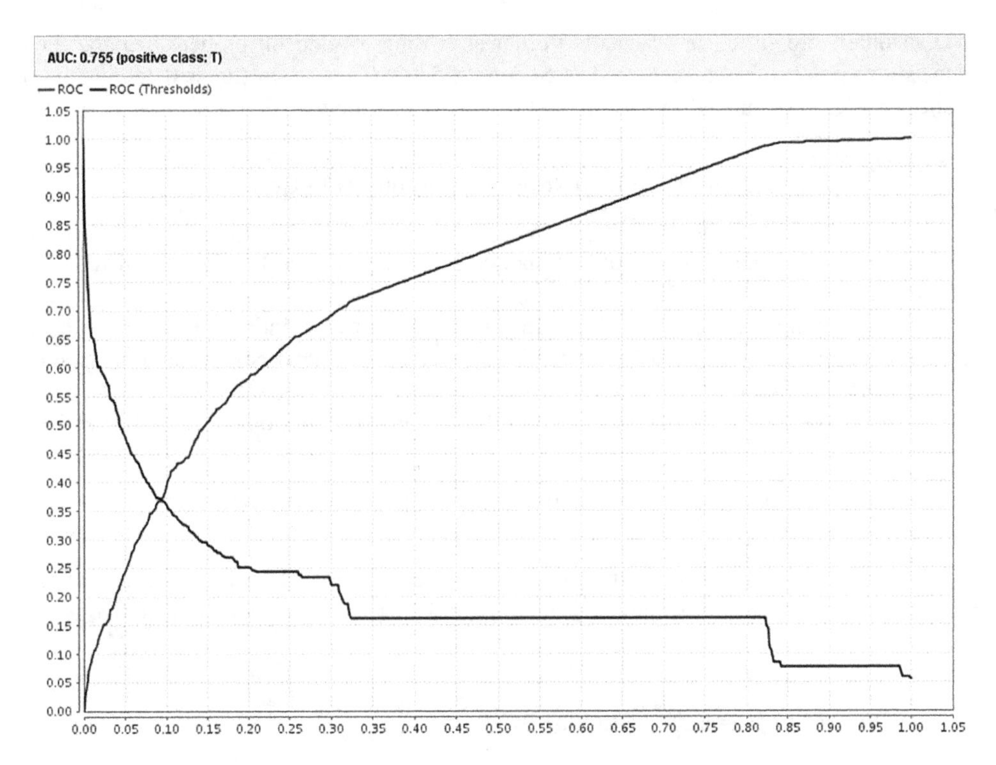

Figure 2.46 Example of a ROC curve

chart for comparison purposes. The operator also applies an n-fold cross-validation technique for better evaluation of the models.

An ROC curve for a binary classifier using RapidMiner is presented in Figure 2.46.

2.5.1.4 Profit/ROI charts

IV.1. Thorough evaluation of the model accuracy	*IV.1.4. Profit/ROI charts*

Profit and ROI charts are extensions of the Gains/Response charts which incorporate cost and revenue information to help marketers decide their target lists based on estimated revenue.

The model estimated target probabilities are combined with expected cost and revenue information to calculate the probabilistic Profit and/or ROI for the model percentiles.

Marketers must specify:

Cost: The estimated cost per offer (for each customer included in the campaign)

Revenue: The anticipated revenue associated with each hit, that is, for each customer accepting the offer

Customers are sorted in descending order according to their hit propensities and then binned into percentiles as presented in Section 2.5.1.2. Then, the estimated (cumulative) profit per offer is calculated for each model percentile as follows:

$$\text{Profit per offer} : (\text{Revenue} \times \text{Response \%}) - \text{Cost}$$

Obviously, revenues concern only responders (hits), while costs apply to all records/offers.

By multiplying the profit per offer with the number of customers of the tile, we have the total (cumulative) profit for the tile. In case of an overhead cost, it should be subtracted from the total profit.

The estimated ROI per offer, expressed as the percentage return on cost, is calculated as

$$\text{ROI \% per offer} = (\text{estimated Profit per offer / Cost}) \%$$

Hence, the ROI per offer is the ratio of the average profit to average cost for each record of the tile. Negative values indicate loss per offer and correspond to negative profit per offer.

@Tech tips for applying the methodology

Profit charts in IBM SPSS Modeler

Revenue and cost information can be specified in the Evaluation plot of IBM SPSS Modeler to build Profit and ROI charts as shown in Figure 2.47.

In this example, a fixed cost of 50$ is anticipated for each offer in which, in the case of acceptance, it is expected to return 200$. The respective (cumulative) total profit for the model quantiles is shown in Figure 2.48.

By studying the Profit chart, we observe the large increase of the sum of profits at the top percentiles. After a point, the profit remains at a high plateau before trailing off on the right side of the chart. This is the general line pattern for a good model. In this hypothetical example, the maximum profit is been reached at about the 40% tile, a candidate selection point for inclusion in the campaign.

An example of an ROI % evaluation chart is shown in Figure 2.49.

This fictitious model doubles each $ spent for each customer belonging in the top 20% tile. The ROI then decreases up to about the 80% tile from which point on each offer is anticipated to be loss-making.

Profit charts in Data Mining for Excel

In Data Mining for Excel, the Profit chart wizard (Figure 2.50) can be applied to build Profit charts for the generated models. Users should specify an individual cost per offer as well as an overhead cost and of course the estimated revenue per offer. Additionally, they can specify the number of customers to be included in the direct marketing campaign.

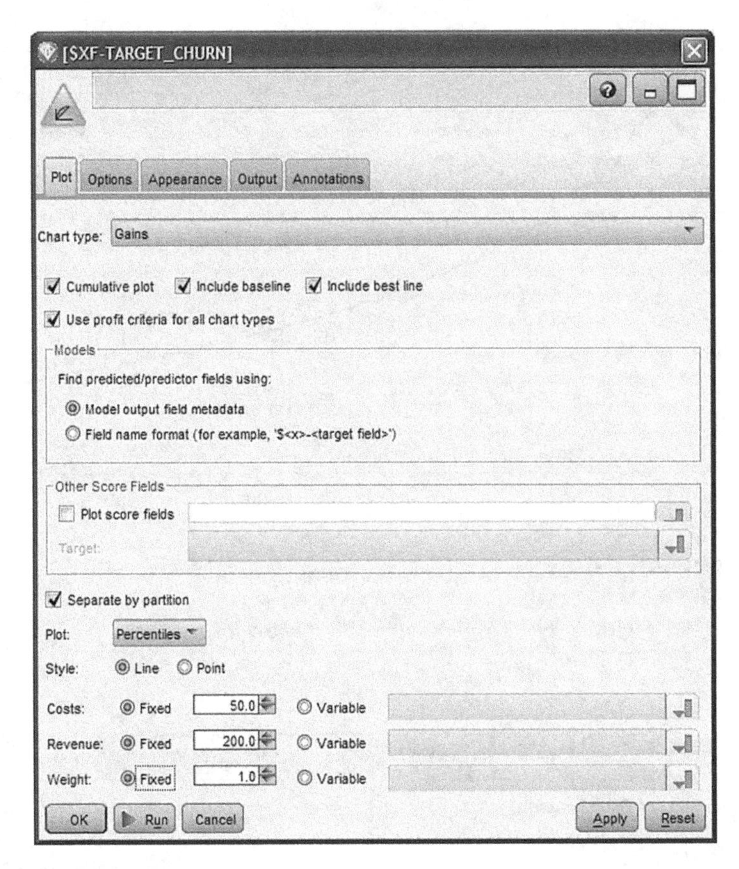

Figure 2.47 The IBM SPSS Modeler Evaluation node for building Profit/ROI charts

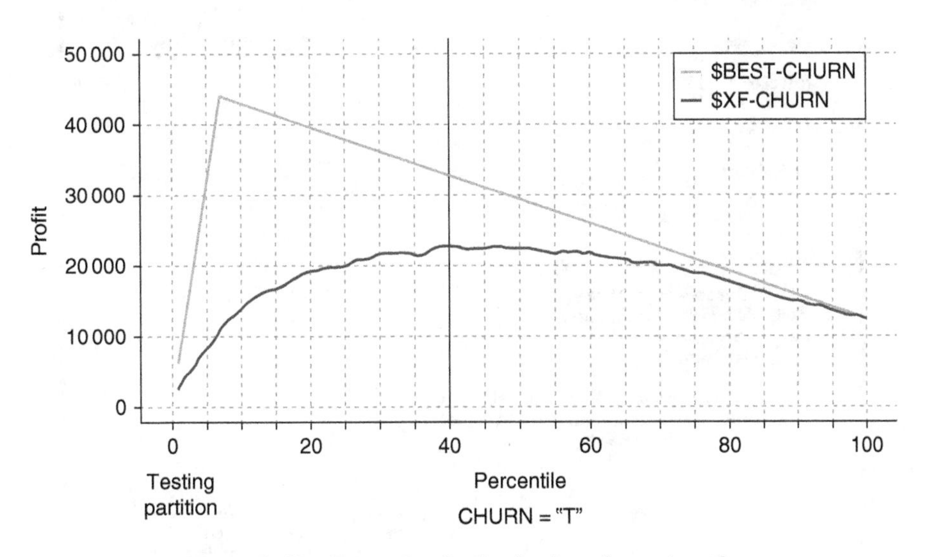

Figure 2.48 Example of a Profit chart for a classifier

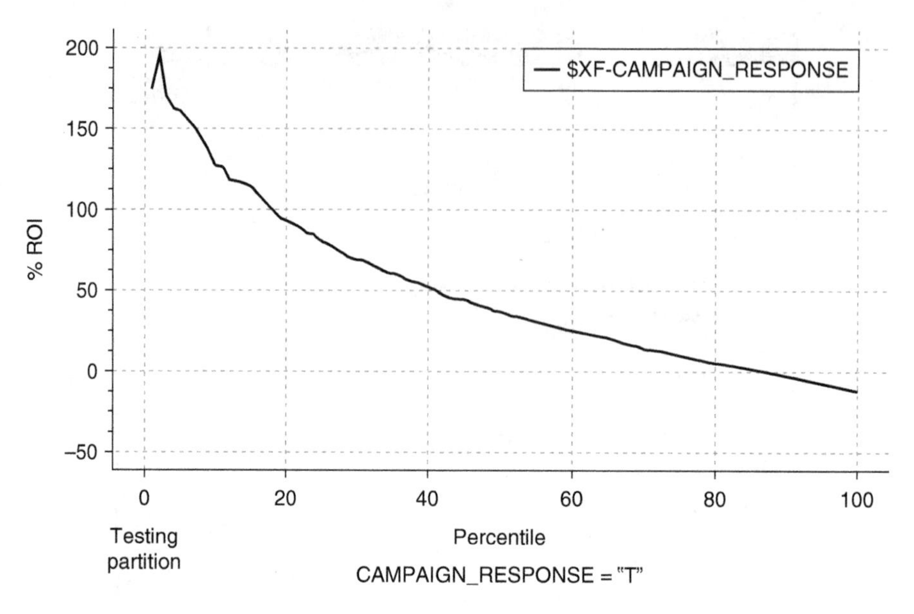

Figure 2.49 Example of a Profit chart for a classifier

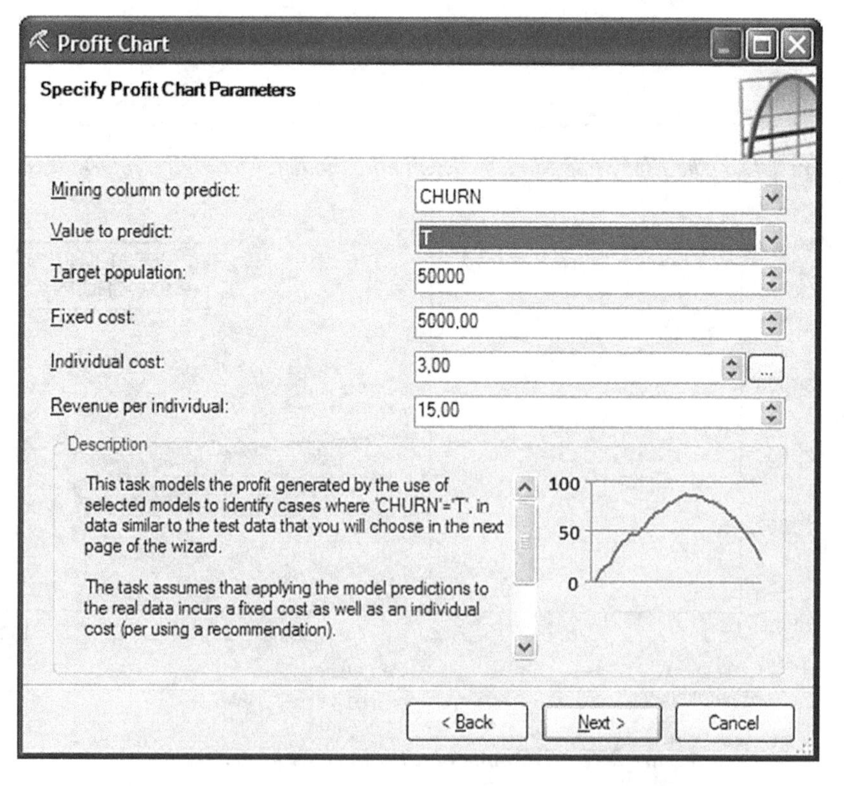

Figure 2.50 The Data Mining for Excel Profit chart wizard

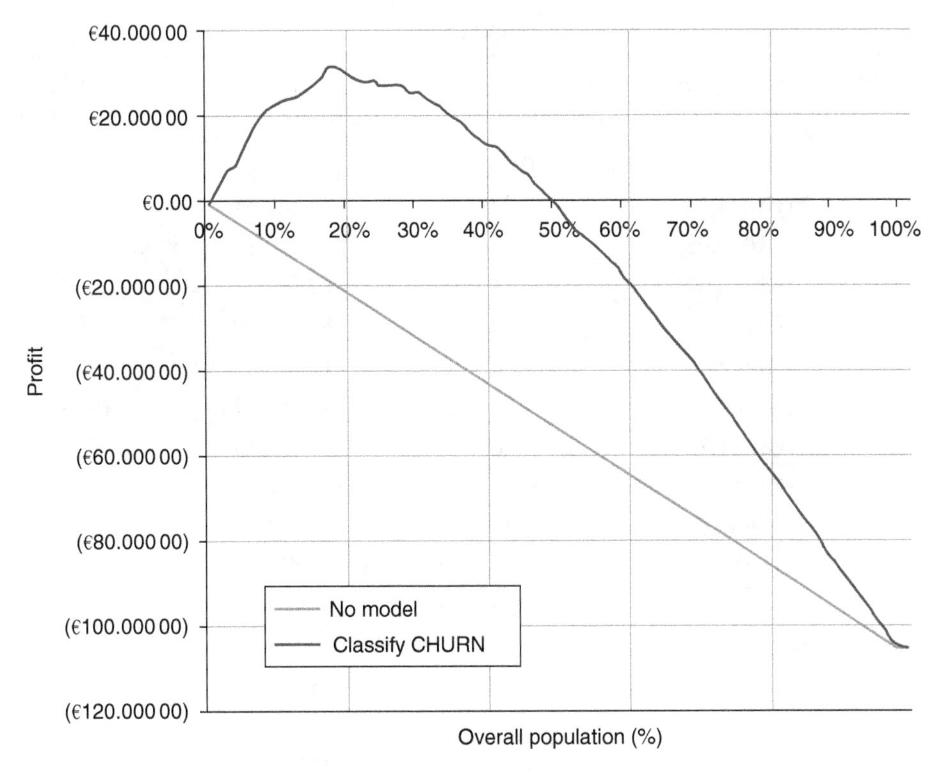

Figure 2.51 A Profit chart generated by Data Mining for Excel Profit chart wizard

The Profit chart is shown in Figure 2.51. The diagonal line corresponds to random guessing.

The chart is supplemented with an informative table report, a sample of which is listed in Table 2.9.

By studying the chart and the report, we can see that the model yields its maximum total profit at around the 18% tile which corresponds to a propensity threshold of 13.94%.

2.5.2 Evaluating a deployed model with test–control groups

IV. Model evaluation	***IV.2. Evaluating a deployed model: test–control groups***

After the rollout of the direct marketing campaign, the model should be reevaluated in terms of actual response. The campaign responses should be stored as they may be used as the training dataset for subsequent, relevant marketing actions. Besides the model predictability, the effectiveness of the marketing intervention and the design of the campaign should also be tested. Hence, all parameters of the campaign (offer, message, and channel) should also be assessed.

A common schema used for testing both the model and the campaign is displayed in Table 2.10. It involves the grouping of customers according to whether they have been selected by the model and whether they have been finally included in the campaign list.

Table 2.9 The Profit report generated by Data Mining for Excel

Percentile (%)	Random guess profit	Classify CHURN population correct (%)	Classify CHURN profit	Classify CHURN probability (%)
0	(€1 000.00)	0.00	(€1 000.00)	100.00
1	(€2 044.54)	3.23	€1 203.61	44.38
2	(€3 089.08)	7.68	€5 170.12	35.27
3	(€4 133.62)	11.11	€7 667.55	26.72
4	(€5 178.16)	13.54	€8 695.90	26.72
5	(€6 222.70)	17.98	€12 662.41	19.49
6	(€7 267.24)	21.82	€15 747.47	19.49
7	(€8 311.78)	25.66	€18 832.53	19.49
8	(€9 356.32)	28.69	€20 742.32	17.75
9	(€10 400.86)	31.72	€22 652.12	17.75
10	(€11 445.40)	33.94	€23 386.66	13.94
11	(€12 489.94)	35.96	€23 827.38	13.94
12	(€13 534.47)	37.98	€24 268.11	13.94
13	(€14 579.01)	40.20	€25 002.64	13.94
14	(€15 623.55)	42.42	€25 737.18	13.94
15	(€16 668.09)	44.85	€26 765.54	13.94
16	(€17 712.63)	48.08	€28 969.15	13.94
17	(€18 757.17)	51.31	€31 172.76	13.94
18	(€19 801.71)	53.54	€31 907.30	13.94
19	(€20 846.25)	54.75	€31 172.76	9.44
20	(€21 890.79)	55.56	€29 850.59	9.44
21	(€22 935.33)	56.57	€28 822.24	9.44
22	(€23 979.87)	57.78	€28 087.70	9.44
23	(€25 024.41)	59.39	€27 940.80	9.44

Table 2.10 The schema used for testing both the model and the offer of a direct marketing campaign

		Selected by model	
		Yes	**No**
Marketing intervention	**Yes**	*Test group: model group—targeted*	*Control group: random group—targeted*
	No	*Model Holdout group*	*Random Holdout group*

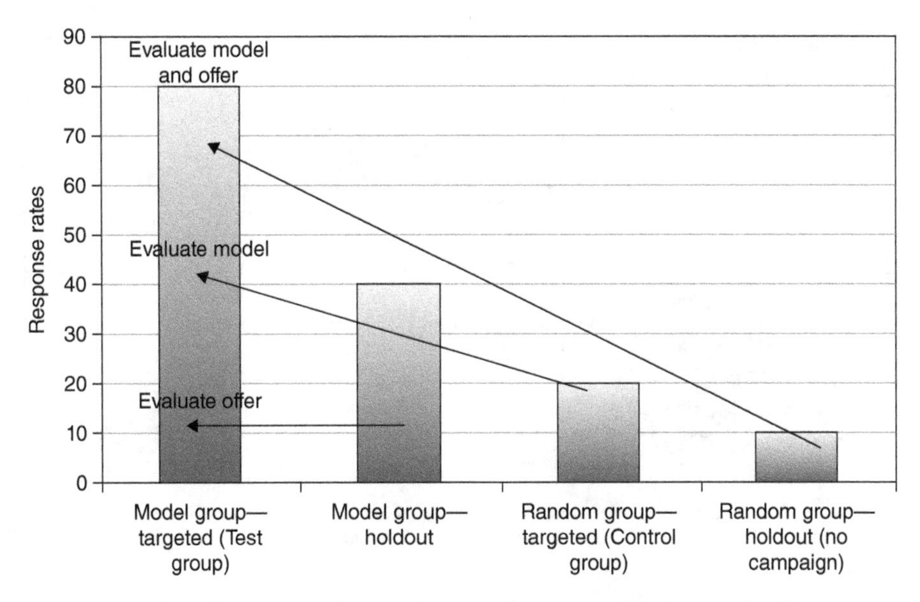

Figure 2.52 The measured response rate of a cross-selling campaign by group

The Test group includes those customers selected by the model to be included in the campaign. So, they are the customers of the top percentiles, with the higher propensities, which were chosen to be targeted.

Normally, the campaign list would not contain anyone else. However, we typically also add a Control group to evaluate the model predictability. The Control group is formed by a random sample of nonselected customers with lower scores.

Finally, two additional holdout groups are formed. These groups are not included in the final campaign list and are not reached by the offer. The Model Holdout group is a sample of model-selected probable responders which was sacrificed for our evaluation purposes. The Random selected holdout group is a sample of customers with lower propensities.

The pairwise comparison of groups allows us to evaluate all aspects of both the model and the marketing intervention as shown in Figure 2.52 which presents the recorded response rate of a hypothetical cross-selling campaign.

The comparison of the Test versus the Control group assesses the model's performance on those who received the offer. The data miner hopes to see substantially higher responses in the Test group.

The comparison of the Test group versus the Model Holdout group shows the intervention's influence/appeal on those scored as probable buyers. The marketer responsible for the campaign design hopes to see a significant higher response percentage in the Test group.

Finally, the comparison of the Test group versus the Random Holdout group reveals the compound effect of both the model and the offer.

The situation is different in the case of a churn model and a retention campaign with offering of incentives to prevent attrition. Churn rates by group are presented in Figure 2.53.

The model, with no marketing intervention, can be evaluated by comparing the Model Holdout with the Random Holdout group. The offered incentive on high-risk customers can be assessed by comparing the Model targeted group with the Model Holdout group.

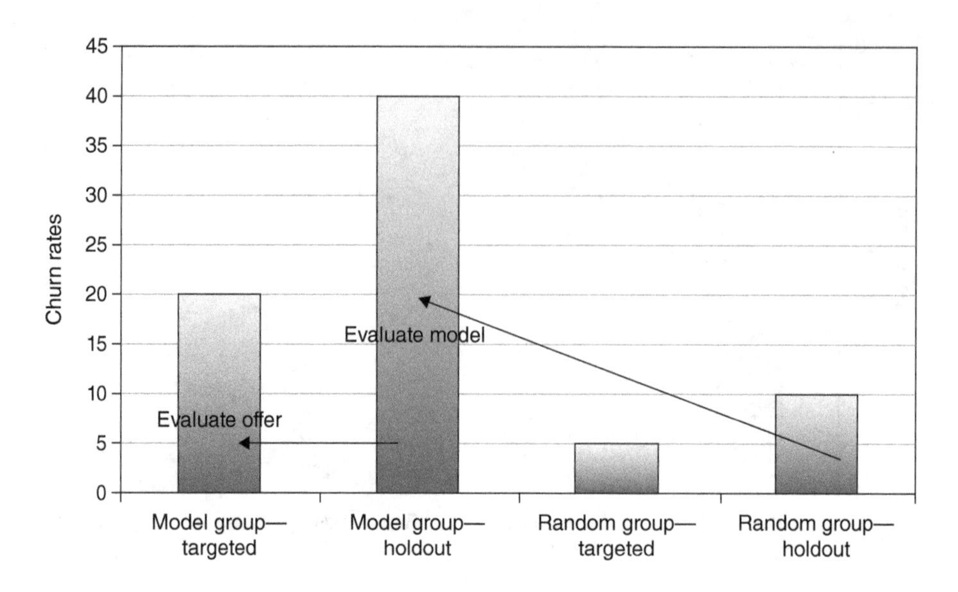

Figure 2.53 The measured churn rate of a retention campaign by group

2.6 Model deployment

The deployment phase is the last phase of propensity modeling for direct marketing campaigns. Each model, apart from offering insight through the revealed data patterns, can also be used as a scoring engine. The evaluated classifier scores new cases and classifies unseen instances according to their input patterns. Customers predicted to belong to the target class are included in the campaign list. Obviously, all model inputs should also be available in the scoring dataset.

2.6.1 Scoring customers to roll the marketing campaign

V. Model deployment	***V.1. Scoring customers to roll the marketing campaign***

When new data are scored by a classifier, the class prediction along with its confidence, the prediction probability, is estimated. Probabilistic classifiers such as logistic regression and Bayesian networks can estimate the probability of belonging at each class. Most other classifiers can also be modified to return class probabilities. For instance, in the case of Decision Tree models, the class probabilities can be estimated by the class distribution of the terminal nodes.

Hence, classifiers are able to estimate the propensity score for each customer/record. The propensity score denotes the likelihood of belonging to the target class. In binary classification problems, it equals the prediction confidence, if the prediction is the target class, or 1 minus the prediction confidence, if the prediction is the negative class.

The ability to estimate propensities provides the great advantage of being able to rank order customers according to their "response" likelihood. Therefore, instead of focusing on the predicted class, we can study the estimated propensities and use them to tailor the final marketing list. The common approach is to target selected top propensity tiles for our marketing campaigns. By examining the Gains/Profit/ROC charts and tables presented in Section 2.5, marketers can choose the tiles to target. The selected tiles correspond to a specific propensity threshold, a propensity cutoff which is the boundary of the selected tiles. The tiles to target can be tailored to the campaign scope and resources, resulting in widened or more focused lists. This procedure is equivalent to "tweaking" the propensity threshold until we reach the desired campaign size.

IBM SPSS Modeler offers a great tool for campaign selection. After studying a Gains/Profit chart, users can insert a vertical line at the selected tile which automatically determines the underlying propensity threshold. This feature enables the generation of a Select node to filter the records with propensity values above the threshold to be included in the target list.

Here is a word of caution on the use and the interpretation of propensities. As noted in Section 2.3.7, when balancing or case weighting has been applied in the model training phase, the propensity values do not correspond to the actual probabilities. However, propensities are still comparable and can be used to rank customers according to their target likelihood.

@Tech tips for applying the methodology

Scoring customers in IBM SPSS Modeler

The IBM SPSS Modeler generated models are scoring engines. When the scoring dataset is passed through the generated model, as shown in Figure 2.54, the prediction and the prediction confidence fields (with a \$ and a \$C prefix, respectively) are generated.

In the case of binary classifiers, the hit propensities can also be requested (derived with a \$RP prefix; Figure 2.55). Besides these raw propensities, users can also estimate the adjusted propensities (with a \$AP prefix). These propensities are based on the testing dataset which hasn't been used in model training and hence are considered more realistic estimates of the actual probabilities. Besides, these propensities are unaffected by the balancing or the case weighting applied in the training dataset.

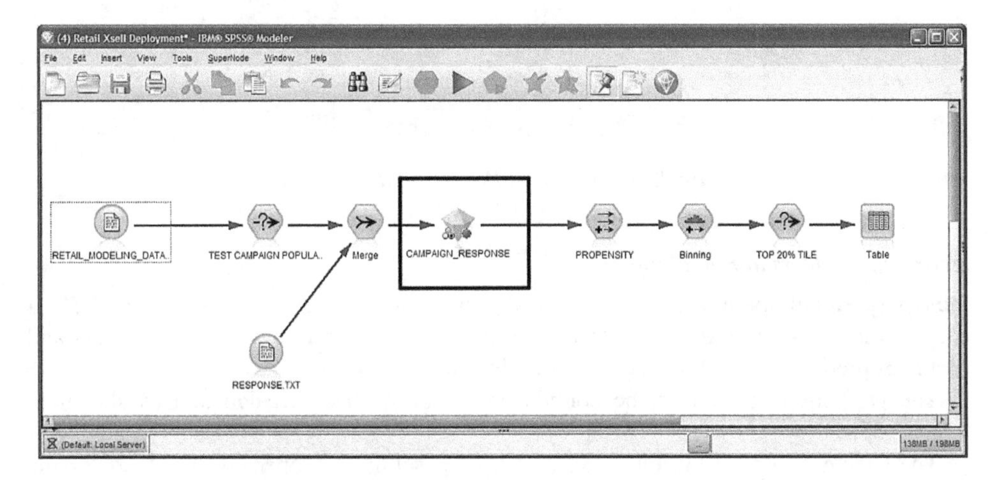

Figure 2.54 Scoring with a generated model in IBM SPSS Modeler

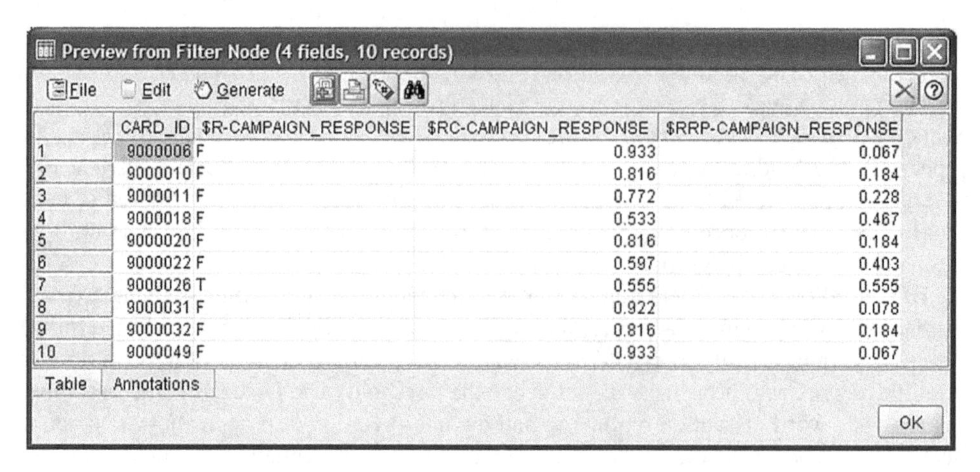

Preview from Filter Node (4 fields, 10 records)

	CARD_ID	$R-CAMPAIGN_RESPONSE	$RC-CAMPAIGN_RESPONSE	$RRP-CAMPAIGN_RESPONSE
1	9000006	F	0.933	0.067
2	9000010	F	0.816	0.184
3	9000011	F	0.772	0.228
4	9000018	F	0.533	0.467
5	9000020	F	0.816	0.184
6	9000022	F	0.597	0.403
7	9000026	T	0.555	0.555
8	9000031	F	0.922	0.078
9	9000032	F	0.816	0.184
10	9000049	F	0.933	0.067

Figure 2.55 The classifier's scores derived by BM SPSS Modeler

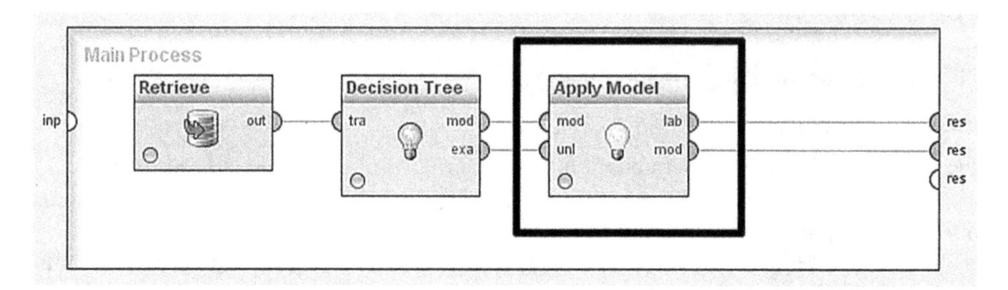

Figure 2.56 The Apply Model RapidMiner operator for scoring with a classifier

ExampleSet (150 examples, 6 special attributes, 4 regular attributes) View Filter (150 / 150): all

Row No.	id	label	confidence(Iris-setosa)	confidence(Iris-versicolor)	confidence(Iris-virginica)	prediction(label)
98	id_98	Iris-versicolor	0	0.979	0.021	Iris-versicolor
99	id_99	Iris-versicolor	0	0.979	0.021	Iris-versicolor
100	id_100	Iris-versicolor	0	0.979	0.021	Iris-versicolor
101	id_101	Iris-virginica	0	0.022	0.978	Iris-virginica
102	id_102	Iris-virginica	0	0.022	0.978	Iris-virginica
103	id_103	Iris-virginica	0	0.022	0.978	Iris-virginica

Figure 2.57 The RapidMiner prediction fields

Scoring customers in RapidMiner

The Apply Model operator is used for scoring in RapidMiner as shown in Figure 2.56. It receives the testing (unlabeled) dataset and the generated or stored model as inputs and calculates predictions and confidences for all instances (it "labels" the data).

The prediction as well as the confidence for each class is calculated as shown in Figure 2.57.

The Create Threshold operator can be used for setting a propensity threshold for the target class as shown in Figure 2.58.

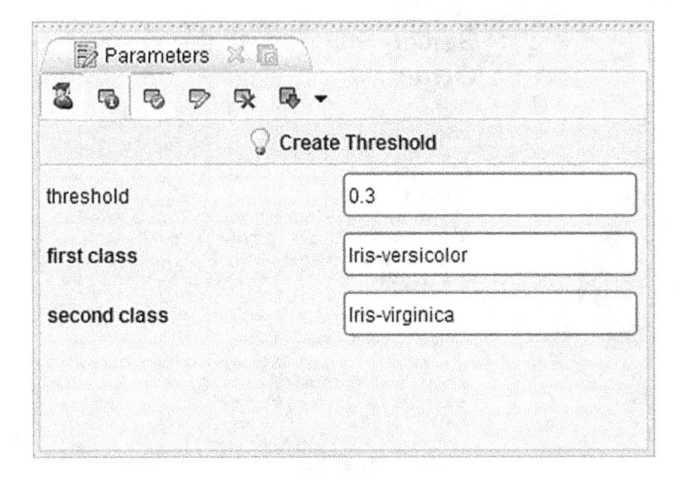

Figure 2.58 The Create Threshold operator for setting a propensity cutoff value

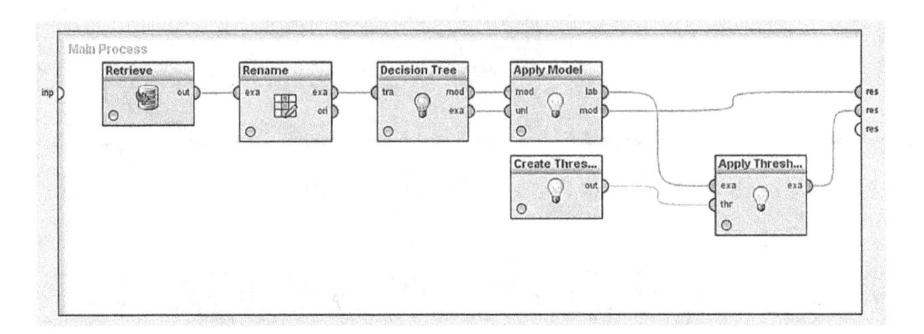

Figure 2.59 Setting a propensity threshold with the Create Threshold operator

A scored record with a hit propensity above the chosen cutoff value is classified in the target class (referred by RapidMiner as the second class, in this example "Iris-virginica"); otherwise, it is classified in the first class (in this example "Iris-versicolor"). The selected threshold is fed through an Apply Threshold operator to the labeled (scored with an Apply Model operator) dataset as shown in Figure 2.59.

Scoring customers in Data Mining for Excel

In Data Mining for Excel, the scoring procedure is implemented using the Query wizard (Figure 2.60).

Users must select the model as well as the scoring dataset which, as mentioned previously, can be the model dataset, an Excel spreadsheet, or even an external data source read with the appropriate query. After the required mapping of the model inputs (the mining inputs) with the dataset fields, users can select the score fields to be estimated, including, among other outputs, the model prediction and its confidence as shown in Figure 2.61.

In binary classification problems, the hit propensity equals the prediction confidence, if the prediction is the target class or 1 minus the prediction confidence otherwise. The scored dataset can be extracted in an Excel worksheet as shown in Figure 2.62.

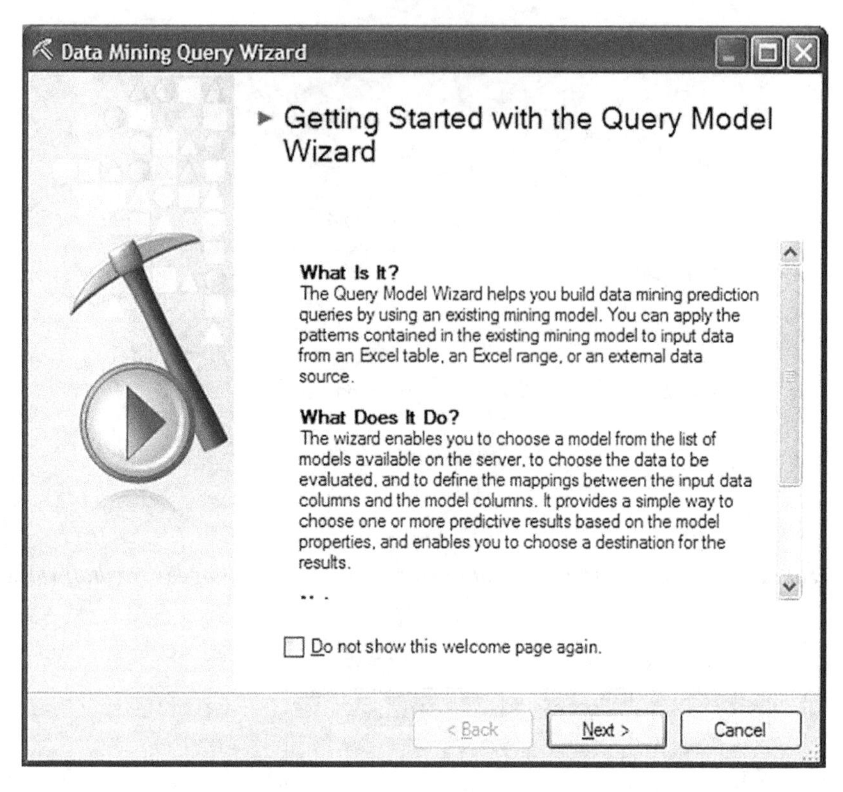

Figure 2.60 The Query wizard used in Data Mining for Excel for model deployment

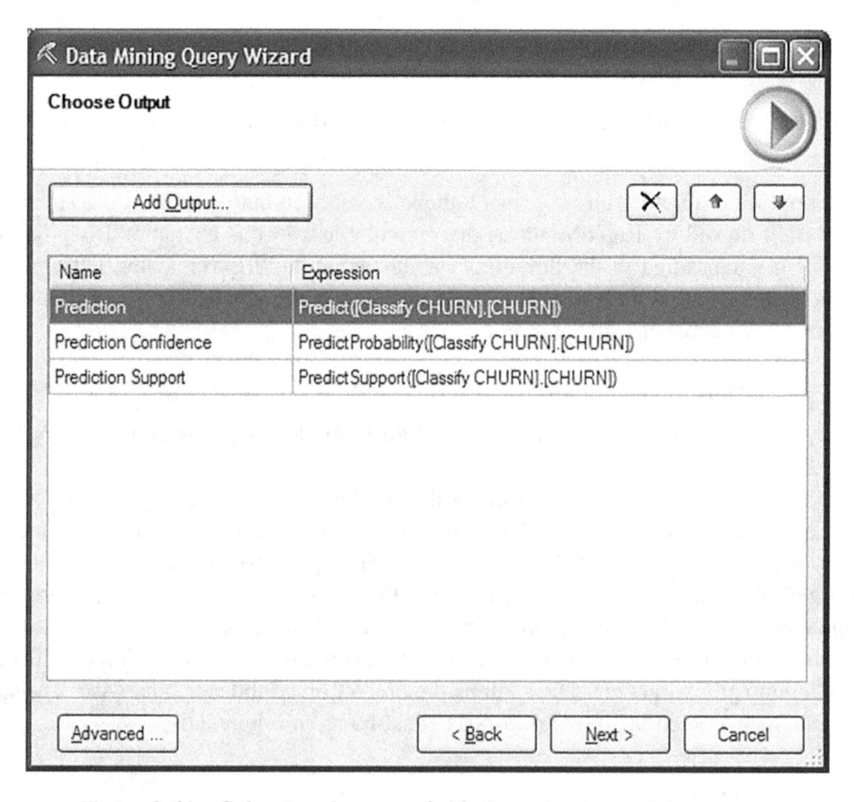

Figure 2.61 Selecting the score fields from the Query Wizard menu

CUSTOMER_ID	Prediction	PredictionConfidence	ChurnPropensity
C100001	F	0.758	0.242
C100002	F	0.975	0.025
C100003	F	0.894	0.106
C100004	F	0.975	0.025
C100005	F	0.861	0.139
C100006	F	0.949	0.051
C100007	F	0.894	0.106
C100008	F	0.975	0.025
C100009	F	0.894	0.106
C100010	F	0.758	0.242
C100011	F	0.941	0.059
C100012	F	0.894	0.106
C100013	F	0.989	0.011

Figure 2.62 Deploying a classifier in Data Mining for Excel

2.6.1.1 Building propensity segments

V.1. Scoring customers to roll the marketing campaign	*V.1.1. Building propensity segments*

Estimated propensities can be used for grouping customers in segments, in propensity pools, according to their target class likelihood. For instance, customers can be divided into groups of low, medium, and high churn likelihood as follows:

1. High Churn Risk
   Comprised of scored customers with churn scores above a selected propensity threshold with business meaning, for example, at least n times higher than the total population churn rate. This segment will be the basic pool for customers to be included in a retention campaign.

2. Medium Churn Risk
   Includes the customers with churn propensities lower than the cutoff value for the high-risk group but higher than the observed overall population churn rate.

3. Low Churn Risk
   Includes the customers with the lowest churn propensities, for example, below the observed overall population churn rate.

If the propensity segmentation is based on specific propensity cutoff values ("hard" thresholds) and it is monitored over time, then possible changes in the distribution of the pools should be tracked, investigated, and explained. If alternatively the grouping is based on frequency bins and propensity percentiles ("soft" thresholds, e.g., top 10%, medium 40%, low 50%), then the boundaries of the pools should be monitored over time to identify changes in the propensity distribution.

Propensity scores and respective segmentations can also be combined with other standard segmentation schemes such as value-based segments. For instance, when value segments are cross-examined with churn probability segments, we have the value-at-risk segmentation, a compound segmentation which can help in prioritizing the retention campaign according to each customer's value and risk of defection.

2.6.2 Designing a deployment procedure and disseminating the results

V. Model deployment	***V.2. Designing a deployment procedure and disseminating the results***

The final stage of propensity modeling includes the design of the deployment procedure and the dissemination of the results and findings. The deployment procedure must be a scheduled, automated, and standardized process which should:

1. Gather the required model inputs from the mining datamart and/or all necessary data sources
2. Prepare the predictors and apply the generated model to refresh the predictions and scores
3. Load the predictions and propensity scores to the appropriate systems of the organization (e.g., data warehouse, campaign management systems) from which they can be accessible and usable for marketing campaigns and insight on the customer base

2.7 Using classification models in direct marketing campaigns

Marketing applications aim at establishing a long-term and profitable relationship with the customers throughout the whole customer lifetime. Classification models can play a significant role in marketing, specifically in the development of targeted marketing campaigns for acquisition, cross-/up-/deep-selling, and retention. Table 2.11 presents a list of these applications along with their business objective.

Table 2.11 Marketing applications and campaigns that can be supported by classification modeling

Business objective	Marketing application
Getting customers	• Acquisition: finding new, profitable customers to increase penetration and to expand the customer base
Developing customers	• Cross-selling: promoting and selling additional products/services to existing customers, for instance, selling investment accounts to savings-only customers • Up-selling: switching customers to premium products. By the term premium products, we refer to products more profitable than the ones they already have. An example is the offering of a hold credit card to holders of a normal credit card • Deep-selling: selling more, increasing usage of the products/services that customers already have. For instance, increasing the balance of existing savings accounts
Retaining customers	• Retention: prevention of voluntary churn, with priority given to presently or potentially valuable customers

All the aforementioned applications can be supported by classification modeling. A classification model can be applied to identify the target population and recognize customers with increased likelihood for churning or additional purchases. In other words, the target event can be identified and an appropriate classifier can be trained to identify the target class. Targeted campaigns can then be conducted with contact lists based on the generated models.

Setting up a data mining procedure for the needs of these applications requires special attention and cooperation between data miners and marketers. In the next paragraphs, we'll try to outline the modeling process for the aforementioned applications by tackling issues such as:

- How to identify the modeling population
- Which mining approach to follow
- How to define the target and the scoring population

2.8 Acquisition modeling

Acquisition campaigns aim at the increase of the market share through the expansion of the customer base with customers new to the market or drawn from competitors. In mature markets, there is a fierce competition for acquiring new customers. Each organization incorporates aggressive strategies, massive advertisements, and discounts to attract prospects.

Predictive models can be used to guide the customer acquisition efforts. However, a typical difficulty with acquisition models is the availability of input data. The amount of information available for people who do not yet have a relationship with the organization is generally limited compared to information about existing customers. Without data, you cannot build predictive models. Thus, data on prospects must be available.

The mining approaches which can be applied for the needs of acquisition modeling include the following.

2.8.1.1 Pilot campaign

Mining approach: This approach involves the training of a classification model on a random sample of prospects. We assume that a list of prospects is available with sufficient profiling information. A test campaign is run on a random sample of prospects; their responses are recorded and analyzed with classification modeling in order to identify the profiles associated with increased probability of offer acceptance. The trained models can then be used to score all prospects in terms of acquisition probability. The tricky part in this method is that it requires the rollout of a test campaign to record prospect responses in order to be able to train the respective models.

Modeling population: The modeling population of this approach is the random sample of prospects included in the pilot campaign.

Target population: The target population includes those who responded in the campaign.

Scoring population: The scoring population is consisted of all prospects who didn't participate in the campaign.

Hints:

- All marketing parameters of the pilot campaign—such as product, message, and channel—must be the same with the ones of the actual designed campaign.
- A problem with pilot campaigns is that you may need large random samples and hence they are expensive. To build a good model, you need at least 100 respondents (positive responses). With an estimated response rate of around 1%, you need at least 10 000 prospects to achieve this.

The modeling phase of this approach is outlined in Figure 2.63.

2.8.1.2 Profiling of high-value customers

Mining approach: An alternative approach, often combined with the one described earlier, is to mine the list of prospects looking for potentially valuable customers. According to this approach, a classifier is trained on existing customers to identify the key characteristics of the high-value customers. The trained model is then deployed on the prospects to discern the ones with similar characteristics. Propensities now indicate similarity to high-value customers and not likelihood to uptake an acquisition offer.

Modeling population: The model training in this approach is based on existing customers.

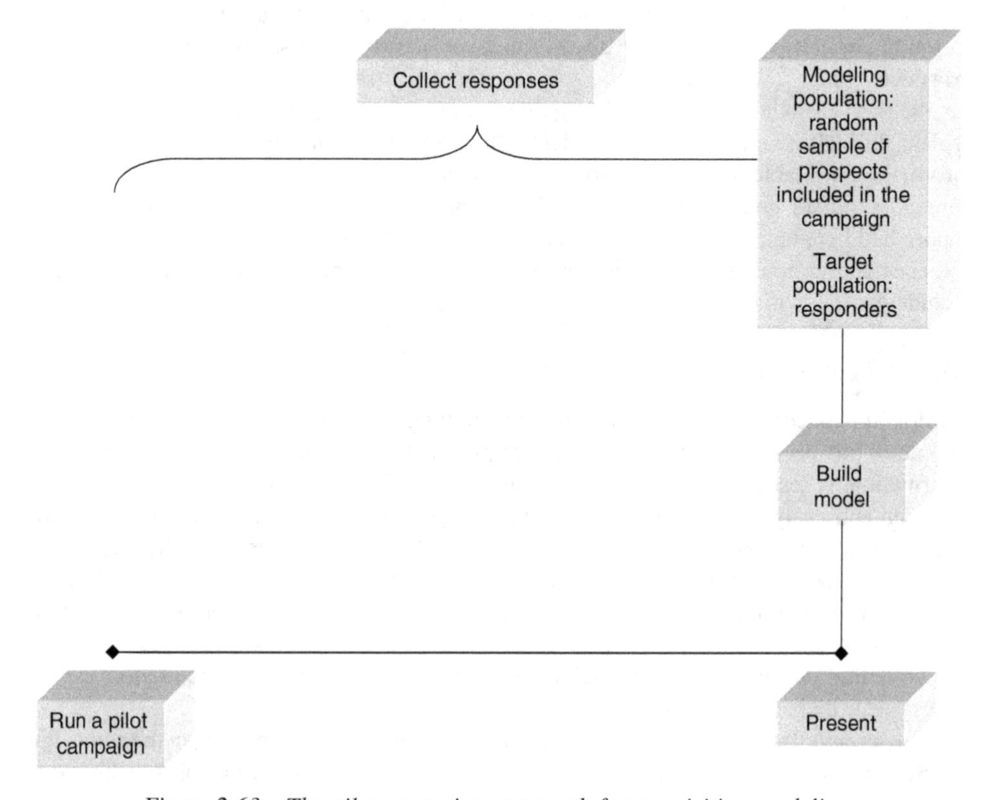

Figure 2.63 The pilot campaign approach for acquisition modeling

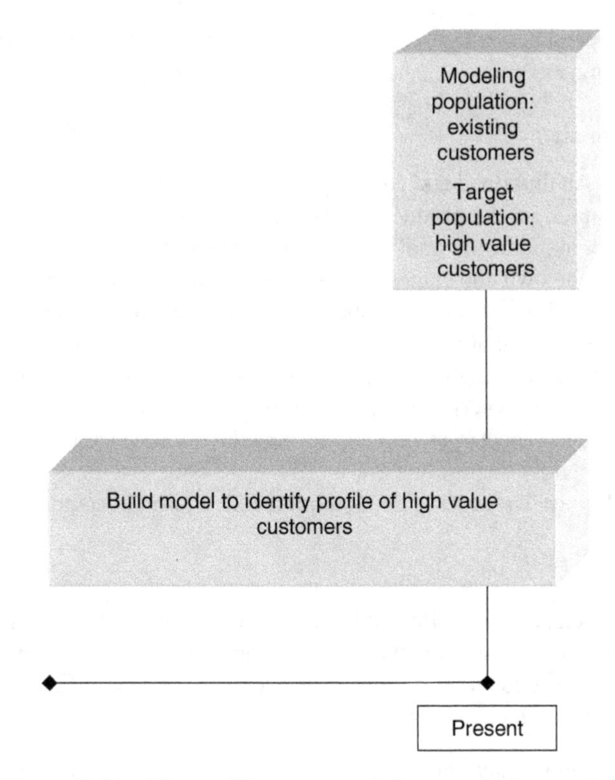

Figure 2.64 The profiling approach for acquisition modeling

Target population: The target population is comprised of high-value customers, for instance, customers belonging to the highest value segments.

Scoring population: The model rules are applied to the list of prospects.

Hints:

- The key to this process is to build a model on existing customers using only fields which are also available for prospects.
- For example, if only demographics are available for prospects, the respective model should be trained only with these data. Acquisition marketing activities could target new customers with the "valuable" profile, and new products related to these profiles could be developed, aiming to acquire new customers with profit possibilities.

The modeling phase of this approach is illustrated in Figure 2.64.

2.9 Cross-selling modeling

Cross-selling campaigns aim at selling additional products to existing customers, for instance, promoting investment products to savings-only bank customers or persuading voice-only callers to start using other mobile phone services. These campaigns can be targeted with the use of classification models. The models can estimate the relevant propensities based on the identified data patterns associated with offer acceptance/product uptaking.

The mining approaches which can be used for the identification of probable buyers include the following.

2.9.1.1 Pilot campaign

Mining approach: An outbound test campaign is rolled out on a random sample of existing customers who do not own the product. Responses are mined with classification modeling and the profiles of responders are analyzed. The generated model is then applied to existing customers who do not own the target product/service and hadn't participated in the test campaign. Customers with increased probability to uptake the offer are identified and included in the large-scale campaign that follows.

Modeling population: The modeling population of this approach is the random sample of existing customers, not owners of the target product, included in the pilot campaign.

Target population: The target population includes those responded positively in the campaign offer and bought the product.

Scoring population: The scoring population includes current customers who do not own the product.

Hints:

- This approach can also be followed in the case of an inbound campaign. During a test period, a test inbound campaign is carried out on a sample of incoming contacts. Their responses are recorded and used for the development of a classifier which then targets incoming cross-selling campaigns.

- As mentioned previously, this approach is demanding in terms of time and resources. However, it is highly effective since it is a simulation of the actual planned campaign, provided of course all aspects—namely, product, message, channel, and direction (inbound/outbound)—are the same with the ones of the designed campaign.

The modeling phase of this approach is outlined in Figure 2.65.

2.9.1.2 Product uptake

Mining approach: Historical data are used and a classifier is trained on customers who did not own the target product at a specific time point in the recent past. The characteristics of the recent "buyers" are identified, and customers with the same outlook who do not currently own the product are selected for inclusion in the campaign.

The "product uptake" approach is effective since it tries to identify the behavioral patterns in the observation period which were followed by the occurrence of the target event, in this case product purchase. However, it requires the building of a customer view in more than one time points. In order to build the model, we must go back and use historical data which summarize the customer behavior at a past time period before the event occurrence. And then we have to move forward in time and use the current customer view in order to score customers with the generated model and estimate their likelihood to uptake the product.

Modeling population: All active customers not owning the target product at the end of the analyzed observation period.

Target population: Those customers who acquired the product in the recent past, within the examined event outcome period.

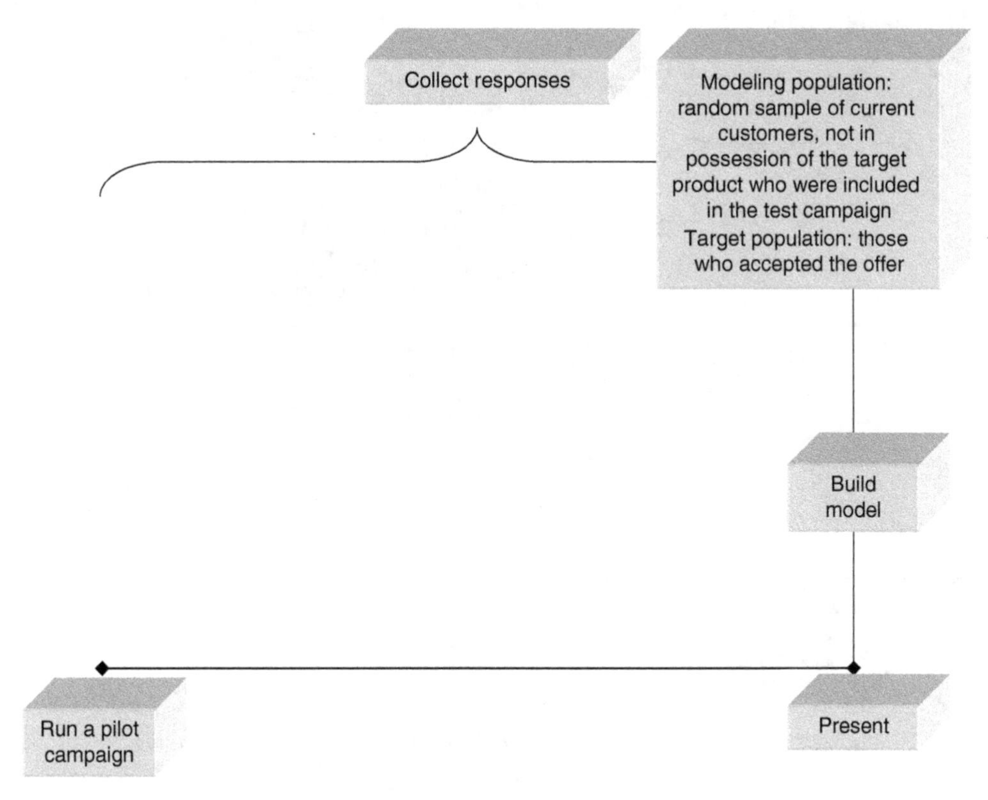

Figure 2.65 The pilot campaign approach for cross-selling modeling

Scoring population: All active customers not currently owning the product.
Hints:

- This approach is effective; however, it is demanding in terms of data preparation since it involves different time frames. In model training, the customer "signature" is built in the observation period, and it is associated with subsequent product purchase. Then, current "signatures" are used for model deployment.
- The model propensities of this approach denote the likelihood of buying the product in the near future and can be used for targeting a planned cross-selling campaign. However, they are not estimates of offer acceptance probabilities as the ones calculated with the pilot campaign method.

The modeling phase of this approach is illustrated in Figure 2.66.

2.9.1.3 Profiling of owners

Mining approach: A model is trained on all active customers and identifies the data patterns and characteristics associated with ownership of the target product. The profile of owners (and preferably heavy users) of the target product is outlined. Customers with the same

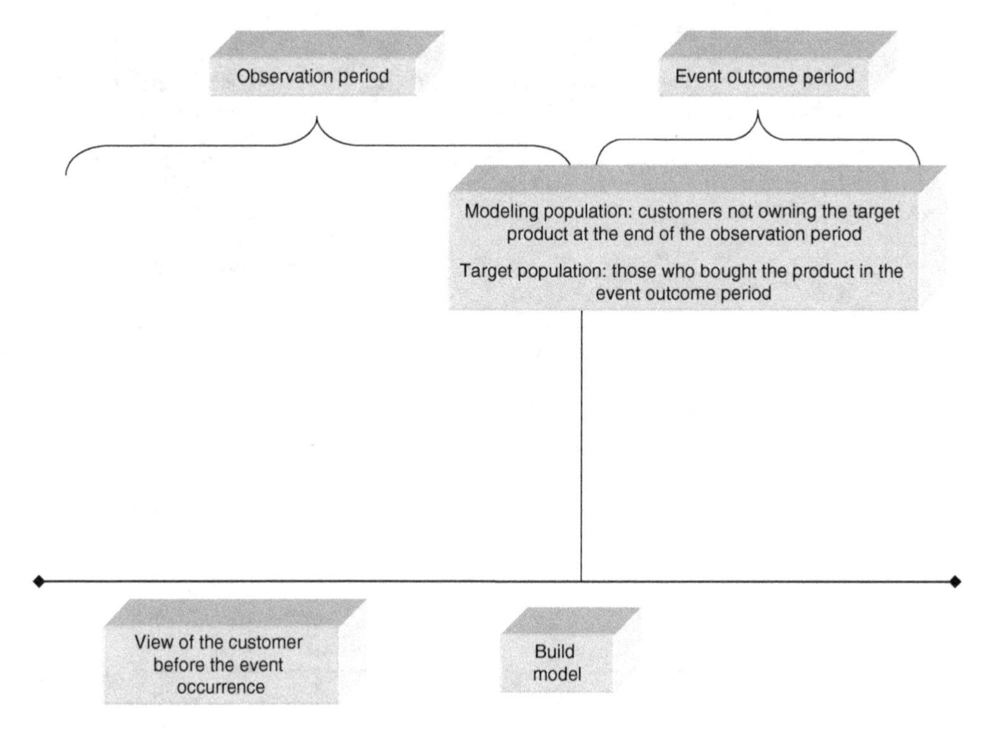

Figure 2.66 The product uptake approach for cross-selling modeling

profile are identified among the population of nonowners, and they are included in the planned cross-selling campaign.

Modeling population: All currently active customers.

Target population: Customers owning (and preferably heavily using) the product to be promoted.

Scoring population: All nonowners of the product.

Hints:

- This approach is appealing due to its straightforwardness and simple data preparation. It does not require distinct observation and outcome windows and the examination of the customer view in different time periods. However, this comes at a cost. The predictive efficiency of this approach is limited since it takes into account the current profile of customers instead of the characteristics prior the purchase event who most likely leaded to the purchase decision.

The modeling phase of this approach is presented in Figure 2.67.

2.10 Offer optimization with next best product campaigns

When planning an outbound or an inbound cross-selling campaign, the marketers of the organization should take into account the customer eligibility for the promoted product. Customers who already have the product or have stated that they are not interested in it will have to be left out. So are customers with a bad risk score.

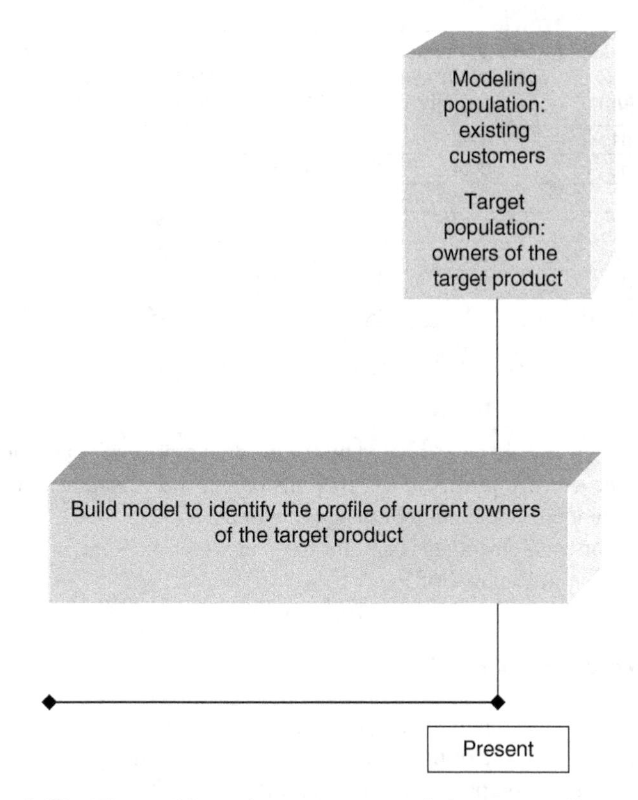

Figure 2.67 The profiling of owners approach for cross-selling modeling

But what happens when customers are eligible for more than one product? How should the organization choose the best product to offer them? The answer is through offer optimization. Offer optimization refers to using analytics and propensity models to identify the products or services that customers are most likely to be interested in for their next purchase. But that's only half the equation. The next best offer strategy should also determine the best offer which will improve the customer lifetime value and will provide value to the organization.

In simple words, each customer should be contacted with the most profitable and most likely to accept product which currently does not own.

This approach requires the development of individual cross-selling models for the different products/product groups. Propensities are combined with the projected net present value (NPV) of each product to estimate the propensity value of each offer as shown in the formula below:

$$\text{Propensity value of the offer} = \text{Product uptake propensity} \times \text{Projected NPV} - \text{Offer cost}$$

The offer cost is the sum of all costs related with the promotion of the product including incentives, cost of the mail piece, etc.

The product uptake propensity is estimated with a respective cross-selling model based on the approach described in Section 2.9.

Customer ID	Offer / product	(model estimated) product uptake propensity	Projected NPV	Cost of the offer	Propensity value of the Offer	Best offer (next best product)
1	Prod1	0.06	$1,000	$2	58	
1	Prod2	0.14	$800	$2	110	x
1	Prod3	0.16	$700	$4	108	
2	Prod1	0.12	$1,000	$2	118	x
2	Prod2	0.14	$800	$2	110	
3	Prod1	0.08	$1,000	$2	78	
3	Prod3	0.17	$700	$4	115	x

Figure 2.68 The next best offer approach

The projected NPV should be estimated by the product manager. A product-level average NPV is based on the average profits of existing customers. It is calculated as the sum of the NPV over n years divided by the number of initial customers.

The best offer for each customer is simply the one which maximizes the estimated propensity value as shown in Figure 2.68.

2.11 Deep-selling modeling

The scope of deep-selling campaigning is to increase the usage of existing products. In other words, they aim at selling more of the same product/service to the customer. As an example, consider the case of a mobile phone network operator which tries to increase the outgoing phone usage (number of minutes of outgoing calls) of its customer base or a bank who targets infrequent users of credit cards, intending to increase their usage. The mining approaches which can be used for the development of deep-selling models are similar to the ones presented in Section 2.9 for cross-selling.

2.11.1.1 Pilot campaign

Mining approach: Customers owning but not heavily using the promoted product are identified, and a random sample of them is drawn to be included in the test campaign. Those selected receive an offer which promotes the usage increase, and their responses are collected and analyzed. Those who increased their usage comprise the target population. The trained model is then deployed on “infrequent” users who were left out of the pilot campaign, and those scored with propensities above the selected threshold are targeted.

Modeling population: The modeling population of this approach is the random sample of owners but “infrequent” users of the target product, included in the pilot campaign.

Target population: The target population includes those who increased their usage after receiving the respective offer.

Scoring population: “Infrequent” users who didn’t participate in the test campaign.

Hints:

- As in the case of cross-selling models, this approach can also be followed in the case of inbound campaigns.

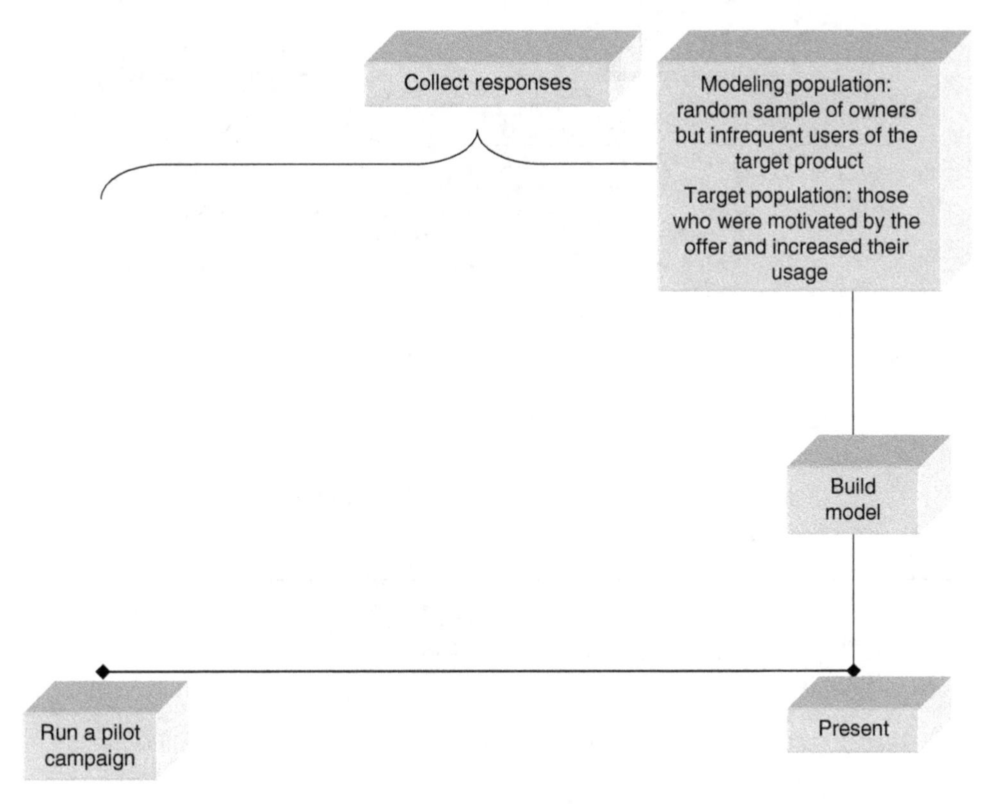

Figure 2.69 The pilot campaign approach for deep-selling modeling

The modeling phase of this approach is outlined in Figure 2.69.

2.11.1.2 Usage increase

Mining approach: A historical view of the customers is assembled, and customers owning but not heavily using the deep-sell product at the end of the observation period are analyzed. Those who substantially increased their usage in the event outcome period are flagged as "positive" cases and form the target class of the classification model. The model is then deployed on current owners/low users of the product. Those scored with high deep-selling propensities are included in the campaign that follows. The "usage increase" approach, as all approaches based on "historical views" of customers, is effective, yet it presents difficulties in its implementation compared to plain profiling approaches.

Modeling population: All owners/low users of the target product at the end of the observation period.

Target population: Those customers who increased their product usage in the outcome period that followed.

Scoring population: All active customers currently owning but infrequently using the deep-sell product.

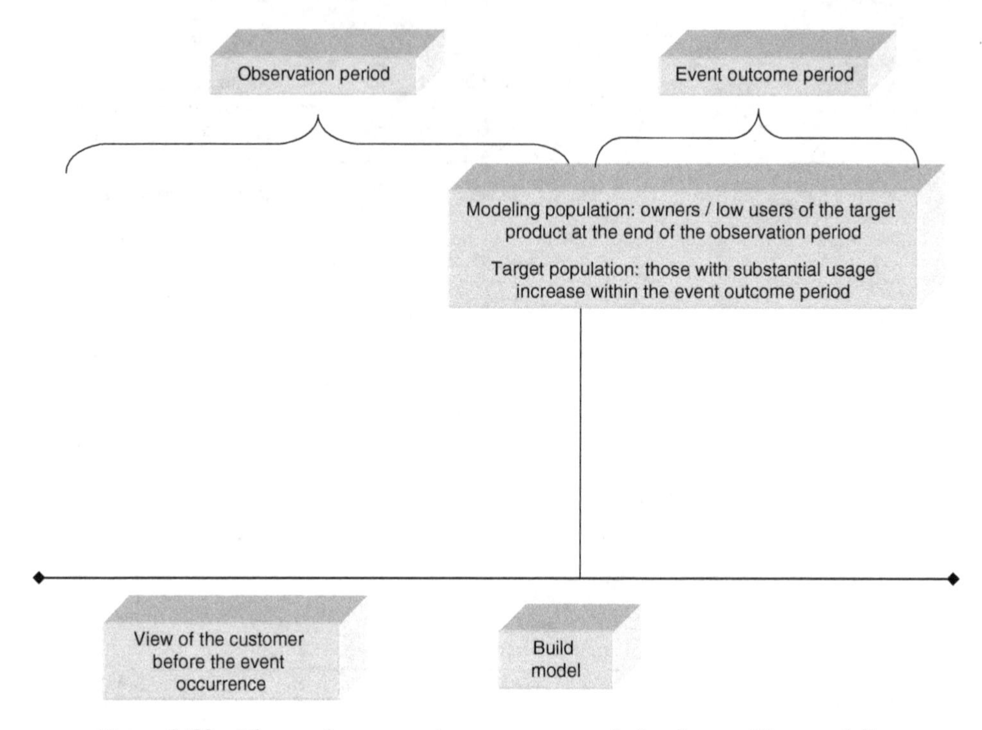

Figure 2.70 The product usage increase approach for deep-selling modeling

Hints:

- An issue which deserves attention and the close collaboration of data miners with the marketers is, as always, the definition of the modeling population and of the target event and/class—in other words, the definition of what constitutes low and heavy usage and what signifies a substantial usage increase. The usage increase can be defined in terms of absolute or relative increase compared to the historical period analyzed.

The modeling phase of this approach is illustrated in Figure 2.70.

2.11.1.3 Profiling of customers with heavy product usage

Mining approach: A model is built on all owners of the deep-sell product which is trained to discern heavy from low users. The drivers of heavy product usage are discovered and clones of heavy users are identified among low users. These customers (low users predicted as heavy users) are targeted by the deep-selling campaign.

Modeling population: All active customers currently owning the target product.
Target population: Those heavily using the product/service to be promoted.
Scoring population: All low users of the product.

The modeling phase of this approach is illustrated in Figure 2.71.

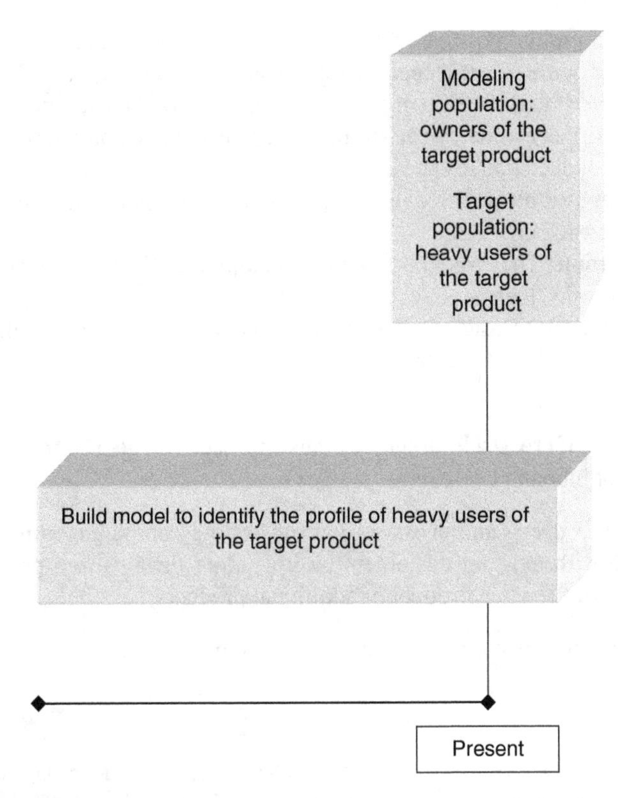

Figure 2.71 The profiling of customers with high usage approach for deep-selling modeling

2.12 Up-selling modeling

The goal of an up-selling campaign is the upgrading of owned products. Existing customers are offered a better and more profitable product than the one they already own. Hence, these campaigns aim at switching customers to "premium" products, and they can be supported by corresponding classification models which estimate the "upgrade" propensities.

For instance, an up-selling campaign of a mobile network operator might address prepaid or basic rate plan customers, trying to promote a postpaid contract or an upgraded rate plan. Ideally, the offered rate plan should be tailored to the customer behaviors and therefore satisfy their needs. At the same time, it should aim at growing and tightening the customer—operator relationship, allowing an immediate or prospect profit increase. Likewise, a gold credit card can be offered to basic cardholders of a bank.

The mining approaches appropriate for the design of up-selling models are once again similar to the ones presented in Section 2.9 for cross-selling.

2.12.1.1 Pilot campaign

Mining approach: A classifier is built on a sample of "basic" customers who were randomly chosen and received an up-selling offer for the target "premium" product. The model is trained on the recorded responses of the test-campaign list. Those who "switched" to the "premium" product are the positive instances and comprise the target population. The data patterns associated with offer acceptance are identified and captured by the generated model which is then deployed on the mass population of "basic" product owners. An "upgrade" propensity is estimated for all the scored customers, and a large-scale up-selling campaign is then conducted using these propensities.

Modeling population: The modeling population of this approach is the random sample of owners of the "basic" product.

Target population: The target population is comprised of those who accepted the offer and agreed to "upgrade" their product.

Scoring population: "Basic" product owners who didn't participate in the test campaign and who obviously don't have the "premium" product.

Hints:

- As in the case of cross-selling models, this approach can also be modified and applied in the case of inbound campaigns.
- As opposed to deep- and cross-selling, up-selling campaigns aim at widening the relationship with existing customers by upgrading their current products instead of selling more of the same product or additional products.

The modeling phase of this approach is outlined in Figure 2.72.

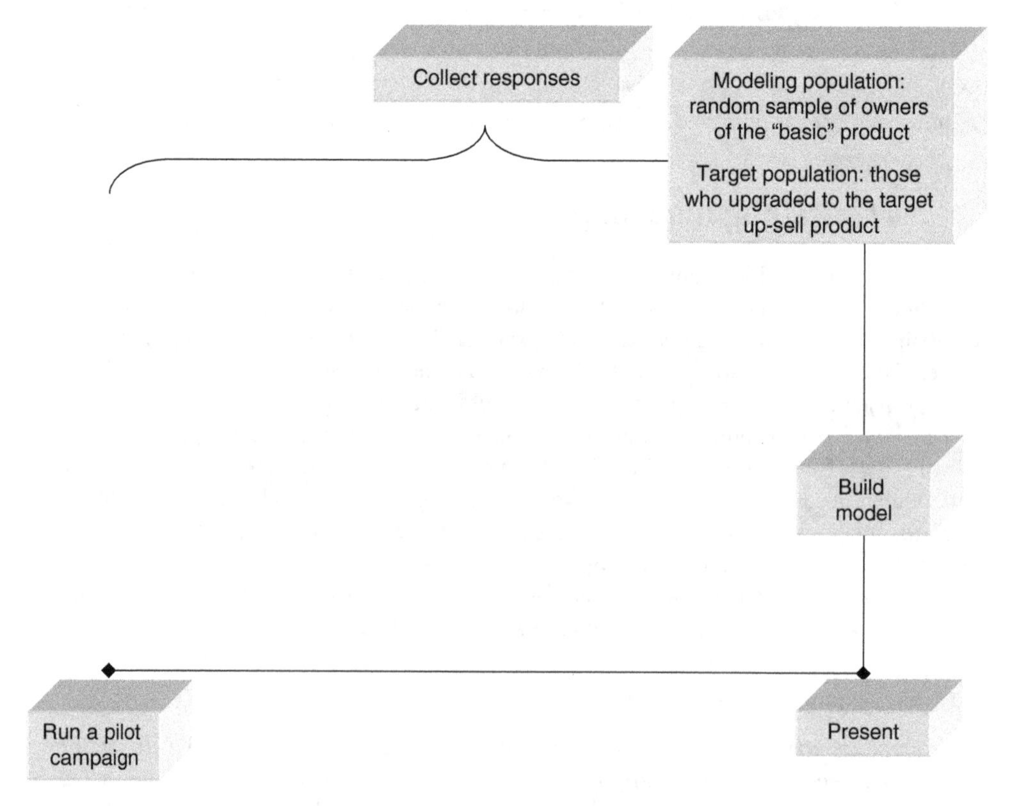

Figure 2.72 The pilot campaign approach for up-selling modeling

2.12.1.2 Product upgrade

Mining approach: This approach requires historical data and a historical view of the customers. A classifier is built on customers who owned the "basic" product/service at a specific time point in the recent past (observation period). The generated model is trained to identify the data patterns associated with upgrading to the premium product during the event outcome period. The scoring phase is based on the current view of the customers. The generated model is deployed on current owners of the entry-level product, and those with relatively high upgrade propensities are included in the campaign list. The "product upgrade" approach is analogous to the "product uptake" and the "usage increase" approaches presented for cross-/deep-selling models.

Modeling population: All owners of the "basic" product/service at the end of the observation period.

Target population: Those customers who upgraded to the "premium" target product within the event outcome period.

Scoring population: All customers who currently own the "basic" but not the "premium" product.

The modeling phase of this approach is illustrated in Figure 2.73.

2.12.1.3 Profiling of "premium" product owners

Mining approach: A classification model is built on the current view of both the "basic" and the "premium" product owners. The generated model discerns the data patterns associated with ownership of the target up-sell product. The scoring population is consisted of all owners

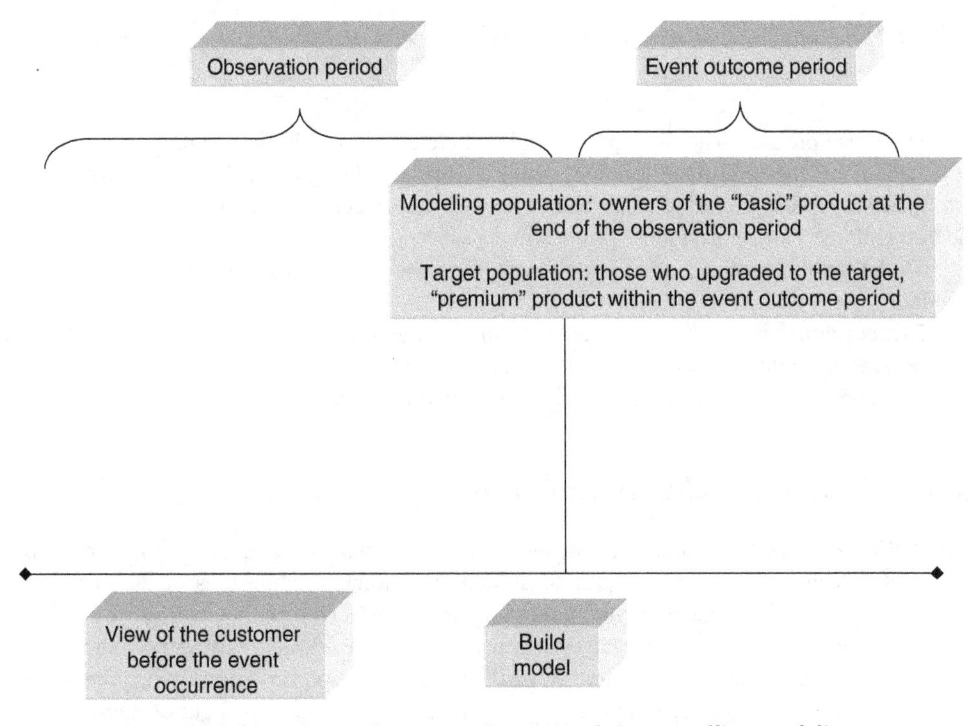

Figure 2.73 The product upgrade approach for up-selling modeling

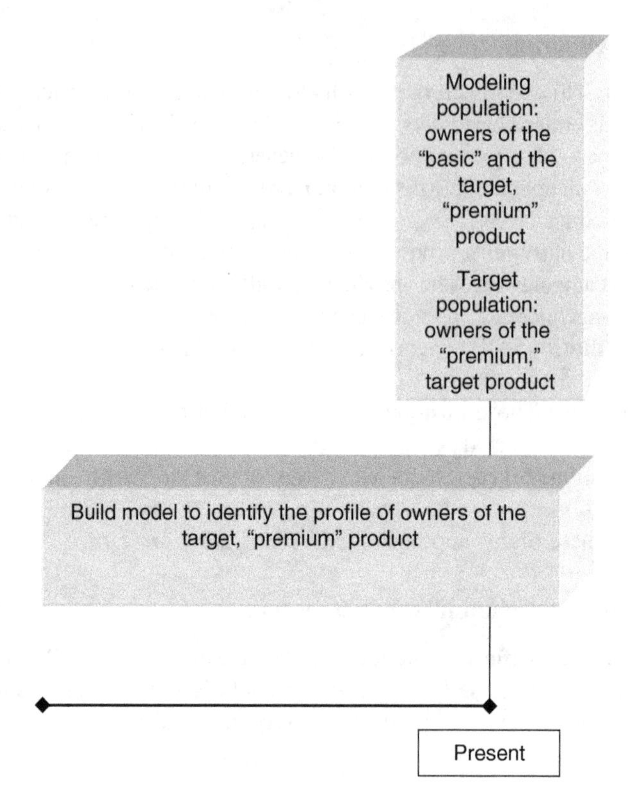

Figure 2.74 The profiling of "premium" product owners approach for up-selling modeling

of the "basic" product. The final campaign list includes all the "basic" product owners who are "clones" of the "premium" product owners. That is, although they do not own the target product, they present similar characteristics with its owners, and consequently, they are scored with increased "premium" product ownership propensities.

Modeling population: All current owners of the "basic" and the "premium" target product.

Target population: Owners of the "premium" target product.

Scoring population: All owners of the "basic" product.

The modeling phase of this approach is presented in Figure 2.74.

2.13 Voluntary churn modeling

The CRM scope is to establish, grow, and retain the organization–customer relationship. Acquisition campaigns cover the part of attracting customers, while cross/deep/up-selling campaigns deal with widening and upgrading the relationship with existing customers. Retention campaigns, on the other hand, target at preventing the leak of customers to the competition. The prevention of voluntary churn is typically focused on currently or potentially valuable customers. These campaigns can be targeted with the development of data mining models, known as churn or attrition models, which analyze the behavior of churned customers and identify data patterns and signals associated with increased probability to leave.

These models should be based on historical data which should adequately summarize the customer behavior and characteristics at an observation period preceding the churn event.

Mining approach: A classification model is trained on customers that were active at the end of the observation period analyzed. These customers comprise the training population. The target population is consisted of those that have voluntary churned (for instance, applied for disconnection) within the event outcome period. The current view of the customers is used in the model deployment phase in which customers presently active are scored according to the model and their churn propensities are estimated.

Modeling population: Customers active at the end of the observation period.

Target population: Those who left within the event outcome period.

Scoring population: All presently active customers.

The modeling phase of this approach is outlined in Figure 2.75.

Hints:

- A more sophisticated approach would limit the training and consequently the scoring population only to customers with substantial value for the organization. With this approach, trivial customers are excluded from the retention campaign.
- Typically, a latency period of 1-3 months is reserved for scoring and campaign preparation as described in 2.2.3.4. Data from the latency period are not used as inputs in the model.

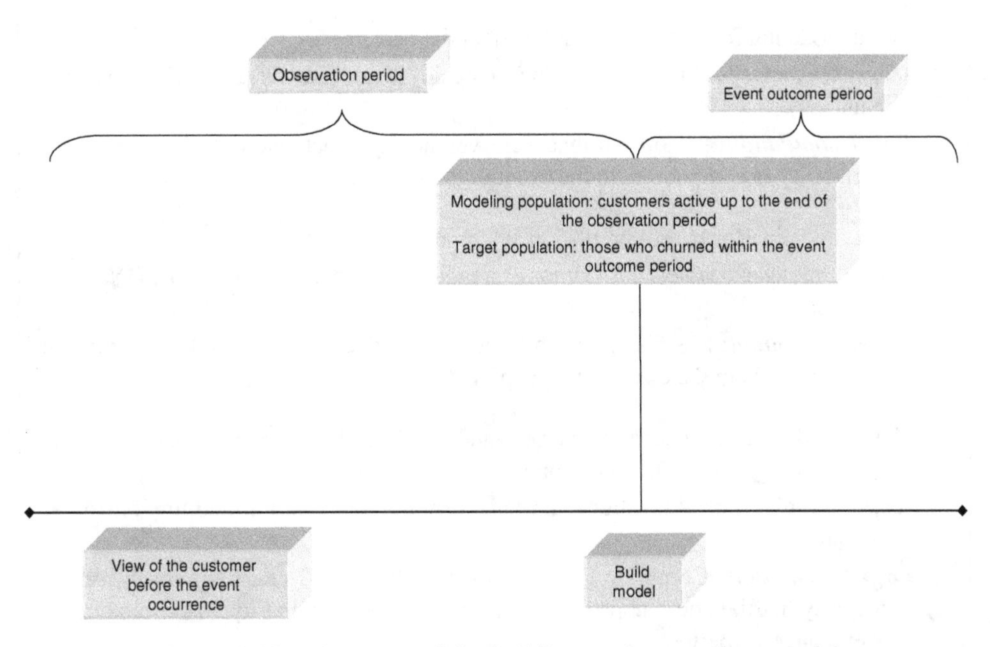

Figure 2.75 The approach for building a voluntary churn model

Examples of voluntary churn modeling for various industries

As stressed many times in this book, one of the most critical points in designing the mining approach for a successful marketing campaign is the definition of the target event. Since it has to make sense from a business point of view, it requires the cooperation of the marketers and the data miners involved. In the following paragraph, we present proposed churn modeling approaches for various major industries. These approaches should be seen as start-up ideas and not as final solutions for model training since each business objective and situation is unique and it requires special consideration.

Churn modeling in banking

Scope: Predict possible churners among bank customers.

Target population: Those who closed all (full churn approach) or the majority (partial churn approach) of their product accounts in the event outcome period.

Credit cards' churn modeling: the termination approach

Scope: Predict voluntary churn among credit cardholders

Target population: Those who voluntarily closed all (full churn approach) or the majority (partial churn approach) of their cards in the event outcome period.

Credit cards' churn modeling: the usage approach.

Scope: Predict voluntary churn among credit cardholders.

Target population: Those who became inactive and made none or trivial spending during the event outcome period.

Churn modeling for telcos, postpaid, residential customers

Scope: Predict possible churn among postpaid, residential (retail) MSISDNs (line numbers).

Target population: MSISDNs that were voluntarily deactivated and/or applied for portability to competitors in the event outcome period.

Churn modeling for telcos, prepaid, residential customers

Scope: Predict possible churn among prepaid, residential (retail) MSISDNs (line numbers).

Target population: MSISDNs which turned to inactive and presented zero or trivial outgoing calls in the event outcome period.

Churn modeling for telcos, corporate, and small/medium enterprise/small office/home office (SME/SOHO) customers

Scope: Predict possible churn among corporate and small/medium customer accounts.

Target population: Corporate accounts for which all (full churn approach) or the majority (partial churn approach) of their lines were voluntarily deactivated in the event outcome period.

Scoring population: Corporate customer accounts active at present.

2.14 Summary of what we've learned so far: it's not about the tool or the modeling algorithm. It's about the methodology and the design of the process

Classification modeling requires more than a good knowledge of the respective algorithms. Its success strongly depends on the overall design instead of the selected software and algorithm. The conclusion after experimenting with the software tools presented in this book (IBM SPSS Modeler, Data Mining for Excel, and RapidMiner) is that it's mainly a matter of the roadmap than of the selected vehicle. All three tools provide excellent modeling algorithms to accurately classify unseen cases, but in the end, classification modeling, and analytics in general, is more than identification of data patterns. It's about the design of the process.

That's why in this chapter we've tried to present a concise and clear step-by-step guide on designing a classification modeling process. We outlined the significance of selecting the appropriate mining approach and focused on the design of the modeling process, dealing with critical issues such as selecting the modeling/analysis level, the target event and population, and the modeling time frames. We provided an overview of the data management tasks typically required to prepare the data for modeling. And we explained the three modeling phases, namely, model training, evaluation, and deployment.

But above all, since this book is mainly addressed to data mining practitioners who want to use analytics for optimizing their everyday business operations, we've concluded this chapter by linking the proposed methodology with real-world applications. We tried to show how this proposed methodology can be applied in real business problems by providing examples for specific direct marketing applications.

3

Behavioral segmentation methodology

3.1 An introduction to customer segmentation

Customer segmentation is the process of dividing customers into distinct, meaningful, and homogeneous subgroups based on various attributes and characteristics.

It is used as a differentiation strategic marketing tool. It enables organizations to understand the customers and build differentiated strategies, tailored to their characteristics.

There are various segmentation types according to the specific segmentation criteria used. Specifically, customers can be segmented according to their value, sociodemographical and life stage information, behavioral, needs/attitudinal, and loyalty characteristics. The type of segmentation to be used depends on the specific business objective. The following types of segmentation are widely used in consumer markets (Source: Tsiptsis and Chorianopoulos (2009). Reproduced with permission from Wiley).

1. **Value based**: In value-based segmentation, customers are grouped according to their value. It is one of the most important segmentation types since it can be used for identifying the most valuable customers and for tracking value and value changes over time. It is also used for differentiating the service delivery strategies and for optimizing the allocation of resources in marketing initiatives.

2. **Behavioral**: A very efficient and useful segmentation type. It is also widely used since it presents minimum difficulties in terms of data availability. The required data include product ownership and utilization data which are usually stored and available in the organization's databases. Customers are divided according to their identified behavioral and

Effective CRM using Predictive Analytics, First Edition. Antonios Chorianopoulos.

Companion website: www.wiley.com/go/chorianopoulos/effective_crm

usage patterns. This type of segmentation is typically used for developing customized product offering strategies, for new product development and design of loyalty schemes.

3. **Propensity based**: In propensity-based segmentation, customers are grouped according to propensity scores, such as churn scores, cross selling scores, etc., which are estimated by respective classification (propensity) models. Propensity scores can also be combined with other segmentation schemes to better target marketing actions. For instance, the value at risk segmentation scheme is developed by combining churn propensities with value segments to prioritize retention actions.

4. **Loyalty based**: Loyalty segmentation involves the investigation of the customers' loyalty status and the identification of loyalty-based segments such as "loyal" and "migrator" segments. Retention actions can then be focused on high-value customers with a nonloyal profile while product offering on prospectively loyal customers.

5. **Sociodemographical**: Reveals different customer groupings based on sociodemographical and/or life stage information such as age, income, and marital status. This type of segmentation is appropriate for promoting specific life stage-based products as well as supporting life stage marketing.

6. **Needs/attitudinal**: This segmentation type is typically based on market research data and identifies customer segments according to their needs, wants, attitudes, preferences, and perceptions pertaining to the company's services and products. It can be used for supporting new product development and for determining the brand image and the key product features to be communicated.

A detailed methodological approach for behavioral segmentation is presented in the following paragraphs.

3.2 An overview of the behavioral segmentation methodology

A segmentation project starts with the definition of the business objectives and ends with the delivery of differentiated marketing strategies for the segments. In this chapter, we'll focus on behavioral segmentation and present a detailed methodological approach for the effective implementation of such projects. The proposed methodology for behavioral segmentation is based on the industry standard CRISP DM methodology, and it includes the following main phases.

1. Business understanding and design of the segmentation process
2. Data understanding, preparation, and enrichment
3. Identification of the segments with cluster modeling
4. Evaluation and profiling of the revealed segments
5. Deployment of the segmentation solution, design, and delivery of differentiated strategies

The sequence of the phases is not strict. Lessons learned in each step may lead analysts to review previous steps. A roadmap of the segmentation procedure is presented in Table 3.1. The phases of the procedure are presented in detail in the following paragraphs.

Table 3.1 The behavioral segmentation methodology

I. *Business understanding and design of the segmentation process*	II. *Data understanding, preparation, and enrichment*	III. *Identification of the segments with cluster modeling*	IV. *Evaluation and profiling of the revealed segments: selecting the optimal cluster solution*	V. *Deployment of the segmentation solution: design and delivery of differentiated strategies*
I.1. Understanding of the business situation, definition of the business objective	**II.1. Investigation of data sources**	**III.1. Trying different models and parameter settings**	**IV.1. "Technical" evaluation of the clustering solution**	**V.1. Building the customer scoring model for updating the segments**
I.2. Design of the modeling process	**II.2. Selecting the data to be used**	**III.2. Proceed to a first rough evaluation of the derived clusters**	**IV.2. Profiling of the revealed segments**	*V.1.1. Using a Decision Tree for scoring: fine-tuning the segments*
I.2.1. Selecting the segmentation population	**II.3. Data integration and aggregation**		**IV.3. Using marketing research information to evaluate the clusters and enrich their profiles**	**V.2. Distribution of the segmentation information**
I.2.2. Selection of the segmentation dimensions	**II.4. Data exploration, validation, and cleaning**		**IV.4. Selecting the optimal cluster solution and labeling the segments**	**V.3. Design and deliver of differentiated strategies**
I.2.3. Determination of the segmentation level	**II.5. Data transformations and enrichment**			
I.2.4. Selecting the observation time window	**II.6. Input set reduction**			

3.3 Business understanding and design of the segmentation process

This phase starts with the understanding of the project requirements from a business perspective. It involves knowledge-sharing meetings and close collaboration between the data miners and the marketers involved in the project to assess the situation, clearly define the specific business goal, and design the whole data mining procedure. In this phase, some crucial questions must be answered, and decisions on some very important methodological issues should be drawn. Tasks of this phase include the following.

3.3.1 Definition of the business objective

I. Business understanding and design of the process	***I.1. Understanding of the business situation, definition of the business objective***

Following the understanding of the business situation, the business objective should be clearly set as it will define all the next steps of the behavioral segmentation procedure. Then it should be translated to a data mining goal. The translation of a marketing objective to a data mining goal is not always simple and straightforward. However, it is one of the most critical issues, since a possible misinterpretation can result in the failure of the entire data mining project.

It is very important for the analysts involved to explain from the beginning to everyone involved in the data mining project the anticipated final deliverables and to make sure that the relevant outcomes cover the initially set business requirements.

3.3.2 Design of the modeling process

I. Business understanding and design of the process	***I.2. Design of the modeling process***

In this phase, crucial decisions should be made, concerning, among other issues, the segmentation perimeter and level as well as the appropriate segmentation dimensions by which customers will be grouped.

3.3.2.1 Selecting the segmentation population

I.2. Design of the modeling process	***I.2.1. Selecting the segmentation population***

This task involves the selection of the customer population to be segmented. Most of the times, the organization will decide to focus on a specific core industry segment (e.g., corporate or retail customers) which needs further analysis and subsegmentation. Core segments, since they have inherent and apparent differences, are typically handled with separate segmentations.

Similarly, customers belonging in obvious segments, such as inactive customers, should be set apart and filtered out from the segmentation procedure in advance. Otherwise, the large differences between active and inactive customers may dominate the solution and inhibit the identification of the existing differences between active customers.

If the size of the selected population is large, a representative sample could be selected and used for the model training. In that case though, a deployment procedure should be designed, for instance, through the development of a relevant classification model, that will enable the scoring of the entire customer base.

3.3.2.2 Selection of the appropriate segmentation criteria

I.2. Design of the modeling process	*I.2.2. Selecting the segmentation dimensions*

One of the key questions to be answered before starting the behavioral segmentation is what attributes should be used for customer grouping. The selection of the appropriate segmentation dimensions depends on the specific business issue that the segmentation model is about to address. The business needs imply, if not impose, the appropriate inputs. Usually, people with domain knowledge and experience can provide a suggestion on the key attribute entities related with the business goal of the analysis. All relevant customer attributes should be identified, selected, and included in the segmentation process. Information not directly related with the behavioral aspects of interest should be omitted.

For instance, if a mobile telephony operator wants to group its customers according to their use of services, all relevant fields, such as the number and volume/minutes of calls by call type, should be included in the analysis. On the contrary, customer information related with other aspects of customer behavior, such as payment behavior or revenue information, should be excluded from the segmentation.

3.3.2.3 Determining the segmentation level

I.2. Design of the modeling process	*I.2.3. Determination of the segmentation level*

The segmentation level defines what groupings are about to be revealed, for instance, groups of customers, groups of telephone lines (MSISDNs in mobile telephony), etc. The selection of the appropriate segmentation level depends on the subsequent marketing activities that the segments are about to support. It also determines the granularity level of the training dataset that will be built for modeling purposes.

3.3.2.4 Selecting the observation window

I.2. Design of the modeling process	*I.2.4. Selecting the observation window*

The goal of the segmentation is usually to outline stable and current behaviors, avoiding inconsistent usage patterns. A "narrow" observation period based on a data snapshot of a few days or even weeks might just capture a random twist in the customer's behavior. Since the objective is to develop a segmentation scheme that would reflect typical behaviors, the observation period must be "wide" enough to avoid behavioral fluctuations but on the other hand not too "wide" to also take into account outdated behaviors.

A general recommendation is to use an observation period of at least 6 months and up to 12 months (especially if we want to take into account yearly, seasonal events).

3.4 Data understanding, preparation, and enrichment

The investigation and the assessment of the available data sources are followed by data acquisition, integration, and processing for the needs of segmentation modeling. The data understanding and preparation phase is probably the most time-consuming phase of the project. Its steps are presented in detail in the following paragraphs.

3.4.1 Investigation of data sources

II. Data understanding, preparation, and enrichment	***II.1. Investigation of data sources***

The available data sources should be evaluated in terms of accessibility and validity. This phase also includes initial data collection and exploration in order to understand the available data.

3.4.2 Selecting the data to be used

II. Data understanding, preparation, and enrichment	***II.2. Selecting the data to be used***

The next step of the procedure involves the definition of the data to be used for the needs of the analysis.

The selected data should cover all the segmentation dimensions as all the additional customer information that will be used for profiling the revealed segments.

Cluster inputs can be numeric as well as categorical. The categorical inputs are handled by the cluster algorithms with appropriate distance measures or internal preprocessing encoding. However, in behavioral segmentation, a general recommendation is to avoid using inputs of mixed scale in the same model as categorical inputs can be overweighted in the formation of the clusters.

Likewise, demographical variables should be avoided in behavioral segmentation projects. Mixing behavioral and demographical information may result in unclear and ambiguous behavioral segments since two customers with identical demographical profile may have completely different behaviors.

For example, imagine the case of a father that has activated a mobile phone line for his teenage son. In a behavioral segmentation, based only on behavioral data, this line would most likely be assigned to the "young—SMS users" segment, along with other teenagers and young techno-fun users. Therefore, we might expect some ambiguities when trying to examine the demographical profile of the segments. In fact, this hypothetical example also outlines why the usage of demographical inputs should be avoided when the main objective is behavioral separation.

Only variables relevant with the specific business objective should be included in the model. Mixing all available attributes in an attempt to build a "silver-bullet" total segmentation that will cover all the aspects of a customer relationship with the organization (e.g., usage and payment behavior) usually leads to unclear segments with poor actionability. It is suggested to develop different segmentation schemes independently and then combine them to build differentiated marketing strategies.

3.4.3 Data integration and aggregation

II. Data understanding, preparation, and enrichment	***II.3. Data integration and aggregation***

The initial raw data should be consolidated for the creation of the final modeling dataset that will be used for the identification of the segments. This task typically includes the collection, filtering, merging, and aggregation of the raw data based on the modeling process designed.

For behavioral segmentation applications, a recent "view" of the customers' behavior should be constructed and used. This "view" should summarize the behavior of each customer by using at least 6 months of observation data.

As outlined previously, the aggregation level of the modeling dataset should correspond to the wanted segmentation level. If the goal, for instance, is to segment bank customers, then the final dataset should be at a customer level. If the goal is to segment telephone lines (MSISDNs), the final dataset should be at a line level. To put it in a simple way, clustering techniques reveal natural groupings of records. So no matter where we start from, the goal is the construction of a final, one-dimensional, flat table, which summarizes behaviors at the selected granularity.

This phase is concluded with the retrieval and consolidation of data from multiple data sources (ideally from the organization's mining datamart and/or Marketing Customer Information File) and the construction of the modeling dataset.

3.4.4 Data exploration, validation, and cleaning

II. Data understanding, preparation, and enrichment	***II.4. Data exploration, validation, and cleaning***

A critical issue for the success of any data mining project is the validity of the used data. The data exploration and validation process includes the use of simple descriptive statistics and charts for the identification of inconsistencies, errors, missing values, and outlier (abnormal) cases. Outliers are cases that do not conform to the patterns of "normal" data. Various statistical techniques can be used in order to fill in (impute) missing or outlier values.

Outlier cases in particular require extra caution. Outliers are records with extreme values and unusual data patterns. They can be singled out by examining data through simple descriptive statistics or specialized algorithms. Outliers deserve special investigation as in many cases they are what we are looking for: exceptionally good customers or, at the other end, fraudulent cases. But they may have a negative impact on clustering models.

They can misguide the clustering algorithm and lead to poor and distorted results. In many cases, the differences between "outlier" and "normal" data patterns are so large that they may mask true differences among the majority of the "normal" cases. As a result, the algorithm can be misled to a poor solution that merely separates outliers. Consequently, the clustering model may come up with a poor solution consisting of one large cluster of "normal" behaviors and many very small clusters representing the unusual data patterns. This analysis may be useful, for instance, in fraud detection, but certainly, it is not appropriate in the case of general purpose segmentation. A recommended approach for an enriched general purpose solution would be to identify and exclude outliers in advance. Another approach could be to run an initial clustering solution, identify small outlier clusters, and then rerun and fine-tune the analysis after excluding the outlier clusters.

@Tech tips for applying the methodology

Data exploration and cleaning in RapidMiner

RapidMiner includes a complete toolbox for data cleaning. The Replace Missing Values operator fills in missing values with statistical measures such as the average for scale attributes and the mode for nominal ones (Figure 3.1).

It also offers four different algorithms for outlier detection. The operator of the algorithm which is based on *k*-nearest neighbors is shown in Figure 3.2.

Data exploration and cleaning in IBM SPSS Modeler

IBM SPSS Modeler offers a handy and automated tool for data preparation, the Auto Data Prep procedure (Figure 3.3). Among other useful functions, it supports missing value replacement with the mean, the mode, or the median.

The anomaly detection algorithm, based on cluster analysis, can be applied in Modeler for detection of outlier instances (Figure 3.4).

Data exploration and cleaning in Data Mining for Excel

The Clean Data wizard of Data Mining for Excel can also be used for handling of outliers. Analysts can study the distribution of an attribute (Figure 3.5) and set the range of acceptable values as well as replacement rules for possible outliers (Figure 3.6).

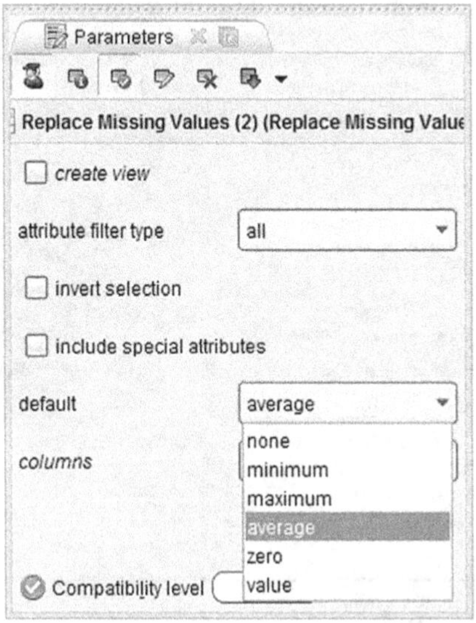

Figure 3.1 The Replace Missing Values operator of RapidMiner for handling missing values

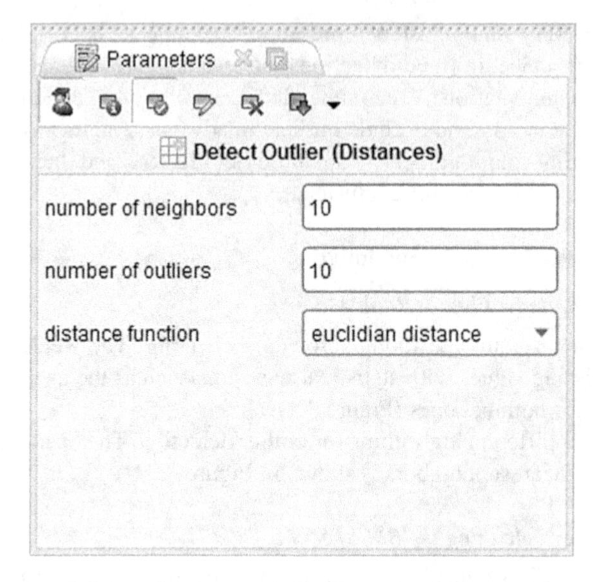

Figure 3.2 The Detect Outlier operator of RapidMiner

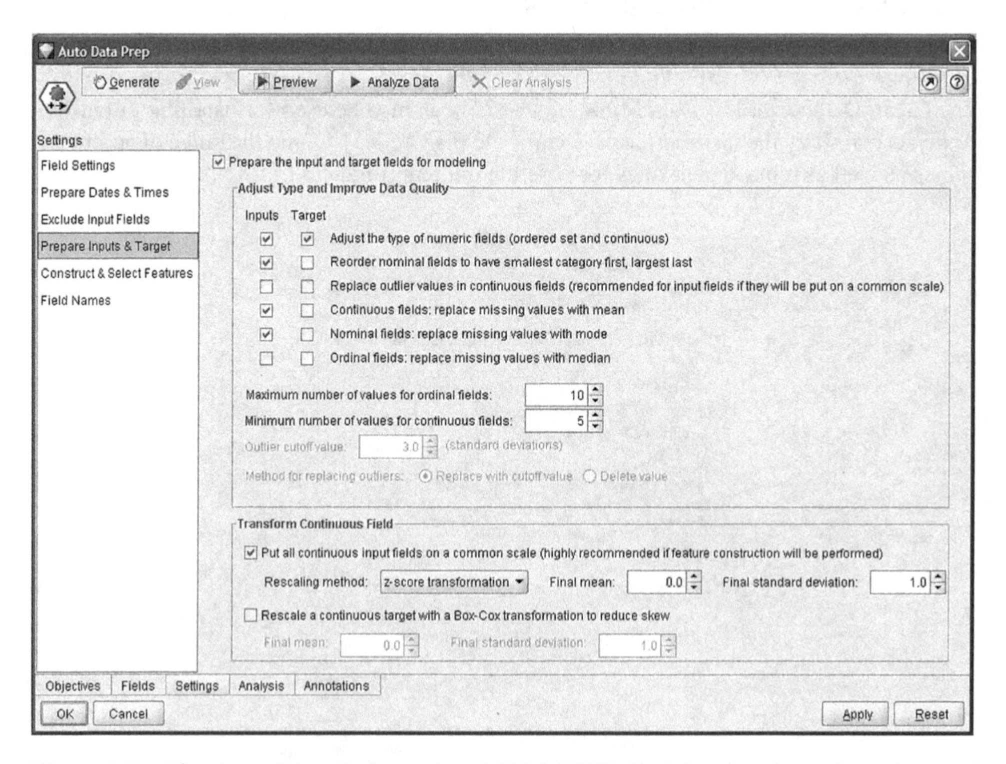

Figure 3.3 The Auto Data Prep node of IBM SPSS Modeler for data cleansing and preparation

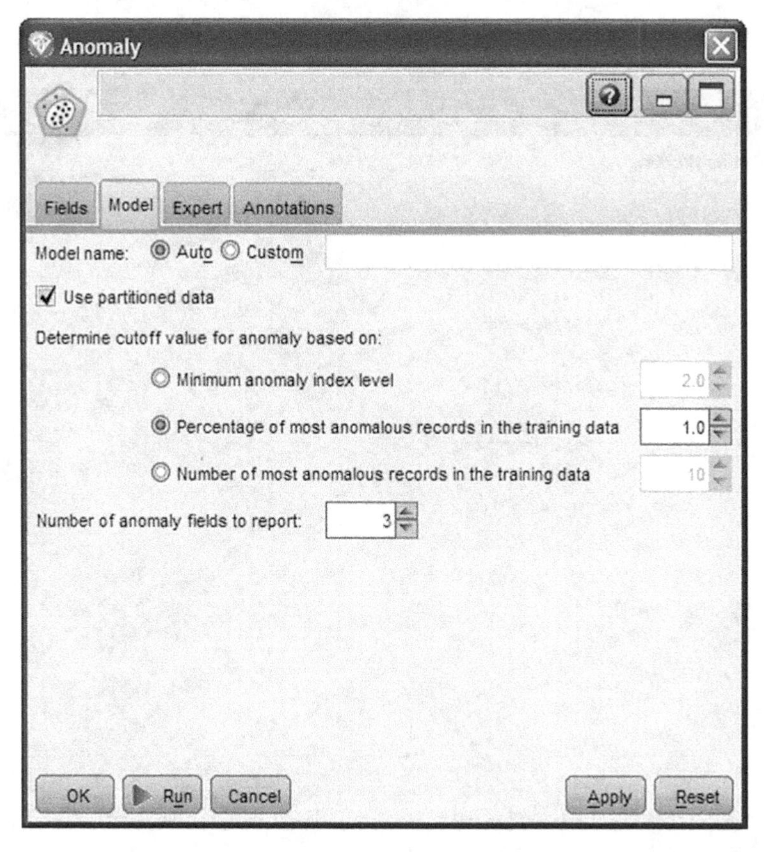

Figure 3.4 The Anomaly node of IBM SPSS Modeler for outlier detection

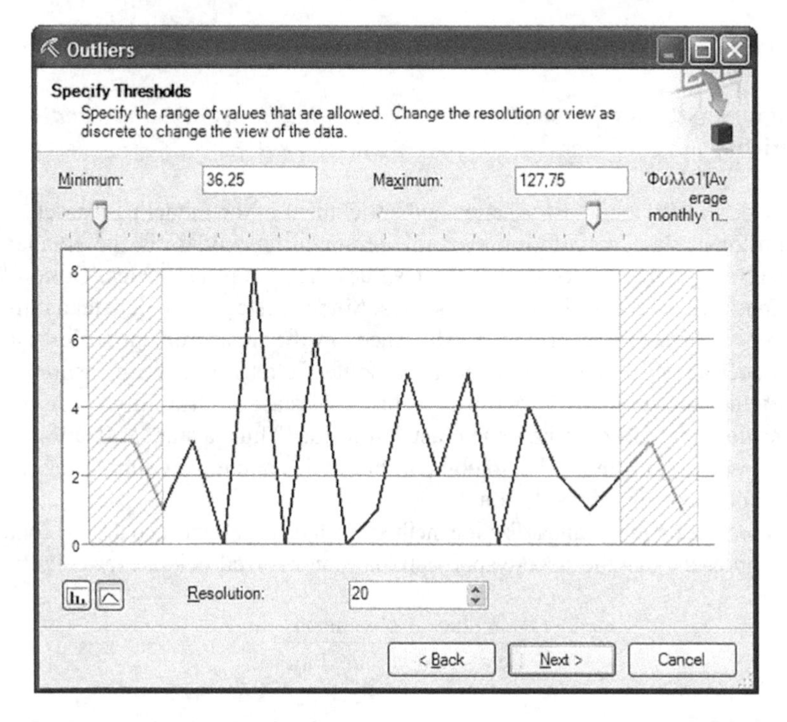

Figure 3.5 Setting the thresholds of acceptable values in the Clean Data wizard of Excel

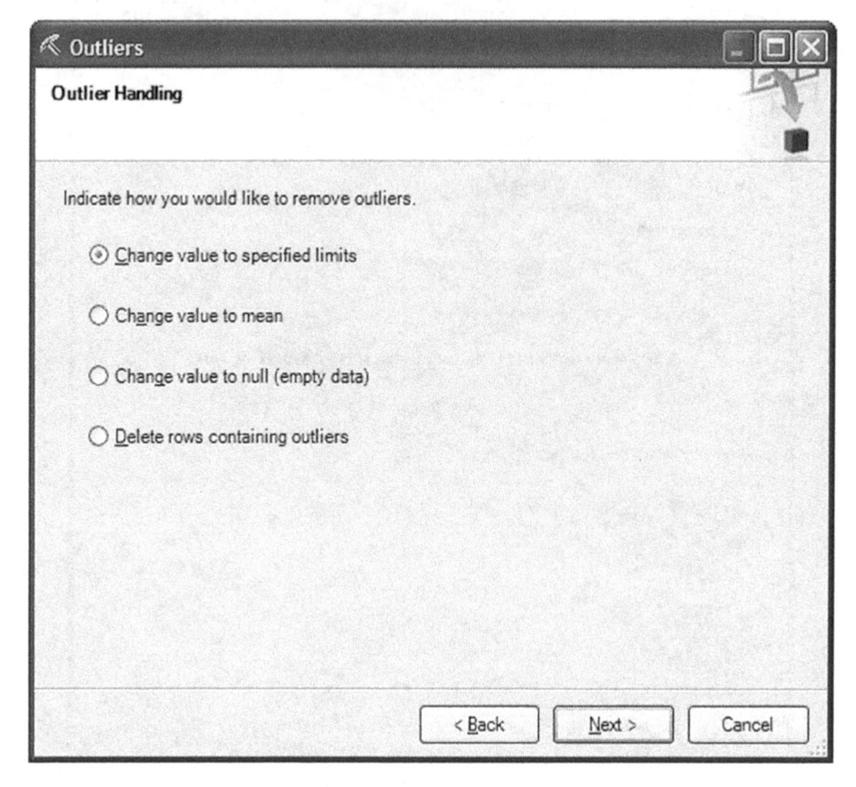

Figure 3.6 Handling outliers with the Clean Data wizard of Excel

3.4.5 Data transformations and enrichment

II. Data understanding, preparation, and enrichment	***II.5. Data transformations and enrichment***

This phase deals with the transformation and enrichment of the modeling dataset.

A significant data preparation step in segmentation models is the standardization (normalization) of the inputs so that their values are in similar scales. Quite often, the cluster inputs are measured in different scales. Since clustering models take into account the differences between records, the differences in the measurement scales can lead to biased clustering solutions simply because some fields might be measured in larger values. Fields measured in larger values have increased variability. If used in their original scale, they will dominate the cluster solution. Thus, a standardization process is necessary in order to bring fields to comparable scales and ensure that fields with larger values do not determine the solution.

The two most common standardization methods include the *z*-score and the 0–1 (or min–max or range) approaches. In the *z*-score approach, the standardized field is created as below:

$$\frac{\text{Value} - \text{field's mean}}{\text{field's standard deviation}}$$

The derived field has a mean of 0 and a standard deviation of 1.

The min–max (typically 0–1) approach rescales all values in the specified range as follows:

$$\frac{\text{Value} - \text{Minimum value of field}}{\text{Maximum value of field} - \text{Minimum value of field}}$$

By the term data enrichment, we refer to the construction of informative indicators, key performance indicators (KPIs), based on the original inputs. The original inputs are combined, and functions such as ratios, percentages, and averages are applied to derive fields which better summarize the customer behavior and convey the differentiating characteristics of each customer.

This is a critical step that heavily depends on the expertise, experience, and business "imagination" of the project team since the construction of an informative list of inputs can lead to richer and more refined segmentations.

@Tech tips for applying the methodology

Normalization in RapidMiner

The Normalize operator of RapidMiner (Figure 3.7) offers the *z*-score as well as the min–max normalization methods.

Normalization in IBM SPSS Modeler

In IBM SPSS Modeler, the normalization functions are part of the Auto Data Prep node (Figure 3.3).

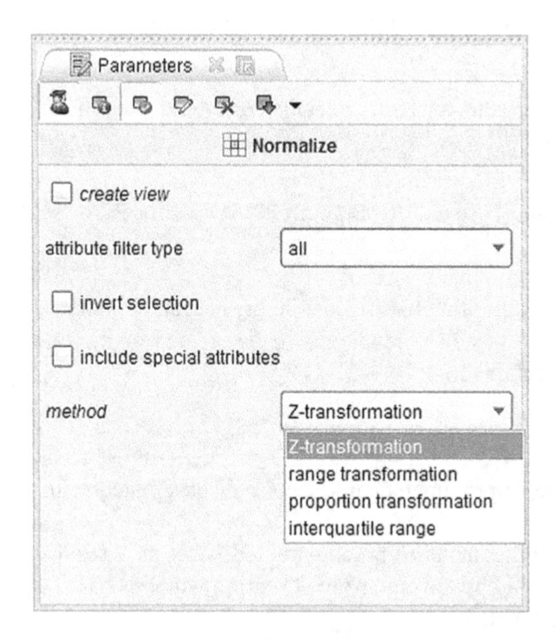

Figure 3.7 The Normalize operator of RapidMiner

Fortunately, many clustering algorithms, such as Modeler's TwoStep cluster or Excel's K-means, offer integrated standardization methods to adjust for differences in measurement scales. Similarly, the application of a data reduction technique like Principal Components/ Factor Analysis also provides a solution since the generated components/factors have standardized values. However, for cluster models without an integrated normalization step, such as Modeler's and RapidMiner's K-means, the standardization of numerical inputs is recommended.

3.4.6 Input set reduction

II. Data understanding, preparation, and enrichment	***II.6. Input set reduction***

The data preparation stage is typically concluded with the application of an unsupervised data reduction technique such as PCA/Factor Analysis. These techniques reduce the dimensionality of the inputs by effectively replacing a typically large number of original inputs with a relatively small number of compound scores, called factors or principal components. They identify the underlying data dimensions by which the customers will be segmented. The derived scores are then used as inputs in the clustering model that follows. The advantages of using a data reduction technique as a data preprocessing step include:

- Simplicity and conceptual clarity. The derived scores are relatively few, interpreted, and labeled. They can be used for cluster profiling to provide the first insight on the segments.
- Standardization of the clustering inputs, a feature important in order to yield an unbiased solution.
- Ensuring that all the data dimensions contribute equally to the formation of the segments.

@Tech tips for applying the methodology

PCA in RapidMiner

The RapidMiner PCA algorithm for dimensionality reduction is shown in Figure 3.8. Analysts can set the desired number of components to be created or the minimum variance of the original inputs to be retained.

PCA in IBM SPSS Modeler

In IBM SPSS Modeler principal components can be created with the PCA/Factor node (Figure 3.9).

The number of components to be constructed is set as a fixed value by the analyst or determined automatically by the eigenvalue criterion which is based on the information conveyed by each component.

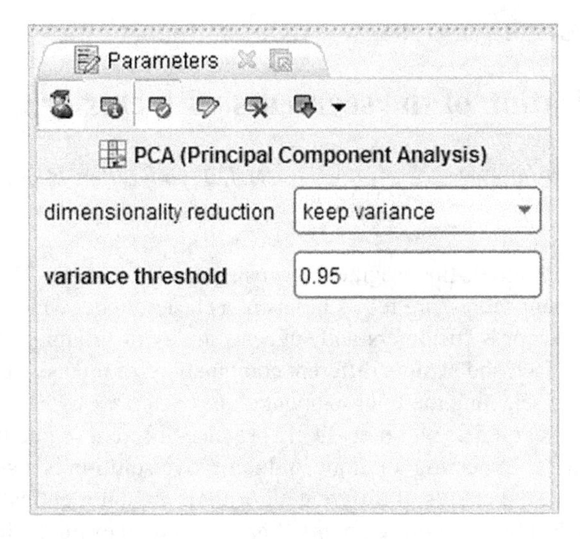

Figure 3.8 The PCA operator of RapidMiner

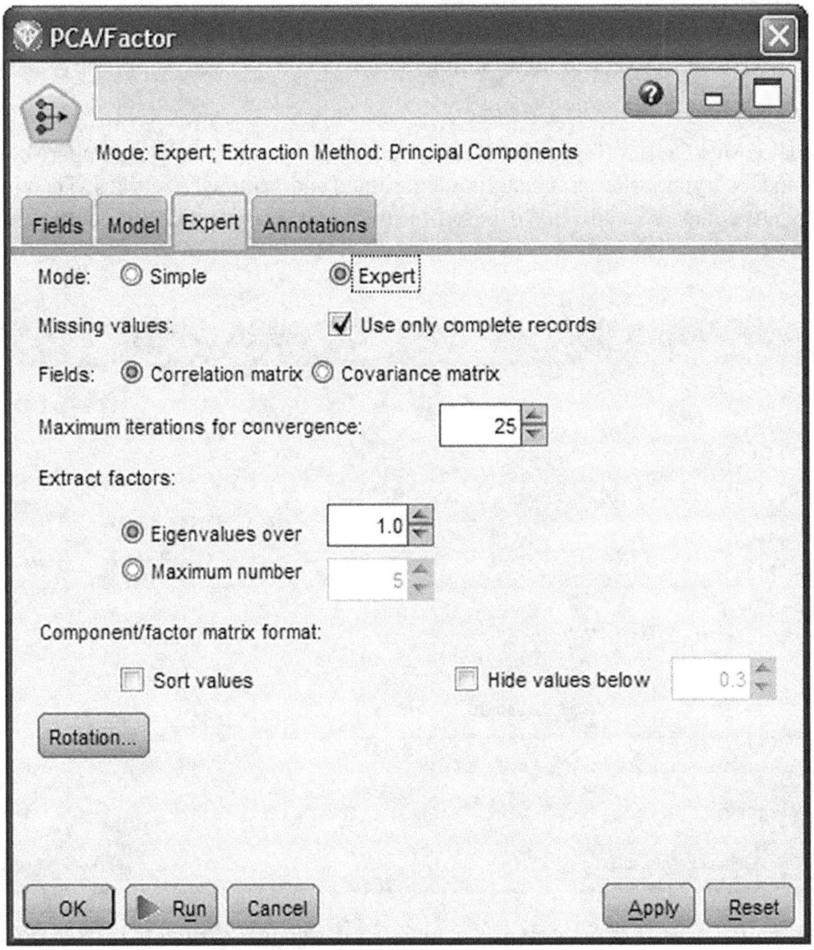

Figure 3.9 The PCA/Factor node of Modeler

3.5 Identification of the segments with cluster modeling

III. Classification modeling	*III.1. Trying different models and parameter settings*

Customers are divided into distinct segments by using cluster analysis. The clustering fields, typically the component scores, are fed as inputs in a cluster model which assesses the similarities between the records/customers and suggests a way of grouping them. Data miners should try a test approach and explore different combinations of inputs, different models, and model settings before selecting the final segmentation scheme.

Different clustering models will most likely produce different segments, and this should not come as a surprise. Expecting a unique and definitive solution is a sure recipe for disappointment. Usually, the results of different algorithms are not identical but similar. They seem to converge to some common segments. Analysts should evaluate the agreement level of the different models and examine in which aspects disagree. In general, a high agreement level between many different cluster models is a good sign for the existence of discern groupings.

@Tech tips for applying the methodology

Cluster modeling in IBM SPSS Modeler

The Modeler Auto Cluster (Figure 3.10) node enables analysts to train and compare numerous cluster models, trying different clustering algorithms and parameter settings. The clustering solutions are compared and ranked based on measures such as the Silhouette coefficient

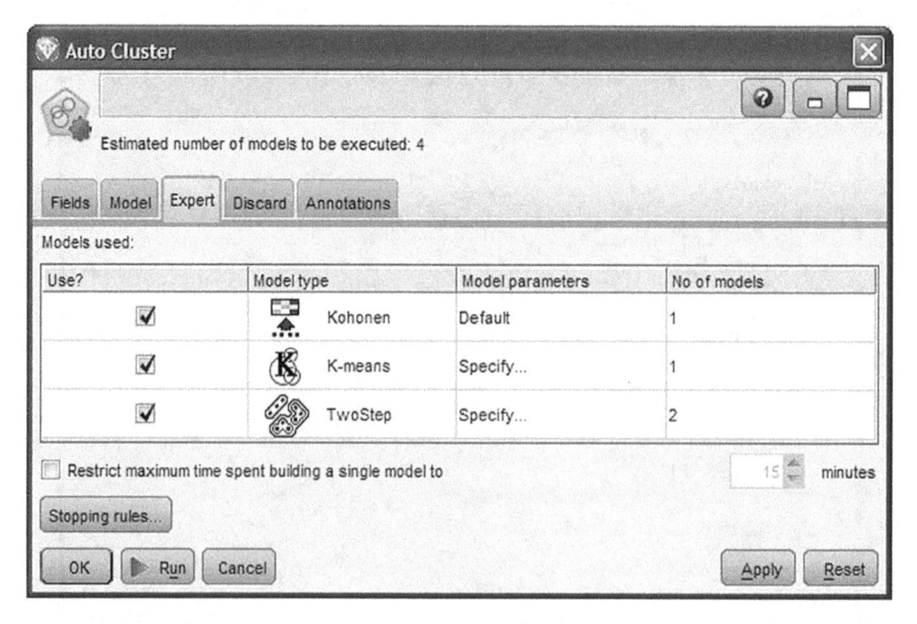

Figure 3.10 The IBM SPSS Modeler Auto Cluster node for training cluster models

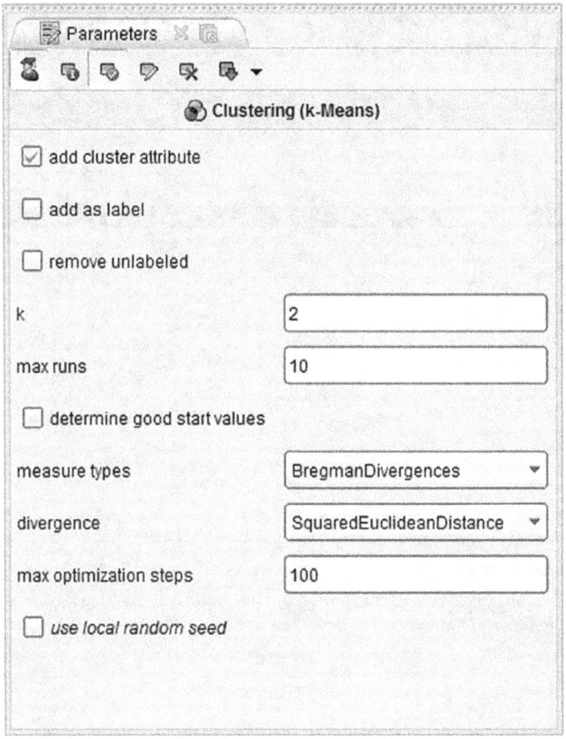

Parameters

Clustering (k-Means)

add cluster attribute

add as label

remove unlabeled

k	2
max runs	10

determine good start values

measure types	BregmanDivergences
divergence	SquaredEuclideanDistance
max optimization steps	100

use local random seed

Figure 3.11 The RapidMiner operator for K*-means clustering*

which will be discussed in the next paragraph. Available algorithms include K-means, TwoStep, and SOM (Kohonen Network/Self-Organizing Map).

Cluster modeling in RapidMiner

The RapidMiner algorithms for clustering are available in the Clustering and Segmentation group of operators. The operator for K-means clustering and its tunable parameters are shown in Figure 3.11. Apart from K-means and K-medoids, the Expectation Maximization (EM) and *X*-means (a variant of K-means which automatically detects the optimal number of clusters) clustering algorithms are supported.

Cluster modeling in Data Mining for Excel

In Data Mining for Excel, a cluster model is built by using the Cluster wizard (Figure 3.12). In the first steps of the wizard, analysts select the training dataset and the inputs. Then they choose among the supported clustering algorithms (EM and K-means) and set the model parameters.

The identified clustering solution is initially examined with basic checks based on simple reporting and descriptive statistics before thoroughly evaluated for acceptance and deployment. But this gets us to the next stage of the behavioral segmentation procedure.

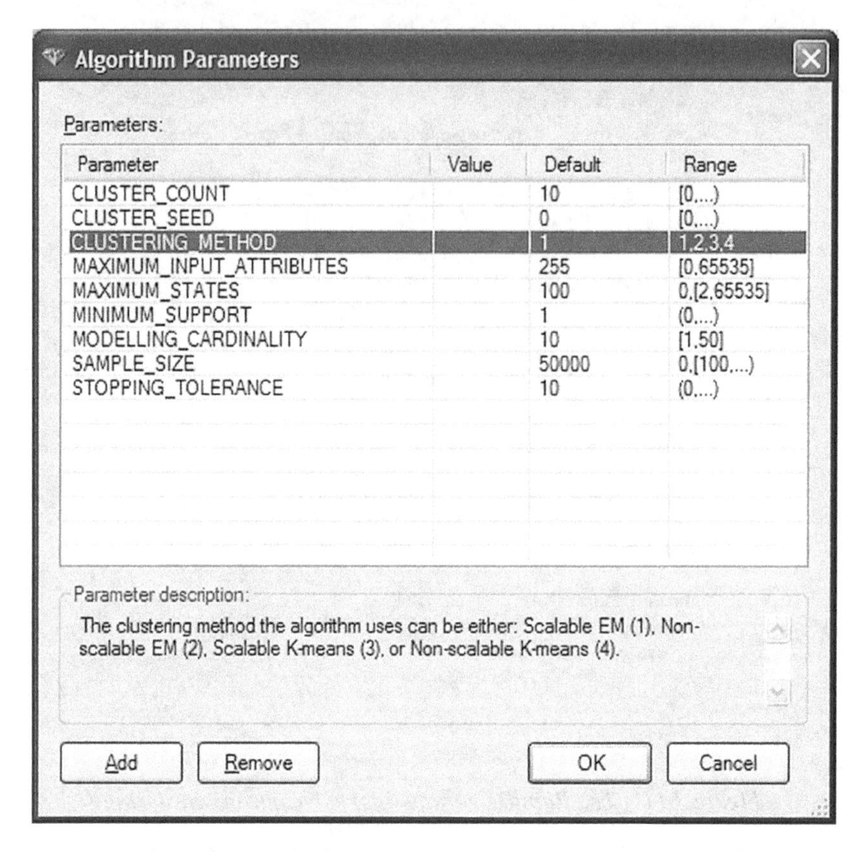

Figure 3.12 The Cluster wizard for building cluster models in Data Mining for Excel

3.6 Evaluation and profiling of the revealed segments

In this phase, the modeling results are evaluated, and the segmentation scheme that best addresses the needs of the organization is selected for deployment. Data miners should not trust blindly the solution suggested by one algorithm. They should explore different solutions and always seek guidance from the marketers for the selection of the most effective segmentation. After all, they are the ones who will use the results for segmented marketing, and their opinion on the future benefits of each solution is critical. The selected solution should provide distinct and meaningful clusters that can provide profit opportunities. Tasks of this phase include the following.

3.6.1 "Technical" evaluation of the clustering solution

IV. Evaluation and profiling of the revealed segments	***IV.1. "Technical" evaluation of the clustering solution***

The internal cohesion and the separation of the clusters should be assessed with the use of descriptive statistics and specialized technical measures (such as standard deviations, interclass and intraclass distances, Silhouette coefficient, etc.). Additionally, data miners should also examine the distribution of the customers in the revealed clusters as well as the consistency of the results in different datasets. All these tests assess the segmentation solution in terms of "technical" adequacy. Additionally, the segments should also be assessed from a business perspective in terms of actionability and potential benefits. To facilitate this evaluation, a thorough profiling of the segments' characteristics is needed.

Analysts should examine and evaluate the revealed cluster solution and assess, among others, the number and the relative size of the clusters and their cohesion and separation. A good clustering solution contains tightly cohesive and highly separated clusters. More specifically, the solution, through the use of descriptive statistics and specialized measures, should be examined in terms of:

1. **The number of clusters and the size of each cluster**
   The cluster examination normally begins by looking at the number of revealed clusters. In general purpose clustering, analysts typically expect a rich but manageable number of clusters. Analysts should look at the number of records assigned to each cluster. A large, dominating cluster which concentrates most of the records may indicate the need of further segmentation. On the other end, a small cluster with few records merits special attention. If considered as an outlier cluster, it could be set apart from the overall clustering solution and studied separately.

2. **Cohesion of the clusters**
   A good clustering solution is expected to be comprised of dense concentrations of records around their centroids. Large dispersion values indicate nonhomogeneous groupings and suggest further partitioning of the dataset. A number of useful statistics can be calculated to summarize the concentration and the level of internal cohesion of the revealed clusters such as:

   - Standard deviations and pooled standard deviations of the clustering fields. Data miners should start by examining the standard deviations of the clustering fields for each cluster, hoping for small values which indicate a small degree of dispersion. The pooled standard deviation of a clustering field is the weighted (according to each cluster's size) average of the individual standard deviations for all clusters. Once again, we anticipate low variability and small values which denote increased cohesion.
   - Maximum (Euclidean) distance from the cluster center (centroid). Another statistic that summarizes the degree of concentration of each cluster is the maximum distance from the cluster center, the cluster radius. In a way, it represents the range of each cluster since it denotes how far apart lies the remotest member of the cluster.
   - Analysts should also evaluate the within-cluster distances (within-cluster variation): the distances between the members of a cluster and their cluster centroid. They could average these distances over all members of a cluster and look for clusters with disproportionately large average distances. These clusters are candidate for further segmentation. A technical cluster cohesion measure which is based on the (squared Euclidean) distances between the data points and their centroids is the sum of squares error (SSE). In order to compare between models, we can use the average SSE calculated as follows:

$$\text{Average SSE} = \frac{1}{N}\sum_{i \in C}\sum_{x \in C_i}\text{dist}\left(c_i, x\right)^2$$

where c_i is the centroid of cluster i, x a data point/record of cluster i, and N the total cases. A solution with smaller SSE is preferred.

3. **Separation of the clusters**
   Analysts also hope for well-separated (well-spaced) clusters.
   A good way to quantify the cluster separation is by constructing a proximity matrix with the distances between the cluster centroids. The minimum distance between clusters should be identified and assessed since this distance may indicate similar clusters that may be merged.
   Analysts may also examine a separation measure named sum of squares between (SSB) which is based on the (squared Euclidean) distances of each cluster's centroid to the overall centroid of the whole population. In order to compare between models, we can use the average SSB calculated as follows:

$$\text{Average SSB} = \frac{1}{N}\sum_{i \in C} N_i * \text{dist}\left(c_i, c\right)^2$$

   where c_i is the centroid of cluster i, c the coverall centroid, N the total cases, and N_i the number of cases in cluster i. The SSB measures the between-cluster distances. The higher the SSB, the more separated the derived clusters are.

A combined measure that assesses both the internal cohesion and the external separation of a clustering solution is the Silhouette coefficient which is calculated as follows:

1. For each record i in a cluster, we calculate $a(i)$ as the average (Euclidean) distance to all other records in the same cluster. This value indicates how well a specific record fits a cluster. To simplify its computation, $a(i)$ calculation may be modified to record the (Euclidean) distance of a record from its cluster centroid.

2. For each record i and for each cluster not containing i as a member, we calculate the average (Euclidean) distance of the record to all the members of the neighboring cluster. After doing this for all clusters that i is not a member of, we calculate $b(i)$ as the minimum such distance in terms of all clusters. Once again to ease computations, the $b(i)$ calculation can be modified to denote the minimum distance between a record and the centroid of every other cluster.

3. The Silhouette coefficient for the record i is defined as

$$\text{Si} = \frac{b(i) - a(i)}{\max\left(a(i), b(i)\right)}$$

   The Silhouette coefficient varies between −1 and 1. Analysts hope for positive coefficient values, ideally close to 1, as this would indicate $a(i)$ values close to 0 and perfect internal homogeneity.

4. By averaging over the cases of a cluster, we can calculate its average Silhouette coefficient. The overall Silhouette coefficient is a measure of the goodness of the clustering solution, and it can be calculated by taking the average over all records/data points. An average Silhouette coefficient greater than 0.5 indicates adequate partitioning, whereas a coefficient less than 0.2 denotes a problematic solution.

Other popular cluster performance measures include the Akaike Information Criterion (AIC) and the Bayesian Information Criterion (BIC). The BIC measure is also used by Modeler's TwoStep algorithm for determining the optimal number of clusters. It assesses the goodness of fit of the solution, that is, how well the specific data are fit by the number of clusters.

Technical measures like the ones presented in this paragraph are useful, but analysts should also try to understand the characteristics of each cluster. The understanding of the clusters through profiling is required to take full advantage of them in subsequent marketing activities. Moreover, it is also a required evaluation step before accepting the solution. This process is presented in the next paragraph.

@Tech tips for applying the methodology

"Technical" evaluation of the clusters in IBM SPSS Modeler

The Modeler Auto Cluster estimates the Silhouette coefficient for the derived cluster solutions (Figure 3.13). The clustering solutions can be compared by this performance measure (Figure 3.14).

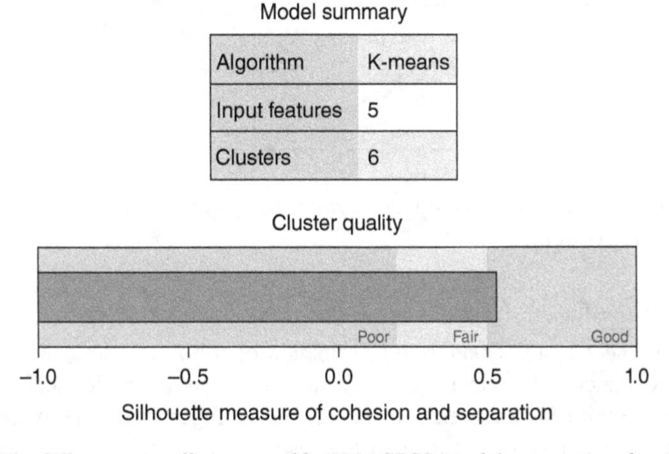

Figure 3.13 The Silhouette coefficient used by IBM SPSS Modeler to assess the cluster solution

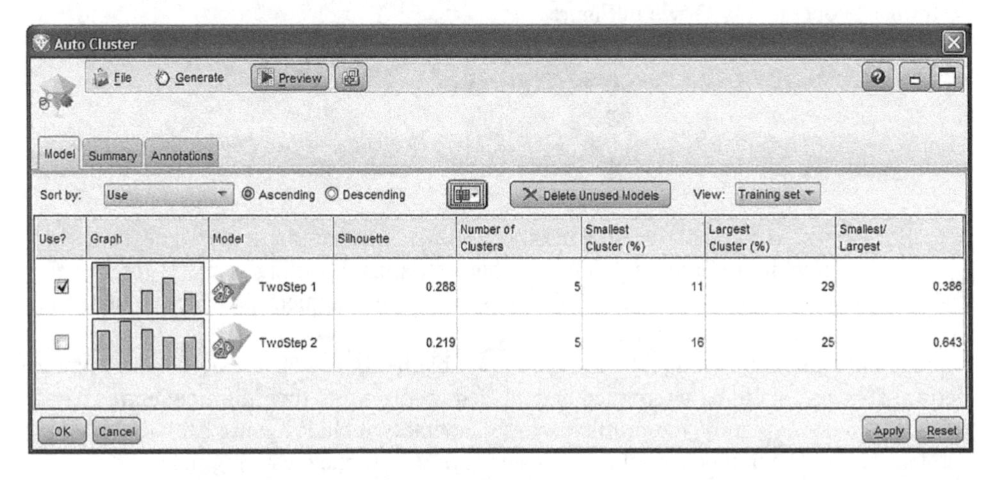

Figure 3.14 Comparing cluster models with the Silhouette coefficient in IBM SPSS Modeler

PerformanceVector

```
PerformanceVector:
Avg. within centroid distance: -0.114
Avg. within centroid distance_cluster_0: -0.055
Avg. within centroid distance_cluster_1: -0.200
Avg. within centroid distance_cluster_2: -0.090
Davies Bouldin: -1.854
```

Figure 3.15 *Assessing the average centroid distances of a cluster solution in RapidMiner using the Cluster Distance Performance*

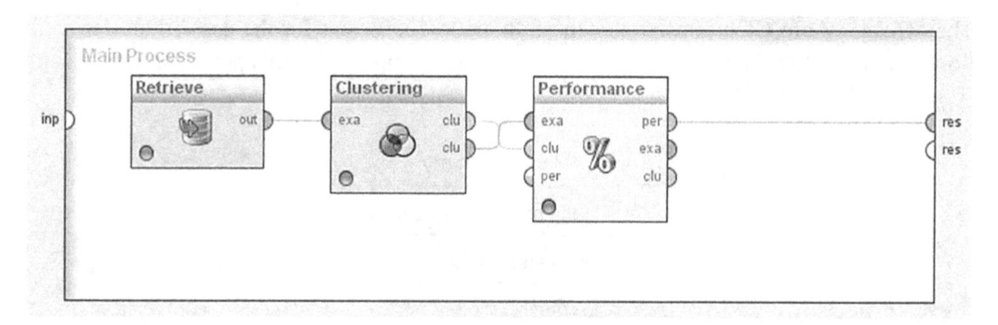

Figure 3.16 *Using the Cluster Distance Performance operator in RapidMiner to assess a cluster solution*

"Technical" evaluation of the clusters in RapidMiner

In RapidMiner, the Cluster Distance Performance operator can be used to evaluate the within-cluster distances of centroid-based cluster models (such as K-means). Each cluster (and the overall solution) is assessed in respect to the average distance between the members of a cluster and their cluster centroid (Figure 3.15).

In RapidMiner, a cluster model is evaluated when passed through a Cluster Distance Performance operator as shown in Figure 3.16.

3.6.2 Profiling of the revealed segments

IV. Evaluation and profiling of the revealed segments	***IV.2. Profiling of the revealed segments***

A profiling phase is typically needed in order to fully interpret the revealed segments and gain insight on their structure and defining characteristics. Profiling supports the business evaluation of the segments as well as the subsequent development of effective marketing strategies tailored for each segment.

Segments should be profiled by using all available fields as well as external information. Derived clusters should be interpreted and labeled according to their differentiating characteristics, and consequently, the profiling phase inevitably includes going back to the inputs and determining the uniqueness of each cluster in respect to the clustering fields.

The description of the extracted segments typically starts with the examination of the centroids' table. A cluster centroid is defined by simply averaging all the input fields over all cluster members. The centroid can be thought of as the prototype or the most typical representative member of a cluster. Cluster centers for the simple case of two clustering fields and two derived clusters are depicted in Figure 3.17.

In the general case of M clustering fields and N-derived clusters, the table summarizing the cluster centers would have the form of Table 3.2.

Analysts should individually check each cluster and compare the means of the input attributes with the grand means, looking for significant deviations from the "overall"/the "typical" behavior. Therefore, they should search for clustering fields with relatively low or high mean values in a cluster.

The profiling phase should also examine the clusters in respect to "external" fields, not directly participated in the cluster formation, including KPIs and demographical information of interest. Normally, the cluster separation is not only limited to the clustering fields, but it is also reflected in other attributes. Therefore, data miners should also describe the clusters

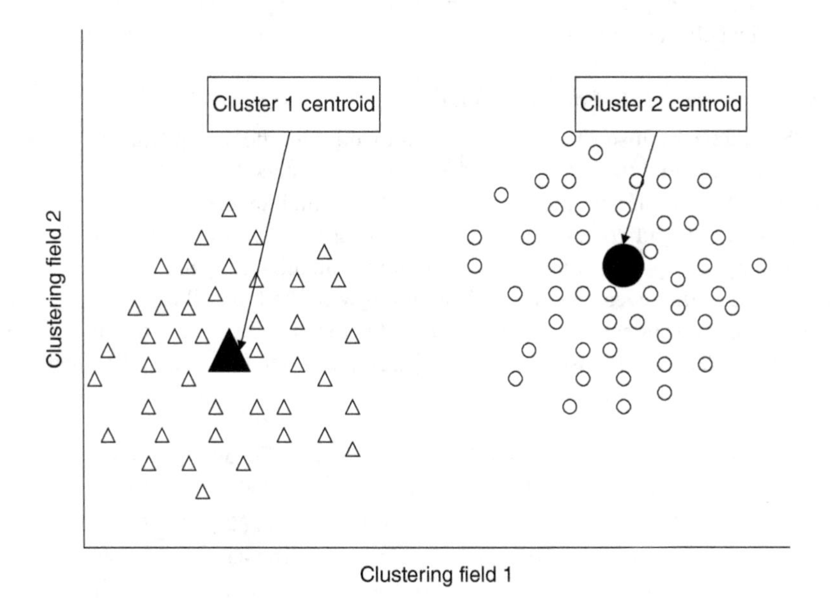

Figure 3.17 Graphical representation of the cluster centers or centroids

Table 3.2 The table of cluster centers

	Cluster 1	Cluster 2	… Cluster N	Overall population: all clusters
Clustering field 1	Mean (1,1)	Mean (1,2)	Mean (1,N)	Mean (1)
Clustering field 2	Mean (2,1)	Mean (2,2)	Mean (2,N)	Mean (2)
…	…	…	…	…
Clustering field M	Mean (M,1)	Mean (M,2)	Mean (M,N)	Mean (M)

by using all important attributes, regardless of their involvement in the cluster building, to fully portray the structure of each cluster and identify the features that best characterize them.

This profiling typically involves the examination of the distribution of continuous profiling fields across clusters. Profiling attributes can be normalized (standardized) in order to diminish the measurement scale differences and more clearly portray the structure of the clusters. For categorical attributes, the procedure would be analogous, involving comparisons of counts and percentages. The scope again is to uncover the cluster differentiation in terms of the categorical fields.

Finally, classification models can augment the reporting and visualization tools in the profiling of the segments. The model should be built with the cluster membership field as the target and the profiling fields of interest as inputs. Decision Trees in particular due to the intuitive format of their results are typically used to describe the segment profiles.

@Tech tips for applying the methodology

Profiling of clusters in RapidMiner

The centroids table is presented in the results pane of RapidMiner's Cluster as shown in Figure 3.18. The clustering inputs in the rows are summarized across the clusters.

Profiling of clusters in IBM SPSS Modeler

IBM SPSS Modeler's cluster models are summarized with the use of Cluster Viewer, a profiling and evaluation tool that graphically describes the clusters. Figure 3.19 presents a cluster-by-inputs grid that summarizes the cluster centroids in a mobile telephony segmentation.

The rows of the grid correspond to the clustering inputs, in this case the five principal components derived by a PCA model trained on mobile telephony data. The columns of the grid correspond to the revealed clusters. The first row of the table illustrates the size of each cluster. The table cells denote the cluster centers: the mean values of the inputs over all members of each cluster. A first insight on the differentiating factors of each cluster is evident.

Attribute	Cluster_0	Cluster_1	Cluster_2	Cluster_3	Cluster_4
pc_1	-0.210	7.146	4.904	-2.306	0.136
pc_2	3.709	1.833	-0.521	-0.208	-0.928
pc_3	0.813	-10.258	0.632	-0.564	0.275
pc_4	-0.008	-0.938	0.263	-0.000	-0.032
pc_5	-0.116	-0.350	0.159	0.105	-0.059
pc_6	0.245	0.496	-0.093	-0.055	-0.032
pc_7	0.127	-1.177	0.330	-0.090	-0.024
pc_8	0.139	1.424	-0.438	-0.189	0.139
pc_9	0.126	0.109	-0.477	-0.345	0.293
pc_10	0.051	0.445	-0.260	-0.211	0.167
pc_11	-0.210	-0.678	0.467	0.374	-0.266
pc_12	0.001	-0.104	0.242	0.076	-0.104

Figure 3.18 The centroids table in RapidMiner

Clusters

Feature importance
1

Cluster	Cluster-1	Cluster-2	Cluster-3	Cluster-4	Cluster-5	Cluster-6
Label	Active users	Voice users	SMSers	Roamers	Tech users	Basic users
Size	1.6% (250)	23.5% (3660)	24.5% (3817)	7.6% (1179)	8.2% (1273)	34.7% (5405)
Features	Component 1- SMS usage 0.52	Component 1- SMS usage −0.15	Component 1- SMS usage 1.83	Component 1- SMS usage 0.03	Component 1- SMS usage 0.24	Component 1- SMS usage −0.43
	Component 2- Voice usage 0.77	Component 2- Voice usage 1.71	Component 2- Voice usage −0.31	Component 2- Voice usage −0.20	Component 2- Voice usage −0.07	Component 2- Voice usage −0.24
	Component 3- Roaming usage 5.58	Component 3- Roaming usage −0.20	Component 3- Roaming usage −0.25	Component 3- Roaming usage 2.48	Component 3- Roaming usage −0.13	Component 3- Roaming usage −0.16
	Component 4- MMS usage 3.25	Component 4- MMS usage −0.11	Component 4- MMS usage −0.26	Component 4- MMS usage −0.13	Component 4- MMS usage 2.44	Component 4- MMS usage −0.12
	Component 5- Internet usage 2.34	Component 5- Internet usage 0.00	Component 5- Internet usage −0.03	Component 5- Internet usage −0.03	Component 5- Internet usage −0.20	Component 5- Internet usage −0.03

Figure 3.19 Modeler's representation of the cluster centroids

According to the centroids of Figure 3.19, cluster 1 seems to include a relatively small number of "superactive" customers with high values on all inputs, thus increased usage of all services. Cluster 2 customers are predominantly voice users. On the contrary, cluster 3 seems to mainly include SMS users as their SMS usage mean is high. Cluster 4 customers show increased roaming usage, while cluster 5 customers show increased usage of the MMS services. The customers in cluster 6 present the lowest mean values and the lowest usage of all services.

The cluster interpretation can be aided by additional Cluster Viewer boxplots such as the one presented in Figure 3.20 which visually compares two clusters on the SMS and voice usage inputs.

The background plot is a boxplot which summarizes the entire training population. The vertical line inside the box represents the population median on the respective clustering field. The median is the 50th percentile, the middle value which separates the population into two sets of equal size. The width of the box is equal to the interquartile range, which is the difference between the 25th and the 75th percentiles and indicates the degree of dispersion in the data. Thus, the box represents the range including the middle 50% of the entire population. Overlaid on the background boxplots are displayed the boxplots for the selected clusters. Point markers indicate the corresponding medians and the horizontal spans the interquartile ranges. As shown in the Figure 3.20, cluster 3 seems to be characterized by relatively high values of SMS usage and lower values of voice usage. On the contrary, cluster 2 shows increased values and preference for voice usage.

Profiling of clusters in Data Mining for Excel

In Data Mining for Excel, the identified clusters are presented in the Cluster Diagram view as connected nodes, as shown in Figure 3.21. The clusters are connected according to their similarities, and the width of the connecting lines denotes the strength of the similarities.

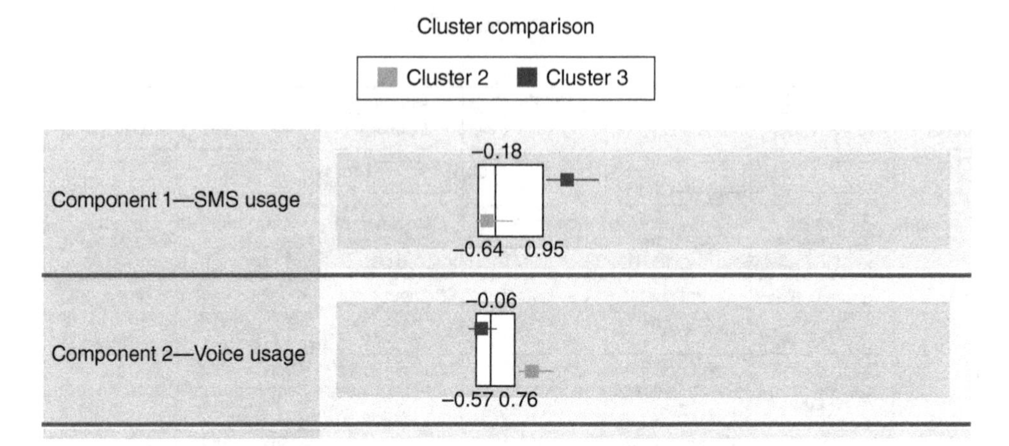

Figure 3.20 Cluster comparison with boxplots in IBM SPSS Modeler

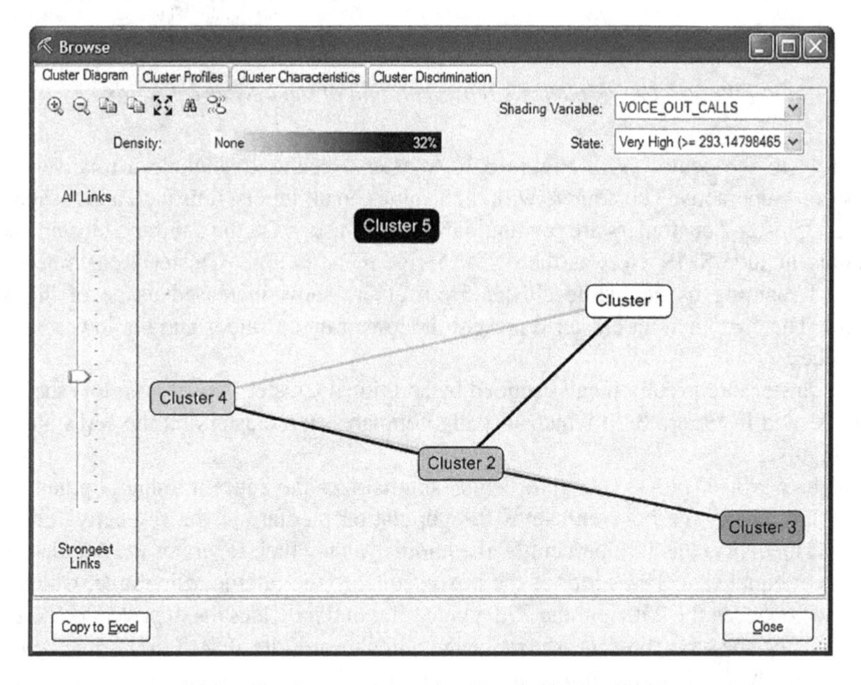

Figure 3.21 The Data Mining for Excel Cluster Diagram view

Figure 3.21 depicts the five clusters identified by a K-means model trained on mobile telephony data. The cluster nodes are shaded according to their voice calls usage. Cluster 5 seems to include an increased percentage (32%) of high voice users with a monthly average of at least 293 voice calls. Cluster 5 also seems different from the rest of identified clusters.

The cluster sizes and centroids are displayed in the Cluster Profiles view of Data Mining for Excel as shown in Figure 3.22.

Cluster Profiles

Variables	States	Population (All)	Cluster 1	Cluster 2	Cluster 3	Cluster 4	Cluster 5
Size		6624	2500	1762	1209	761	392
SMS_OUT_CALLS	Mean	11,48	0,78	12,73	9,16	26,47	49,66
SMS_OUT_CALLS	Deviation	28,51	0,91	13,2	14,17	36,21	84,04
VOICE_OUT_CALLS	Mean	116,66	73,18	131,58	138,46	138,57	205,41
VOICE_OUT_CALLS	Deviation	117,66	54	124,12	146,08	102,42	185,46
OUT_CALLS_ROAMING	Mean	0,7	0	0	3,08	0	2,37
OUT_CALLS_ROAMING	Deviation	2,77			5,17	0,02	4,57
MMS_OUT_CALLS	Mean	0,11	0	0	0	0,4	1,14
MMS_OUT_CALLS	Deviation	0,75			0,02	0,49	2,77
GPRS_TRAFFIC	Mean	1,08	0	0	0	0,51	17,09
GPRS_TRAFFIC	Deviation	21,14		0,01	0,01	0,93	84,91

Figure 3.22 The Cluster Profiles view of Mining for Excel with the cluster sizes and centroids

Cluster Characteristics

Cluster 5

Variables	Values	Probability
VOICE_OUT_CALLS	196,0 - 469,6	44 %
OUT_CALLS_ROAMING	2,6 - 9,0	41 %
SMS_OUT_CALLS	30,7 - 97,0	30 %
MMS_OUT_CALLS	0,6 - 2,4	25 %
GPRS_TRAFFIC	15,3 - 64,5	22 %
VOICE_OUT_CALLS	116,7 - 196,0	16 %
OUT_CALLS_ROAMING	0,7 - 2,6	16 %
VOICE_OUT_CALLS	37,3 - 116,7	13 %
SMS_OUT_CALLS	11,5 - 30,7	9 %
MMS_OUT_CALLS	0,1 - 0,6	7 %
GPRS_TRAFFIC	1,1 - 15,3	7 %
OUT_CALLS_ROAMING	0,0 - 0,7	6 %
VOICE_OUT_CALLS	0,0 - 37,3	5 %
SMS_OUT_CALLS	0,0 - 11,5	5 %
MMS_OUT_CALLS	0,0 - 0,1	2 %
GPRS_TRAFFIC	0,0 - 1,1	1 %

Figure 3.23 The Cluster Characteristics view of Mining for Excel for cluster profiling

The mean and the standard deviation of the input fields (in rows) are presented for the total population and for each cluster (in columns). This view facilitates the comparison of distributions across clusters. By studying the profiles of cluster 5 in Figure 3.22, it is evident that it includes superusers of all type of calls. Cluster 4 seems to group high users with increased preference to SMS usage. Cluster 3 seems to be characterized by increased roaming usage. Cluster 1 includes basic usage customers, while cluster 2 includes typical customers with average usage.

The Cluster Characteristics view presents the top characteristics of the selected cluster as shown in Figure 3.23. Continuous inputs have been discretized, and the bins are sorted in descending probability.

For example, the top characteristic of cluster 5 is the increased voice usage; 44% of its members make more than 196 voice calls per month. However, the defining characteristics of

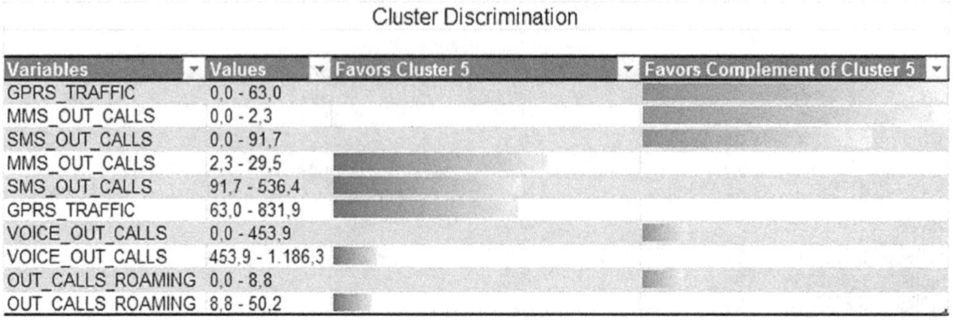

Cluster Discrimination

Variables	Values	Favors Cluster 5	Favors Complement of Cluster 5
GPRS_TRAFFIC	0,0 - 63,0		
MMS_OUT_CALLS	0,0 - 2,3		
SMS_OUT_CALLS	0,0 - 91,7		
MMS_OUT_CALLS	2,3 - 29,5		
SMS_OUT_CALLS	91,7 - 536,4		
GPRS_TRAFFIC	63,0 - 831,9		
VOICE_OUT_CALLS	0,0 - 453,9		
VOICE_OUT_CALLS	453,9 - 1.186,3		
OUT_CALLS_ROAMING	0,0 - 8,8		
OUT_CALLS_ROAMING	8,8 - 50,2		

Figure 3.24 The Cluster Discrimination view of Mining for Excel for comparing the cluster with its complement

each cluster are better identified by studying the Cluster Discrimination view in which each cluster is compared to its complement, everything outside the cluster.

As shown in Figure 3.24, cluster 5 is characterized by increased voice usage (more than 453 voice calls) as well as increased usage of all other call types.

3.6.3 Using marketing research information to evaluate the clusters and enrich their profiles

IV. Evaluation and profiling of the revealed segments	***IV.3. Using marketing research information to evaluate the clusters and enrich their profiles***

Marketing research surveys are typically used to investigate the needs, preferences, opinions, lifestyles, perceptions, and attitudes of the customers. They are also commonly used in order to collect valid and updated demographical information. It is strongly recommended to combine the data mining-driven behavioral segments with market research-driven demographical and needs/attitudinal segments. While each approach helps in the understanding of certain aspects of the customers, combining them provides deeper insight on different customer typologies.

For instance, provided a behavioral data mining segmentation has been implemented, random samples can be extracted from each segment, and through surveys and/or qualitative research and focus group sessions, valuable insight can be gained concerning each segment's needs and preferences. Alternatively, the data mining and the market research approaches can be implemented independently and then cross-examined, not only as a mean for evaluating the solutions but also in order to construct a combined and integrated segmentation scheme which would provide a complete view of the customers.

In conclusion, combining data mining and market research techniques for customer segmentation can enable refined subsequent marketing strategies, based on thorough understanding of customer behavior and needs, as shown in the Figure 3.25.

Consider customers belonging to the same behavioral segment but having diverse needs and perceptions. This information can lead to tailored marketing strategies within each behavioral segment.

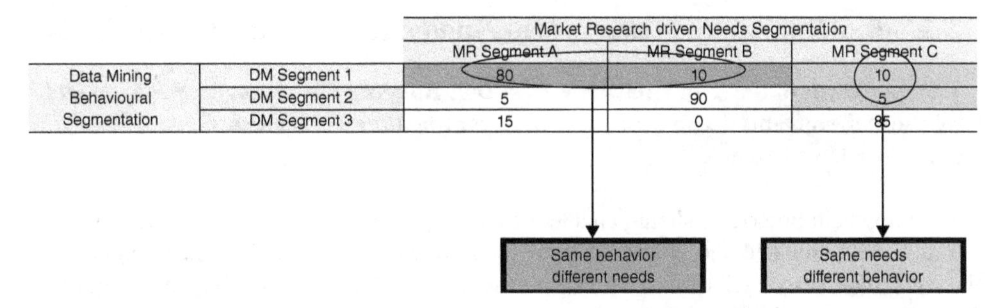

Figure 3.25 Combining data mining and market research-driven segmentations. Source: Tsiptsis and Chorianopoulos (2009). Reproduced with permission from Wiley

3.6.4 Selecting the optimal cluster solution and labeling the segments

IV. Evaluation and profiling of the revealed segments	***IV.4. Selecting the optimal cluster solution and labeling the segments***

Although all the aforementioned evaluation measures make mathematical sense and they can provide valuable help in the identification of the optimal solution, data miners should not solely base their decision on them. Many algorithms, such as Modeler's TwoStep cluster, Excel's clustering, and RapidMiner's X-means, automatically propose an optimal number of clusters based on technical measures and heuristics. Once again, these proposed solutions should not be blindly accepted. A clustering solution is justified only if it makes sense from a business point of view. Actionability, potential business value, interpretability, and easiness of use are factors hard to quantify and, in a way, subjectively measured. However, they are the best benchmarks for determining the optimal clustering solution. The profiling of the clusters and the identification of their defining characteristics are an essential part of the clustering procedure. It should not be considered as a postanalysis task but rather as an essential step for assessing the effectiveness of the solution.

As with every data mining model, data miners should try many different clustering techniques and compare the similarity of the derived solutions before deciding on the one to choose. Different techniques that generate analogous results are a good sign for the identification of a general and valid solution. As in any other data mining model, clustering results should also be validated by applying the model in a disjoint dataset and by examining the consistency of the results.

The profiling and interpretation process ends with the labeling of the identified segments with names that appropriately designate their unique characteristics. Each segment is assigned an informative and revealing name, for instance, "Business Travelers," instead of "Segment 1." The naming of the segments should take into account all the profiling findings. These names will be used for communicating the segments to all business users and for loading them to the organization's operational systems.

3.7 Deployment of the segmentation solution, design and delivery of differentiated strategies

The segmentation project is concluded with the deployment of the segmentation solution and its usage for the development of differentiated marketing strategies and segmented marketing.

3.7.1 Building the customer scoring model for updating the segments

V. Deployment of the segmentation solution: design and delivery of differentiated strategies	***V.1. Building the customer scoring model for updating the segments***

The deployment procedure should enable the customer scoring and update of the segments. It should be automated and scheduled to run on a frequent base to enable the monitoring of the customer base evolvement over time and the tracking of the segment migrations. Moreover, as nowadays markets change very rapidly, it is evident that a segmentation scheme can become outdated within a short time. Refreshment of such schemes should be made quite often. That's why the developed procedure should also take into account the need of possible future revisions.

@Tech tips for applying the methodology

Assigning customers to clusters in IBM SPSS Modeler

In IBM SPSS Modeler, the generated models are scoring engines. Instances passed through the generated model are scored and assigned to the identified clusters as shown in Figure 3.26.

Assigning customers to clusters in RapidMiner

RapidMiner's cluster models are also scoring engines and can be used to segment customers and derive the cluster membership field as shown in Figure 3.27.

Assigning customers to clusters in Data Mining for Excel

In Data Mining for Excel, the scoring procedure is implemented through the Query wizard. Analysts select the cluster model, as well as the scoring dataset. A required step of the scoring

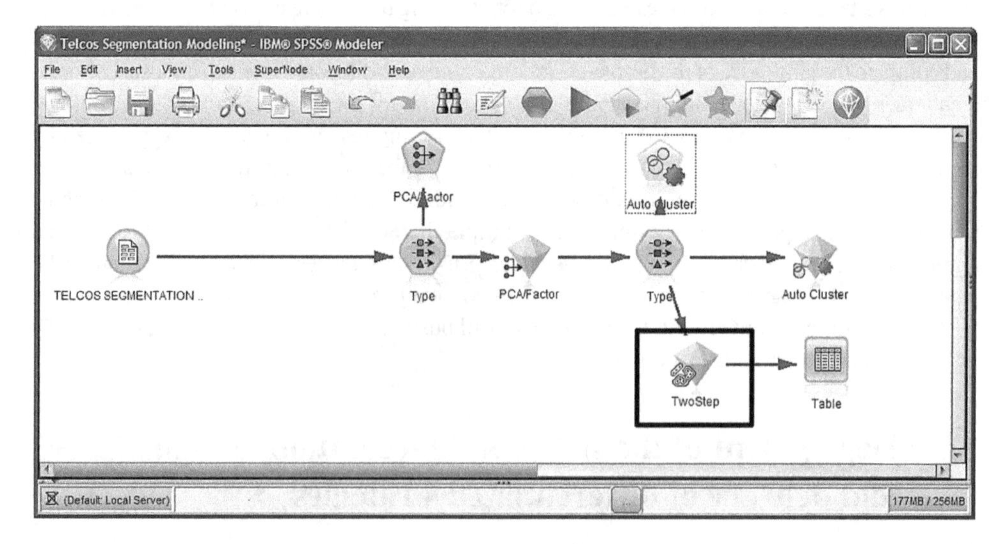

Figure 3.26 Assigning instances into clusters with a generated cluster model in IBM SPSS Modeler

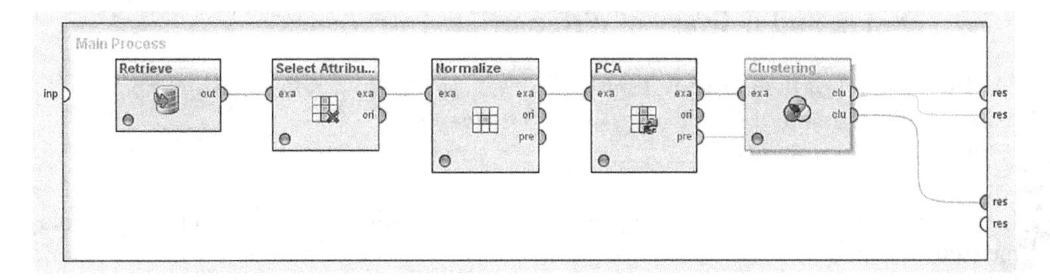

Figure 3.27 Segmenting instances using a trained RapidMiner model

procedure is to designate the cluster inputs by mapping the scoring dataset fields ("Table Columns") with the model inputs ("Mining Columns"). The Query wizard returns the cluster for each instance. Note that in the case of deployment of probabilistic EM cluster models, the wizard along with the predicted primary cluster also returns the probability of belonging to the cluster.

3.7.1.1 Building a Decision Tree for scoring: fine-tuning the segments

V.1. Building the customer scoring model for updating the segments	*V.1.1. Building a Decision Tree for scoring: fine-tuning the segments*

Decision Trees can also be used for scoring and assigning customers to the identified clusters. They can translate the differentiating characteristics of each cluster into a set of simple and understandable rules which can subsequently be applied for classifying new records in the revealed clusters. Each tree's terminal node corresponds to assignment to a specific cluster with an estimated probability. Although this approach also introduces an additional source of estimation errors, it is also a more transparent approach for cluster updating. It is based on understandable, model-driven rules, similar to common business rules, which can more easily be examined and communicated. Additionally, business users can more easily intervene and, if required, modify these rules and fine-tune them according to their business expertise.

3.7.2 Distribution of the segmentation information

V. Deployment of the segmentation solution: design and delivery of differentiated strategies	*V.2. Distribution of the segmentation information*

Finally, the deployment procedure should also enable the distribution of the segmentation information throughout the enterprise and its "operationalization." Therefore, it should cover the upload of the segmentation information in the organization's databases and CRM tools in order to enable customized strategies to be applied across all customers' touch points.

3.7.3 Design and delivery of differentiated strategies

V. Deployment of the segmentation solution: design and delivery of differentiated strategies	***V.3. Design and deliver of differentiated strategies***

In order to make the most of the segmentation, specialization is needed. Thus, it is recommended to assign the management of each segment to a specific team. Each team should build and deliver specialized marketing strategies tailored to the segment's needs, in order to improve the customer handling, develop the relationship with the customers, and grow the profitability of each segment. The responsibilities of the team should also include the tracking of each segment with appropriate KPIs and the monitoring of the competitors' activities for each segment.

3.8 Summary

Customer segmentation is the process of identifying groups that have common characteristics. The main objective of customer segmentation is to understand the customer base and gain customer insight that will enable the design and development of differentiated marketing strategies.

Clustering is a way to identify segments and assign customers into groups not known in advance. The identification of the segments should be followed by the profiling of the revealed customer groupings. The detailed profiling is necessary for understanding and labeling the segments based on the common characteristics of their members.

The decision for the optimal segments should be supported by technical measures of the "separation" of the records; however, it should not solely depend on them. The segmentation scheme is the tool for delivering personalized customer handling. Therefore, the identified segments should be clear and explicable. They should correspond to "actual" and distinct customer typologies in order to enable the passing from mass to personalized marketing.

Part II

The Algorithms

4

Classification algorithms

4.1 Data mining algorithms for classification

This chapter is dedicated to classification algorithms. Specifically, we'll present Decision Tree, Bayesian network, and Support Vector Machine (SVM) models. The goal of these models is the same: to classify instances of an independent dataset based on patterns identified in the training dataset. However, they work in different ways. Bayesian networks are probabilistic models which visually represent the dependencies among attributes. Decision Trees perform a type of supervised segmentation by recursively partitioning the training dataset and identifying "pure" segments in respect to the target attribute, ideally with all instances landing on the same class. SVM models are machine learning models which use nonlinear mappings for classification.

Despite their different mechanisms, these algorithms perform well in most situations. But they also differ in terms of speed and "greediness" with Decision Trees being the lightest and fastest as opposed to SVMs which typically take longer training time and are also quite demanding in terms of hardware resources. Although we'll try to present their pros and cons later in this chapter, the choice of the algorithm to use depends on the specific situation, and it is typically a trial-and-error procedure. In this chapter, we'll focus on the explanation of the algorithms. But remember model training is just a step of the overall classification process. This process is described in detail in Chapter 2.

Effective CRM using Predictive Analytics, First Edition. Antonios Chorianopoulos.

Companion website: www.wiley.com/go/chorianopoulos/effective_crm

4.2 An overview of Decision Trees

Decision Trees belong to the class of supervised algorithms, and they are one of the most popular classification techniques. A Decision Tree is represented in an intuitive tree format, enabling the visualization of the relationships between the predictors and the target attribute.

Decision Trees are often used for insight and for the profiling of target events/attributes due to their transparency and the explanation of the predictions that they provide. They perform a kind of "supervised segmentation": they recursively partition (split) the overall training population into "pure" subgroups, that is, homogeneous subsets in respect to the target attribute.

The training of a Decision Tree model is a top-down, "divide-and-conquer" procedure. It starts from the initial or root node which contains the entire set of the training instances. The input (predictor) that shows the best split according to the evaluation measure is selected for branching the tree. This procedure is recursively iterated, until a stop criterion is met. Through this procedure of successive partitionings, the heterogeneous training population is split into smaller, pure, homogeneous terminal leafs (segments). The procedure is further explained in Section 4.3.

4.3 The main steps of Decision Tree algorithms

1. Initially, all records of the training dataset belong to the root node and obviously to one of the classes of the target attribute. In the theoretical and oversimplified example presented in Figure 4.1, the root node contains eight customers, classified into two known classes (*yes/no*) according to their responses in a marketing campaign.

2. Each Decision Tree algorithm incorporates a grow method which is an attribute selection method to assess the predictive ability of the available inputs and optimally split into subsets. For each split, the effect of all predictors on the target attribute is evaluated, and the population is partitioned based on the predictor that yields the best separation. This attribute selection is repeated for many levels until the tree is fully grown. Starting from the root node, each parent node is split into child nodes. The outcomes of the selected predictor define the child nodes of the split. The child nodes in turn become parent nodes which are further split up to the leaf nodes.

3. Decision Tree algorithms differ in the way they handle predictors. Some algorithms are restrained to binary splits, while others do not have this limitation. Some algorithms collapse categorical inputs with more than two outcomes in an optimal way before the split. Some discretize continuous inputs, while others dichotomize them. Hence, before step 2, a preparatory step is carried out during which all inputs are appropriately transformed before being assessed for the split.

4. In general, the tree growing stops when perfect separation is reached, and all instances of the leaf nodes belong to the same class. Similarly, the tree growth stops when there are no more inputs available for further splitting. Finally, the tree construction can be stopped when specific user-specified terminating criteria have been met.

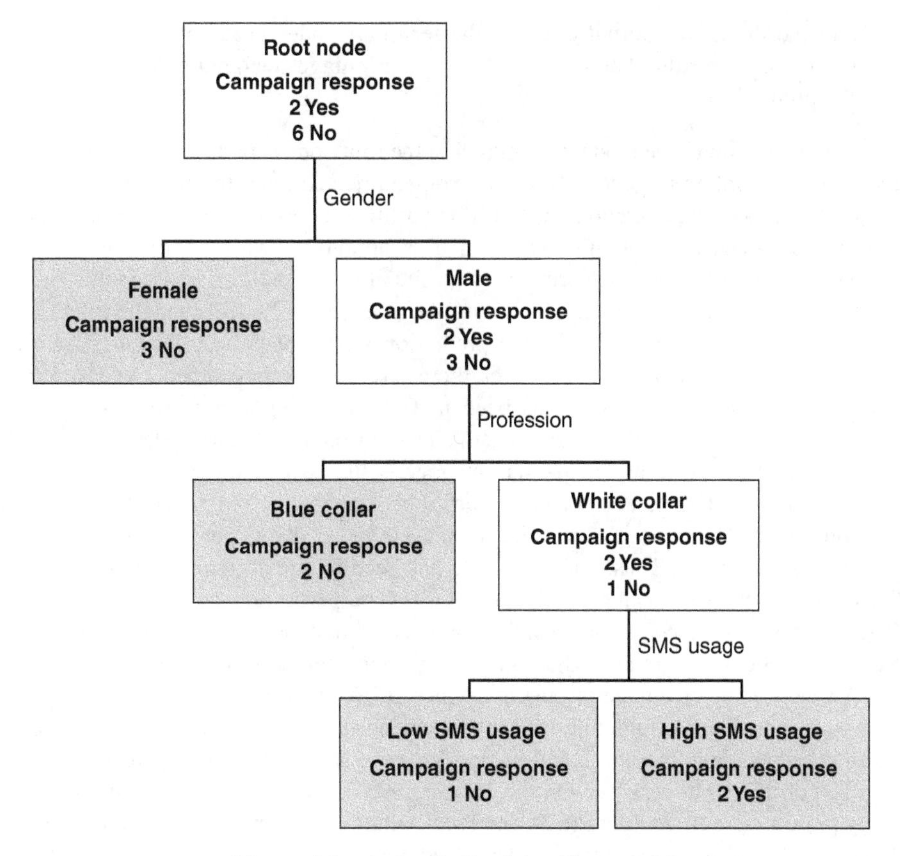

Figure 4.1 A simple Decision Tree model

5. The Decision Tree can be easily converted to decision rules. Specifically, rules of the following format can be extracted from the tree branches:

   IF (PREDICTOR VALUES) THEN (PREDICTION = TARGET CLASS with a specific PREDICTION CONFIDENCE).

   For example:

   IF (Tenure <= 4 years and Number of different products <= 2 and Recency of last Transaction > 20 days) ***THEN*** (PREDICTION = Churner and CONFIDENCE = 0.78)

   The extracted rules are exhaustive and mutually exclusive since in the fully grown tree, each record lands into only one terminal node. Each terminal node comprises a distinct rule which associates the predictors with the target outcomes. All records landing at the same terminal node get the same prediction and the same prediction confidence (probability of prediction). The path of the successive splits from the root node to each terminal node indicates the rule conditions, the combination of input characteristics that leads to a specific outcome. In general (if no misclassification costs have been defined), the majority voting method is applied, and the dominant output class of the terminal node represents the prediction of the respective rule. The dominant class percentage designates the confidence of the prediction.

In the simple Decision Tree model of Figure 4.1, the root node is partitioned into two subsets (child nodes) according to gender. Since all women rejected the offer, the "women" node is terminal due to perfect separation and can't be partitioned any further. Hence, the first rule has been discovered and classifies all women as nonbuyers with a respective prediction confidence of 1.0 and a purchase propensity (probability of the target class, in this case *yes*) of 0. On the other side of the tree, three out of the five men (60%) rejected the offer; thus, if the model had stopped at this level, the prediction would have been no purchase with a respective confidence of 3/5 = 0.6. However, there is room for improvement (we also suppose that none of our stopping criteria has been met), and the algorithm proceeds by further splitting the "men" branch of the tree into smaller subgroups and child nodes.

At this level of the tree, occupation category is the best predictor since it results the optimal separation. The parent branch is partitioned into two child nodes: blue-collar and white-collar men. The node of blue-collar men presents absolute "purity" since it only contains (100%) nonbuyers. Thus, one more rule has been identified which classifies all blue-collar men as nonbuyers with a confidence of 1.0. The percentage of buyers rises at about 67% (2/3) among white-collar men which seem to constitute the target group of the promoted service. In our fictional example, the purity of this node can be further improved if we split on SMS usage. The percentage of buyers reaches 100% among white-collar men with high SMS usage, inducing a confident rule for the identification of good prospects. The previous oversimplified example may be useful in clarifying the way that Decision Trees work, yet it cannot be considered as realistic. On the contrary, we deliberately kept it as simple as possible by presenting only faultless rules and by eliminating uncertainties due to model errors and misclassifications. In real projects, we would not even dare to split nodes with a handful of records. Moreover, the resulting terminal nodes would rarely be so homogeneous. The terminal nodes typically include records of all the outcome categories but with increased purity and decreased diversity in the distribution of the categories of the outcome, compared to the overall population. In these situations, the majority rule (dominant category) defines the prediction of each terminal node. In real-world applications, models make errors. They misclassify cases and they inevitably produce confidence values lower than 1.0. That's why data miners should thoroughly evaluate the performance of the resulted models and assess their accuracy before deployment.

4.3.1 Handling of predictors by Decision Tree models

Decision Tree models can handle both categorical and continuous predictors. Some algorithms, including Classification and Regression Trees (CART), incorporate a *binary attribute selection method* and are restricted to binary splits. They always split a node in two branches. Hence, before assessing a candidate categorical outcome for a split, its outcomes are optimally collapsed in the two subsets that maximize the gain in purity.

Multiway algorithms, including C5.0/C4.5 and CHAID, split a node in as many branches as the categorical input outcomes. However, they also fine-tune this splitting, returning an optimal number of child nodes, in between the binary and the maximum. More specifically, outcomes assessed as homogeneous are merged (regrouped) into fewer categories, and the categorical split is simplified without sacrificing the gained separation.

The CHAID and C5.0/C4.5 algorithms use different approaches for splitting on continuous inputs. The C5.0/C4.5 algorithm dichotomizes continuous predictors. An optimal cut point is selected, and the continuous predictor is turned into a binary one. More specifically, the distinct values of the continuous input are evaluated as potential split points. The value, typically the midpoint between two adjacent observed values, which yields the best separation, is selected as the threshold (less than vs. greater than) for splitting. The dichotomized continuous input is then challenging the rest of predictors for the best split. This approach is shown in the Decision Tree model of Figure 4.1 when the continuous SMS usage attribute is used for the final binary split of the white-collar men node. The CHAID Decision Tree algorithm uses a different method to handle continuous predictors. It discretizes them into bins of the same size and produces ordinal attributes for splitting. Hence, the original continuous fields are converted to ordinal categorical fields with ordered outcomes.

As explained earlier, multiway tree methods can split on more than two categories at any particular level in the tree. Therefore, they tend to create wider trees which reach the target discrimination faster than binary growing methods. More specifically, categorical attributes when used at a particular level of a tree are fully "utilized," delivering their maximum separation at a single, optimal split. On the contrary, binary splits typically lead to bushier trees with more levels. A binary split may fail to yield all the information from an input with a single split. Hence, the same attribute may be used many times at successive levels of the tree, gradually improving the purity of the child nodes.

4.3.2 Using terminating criteria to prevent trivial tree growing

As explained in 4.3, a tree may be forced to stop growing even when perfect separation has not been reached, and there are available inputs for additional splits. Specifically, a tree may be halted when certain user-specified terminating rules have been met. This approach is referred to as prepruning or forward pruning, as opposed to the postpruning procedure applied in some algorithms. The scope is common—to produce smaller, simpler, and more generalizable trees that present increased accuracy when classifying unseen cases. The pruning origin however is opposite. With postpruning, a tree is fully grown and then is trimmed to a smaller tree of comparable accuracy. Prepruning trims in advance. The tree is prevented from further growing because the maximal tree depth has been reached, the additional splits provide trivial purity improvement, or the size of the terminal nodes gets too small to support stable rules.

More specifically, the split-and-grow procedure of a tree can be terminated when one of the following terminating criteria is met:

- **Purity improvement**: A significant predictor cannot be found, and the purity improvement gained by splitting on the available predictors is considered insignificant.

- **Maximal tree depth:** The maximum number of allowable successive splits has been reached. Users can specify in advance the maximum tree depth, the tree levels below the root node, which determines the maximum number of times the training dataset can be recursively split.

- **Minimal node size**: The tree is splitting into nodes of unsafely small sizes. The user-defined stopping criteria concerning the minimum size for splitting (of the parent and child nodes) has been met.

4.3.3 Tree pruning

A crucial aspect to consider in the development of Decision Tree models is their stability and their ability to capture general patterns and not patterns pertaining to anomalies of the training dataset. Overfitting is the unpleasant situation when the generated model represents the outliers and the "noise" of the training instances. In such a case, the model will most probably collapse when used on unseen data.

Prepruning is a way to deal with the pitfall of overfitting by setting relatively prudent and conservative terminating criteria. The number of the tree levels should rarely be set above 6. Additionally, the requested minimum node size should be set large enough to identify rules with acceptable support. Although the respective settings also depend on the total number of records available, it is generally recommended to keep the number of records for the terminal nodes at least above 100 and if possibly between 200 and 300. This will return rules with reasonable coverage and will eliminate the risk of modeling patterns that only apply for the specific records analyzed.

Another way to address overfitting is with tree pruning. Many Decision Tree algorithms incorporate an integrated pruning procedure which trims the tree after full growth and replace specific tree branches with internal nodes. This method is referred to as postpruning or backward pruning. The main concept behind pruning is to end up with smaller and more stable trees, with simpler structure but with improved classification accuracy on independent datasets.

The main criterion for pruning a specific branch is the number of errors. The accuracy measure known as the error rate designates the percentage of the misclassified cases. The trimming is done by comparing the compound error rate of the subtree with the estimate of its parent node. Using the training instances to validate a model leads to biased and optimistic accuracy estimates. A way to tackle this is to estimate the error rates on a holdout dataset. But this would lead to a decrease in the number of the training cases. The C5.0/C4.5 algorithm uses a method called pessimistic pruning to overcome this issue. The error rates of the subtree and the candidate leaf node are estimated on the training dataset, but the inherent bias is adjusted by adding a correction to the estimates. More specifically, the upper limits of the respective confidence intervals are used. If the accuracy estimate of the internal node is less than the one of the subtree, then the subtree is pruned.

CART on the other hand use the method of cost-complexity pruning. This method uses a prune dataset and computes a cost-complexity measure which is a function of error (misclassification) rate as well as complexity (number of terminal nodes). Large impurity and a large number of leaves correspond to a high cost-complexity measure. Subtrees are trimmed so that the highest possible decrease in the cost complexity is achieved. Hence, a large subtree with trivial purity improvement is trimmed to its parent node if the cost complexity of the pruned tree is smaller than the one of the full tree.

4.4 CART, C5.0/C4.5, and CHAID and their attribute selection measures

There are various Decision Tree algorithms. All of them have the same goal of minimizing the total purity by identifying subsegments with a dominant class. However, they use different attribute selection measures and pruning strategies. In this chapter, we'll present the splitting criteria used by the well-established and popular CART, C5.0/C4.5, and CHAID algorithms.

The data listed in Table 4.1 will be the training dataset for a simple Decision Tree example in order to explain the different splitting criteria. The data file records the responses of 50 mobile telephony customers to a test marketing campaign (response to pilot campaign *yes/no*). This information defines the model's target field. Demographical (gender, profession) and behavioral (SMS and voice usage) information is also available for the customers and will be used for predicting responses.

4.4.1 The Gini index used by CART

The CART algorithm produces binary splits based on the Gini impurity measure. Gini is a measure of dispersion that depends on the distribution of the target classes. It ranges from 0 to 1 with larger values denoting greater impurity and a balanced distribution of the target outcomes while smaller values designate lower impurity and a dominant class within the node.

Specifically, the formula of the Gini index used by the CART algorithm is the following:

$$\text{Gini}(N) = 1 - \sum_{i=1}^{m} p_i^2$$

where N is the node N and p_i is the probability of target class i in node N. p_i is calculated as the proportion of cases belonging to target class i in node N. The sum is computed over the m target classes.

For example, in the case of a target field with three classes, a node with a perfectly balanced outcome distribution yields a Gini value of 0.667 ($1-(3\times0.33^2)$). On the contrary, a pure node with all records assigned to a single target class gets a Gini value of 0.

The Gini impurity measure of a binary split S of node N on predictor P is the weighted average of the Gini indexes of the resulting child nodes N_1 and N_2. It is calculated as follows:

$$\text{Gini } S_{(P)} = \text{Gini}(N_1)\frac{n_1}{n} + \text{Gini}(N_2)\frac{n_2}{n}$$

where n_1 and n_2 are the number of cases of the two partitions N_1 and N_2 and n the number of cases of the parent node N.

Before each split, all predictors are optimally dichotomized as explained in 4.3.1. Categorical predictors are collapsed in those subsets that return the minimum Gini index. Analogously, for each continuous input, the best cutoff point is identified that produces the binary split with the smaller Gini index.

Then, all predictors are evaluated for the split on the basis of maximum impurity reduction or equivalently the greatest purity improvement. The purity improvement incurred by the split S of node N on predictor P is calculated as

$$\text{Purity improvement } S_{(P)} = \text{Gini}(N) - \text{Gini } S_{(P)}$$

The Gini indexes of all predictors are compared, and the attribute which yields the greater purity improvement is selected for partitioning the node N.

Table 4.1 The modeling dataset for the Decision Tree example

Customer ID	Gender	Profession	Average monthly number of SMS	Average monthly number of Voice calls	Response to pilot campaign—OUTPUT FIELD
1	Male	White collar	45	100	No
2	Male	Blue collar	32	67	No
3	Male	Blue collar	57	30	No
4	Male	Blue collar	120	150	Yes
5	Female	White collar	87	81	No
6	Male	Blue collar	98	32	No
7	Female	White collar	110	78	No
8	Female	White collar	68	90	Yes
9	Male	White collar	67	90	No
10	Male	Blue collar	23	110	No
11	Female	Blue collar	57	30	No
12	Female	White collar	125	110	Yes
13	Female	White collar	87	81	No
14	Male	Blue collar	56	56	No
15	Female	White collar	56	120	No
16	Female	White collar	90	30	Yes
17	Male	White collar	78	45	No
18	Male	Blue collar	32	87	No
19	Male	Blue collar	57	30	No
20	Male	White collar	110	100	Yes
21	Female	White collar	87	81	No
22	Male	Blue collar	34	45	No
23	Female	White collar	65	32	No
24	Male	White collar	100	90	Yes
25	Male	White collar	21	56	No
26	Male	Blue collar	65	45	No
27	Female	Blue collar	57	30	No
28	Female	Blue collar	100	105	Yes
29	Female	Blue collar	132	81	No
30	Male	Blue collar	110	97	No
31	Female	White collar	112	128	No
32	Male	White collar	100	60	Yes
33	Male	White collar	45	110	No
34	Male	Blue collar	43	67	No
35	Male	Blue collar	57	30	No
36	Female	White collar	143	140	Yes
37	Female	White collar	87	81	Yes
38	Male	Blue collar	90	45	No
39	Male	White collar	130	67	No
40	Female	White collar	98	50	Yes
41	Male	White collar	67	57	No

Table 4.1 (**Continued**)

Customer ID	Gender	Profession	Average monthly number of SMS	Average monthly number of Voice calls	Response to pilot campaign—OUTPUT FIELD
42	Male	Blue collar	32	55	No
43	Female	Blue collar	57	30	No
44	Male	White collar	136	60	Yes
45	Male	White collar	87	81	No
46	Male	Blue collar	67	34	No
47	Male	White collar	118	111	No
48	Female	White collar	113	78	Yes
49	Female	White collar	135	167	Yes
50	Male	Blue collar	23	34	No

As an example, let's consider the 50 cases of Table 4.1 as a root node and try to select the optimal partitioning using the Gini index. To decide the best split, the purity improvement for each possible split must be calculated and compared.

The first step is to identify the best split points for the two continuous inputs, namely, voice and SMS usage. The identified optimal cutoff values are 134 for the voice usage field and 88.5 for the SMS usage field. The two categorical inputs are binary; therefore, they are ready to participate in the split challenge.

The root node contains 50 cases, 14 belonging to the target class of responders and 36 to the class of nonresponders.

The Gini index of the root node is

$$\text{Gini}(\text{root node } N) = 1 - \left(\frac{14}{50}\right)^2 + \left(\frac{36}{50}\right)^2 = 0.4$$

The profession predictor has two outcomes, white and blue collar with 28 and 22 cases each. Twelve of the white-collar and only two of the blue-collar customers responded to the campaign. Hence, the Gini index for the split on profession would be

$$\begin{aligned}\text{Gini}(\text{split } S \text{ on profession}) &= \left(\frac{28}{50}\right)\text{Gini}(\text{white-collar node partition, child node } N_1)\\ &\quad + \left(\frac{22}{50}\right)\text{Gini}(\text{blue-collar node partition, child node } N_2)\\ &= \left(\frac{28}{50}\right)\left(1-\left(\frac{12}{28}\right)^2-\left(\frac{16}{28}\right)^2\right)+\left(\frac{22}{50}\right)\left(1-\left(\frac{2}{22}\right)^2-\left(\frac{20}{22}\right)^2\right) = 0.0347\end{aligned}$$

The purity improvement incurred by splitting on profession is $0.4 - 0.0347 = 0.056$.

Similarly, the Gini indexes for the rest of predictors are calculated and compared. Figure 4.2 presents the purity improvements for all possible splits and predictors as calculated by the Weight by Gini Index operator of RapidMiner.

attribute	weight
Gender	0.039
Profession	0.056
Average monthly number of Voice calls	0.066
Average monthly number of SMS	0.137

Figure 4.2 The reduction of impurity for the four possible splits based on the Gini index

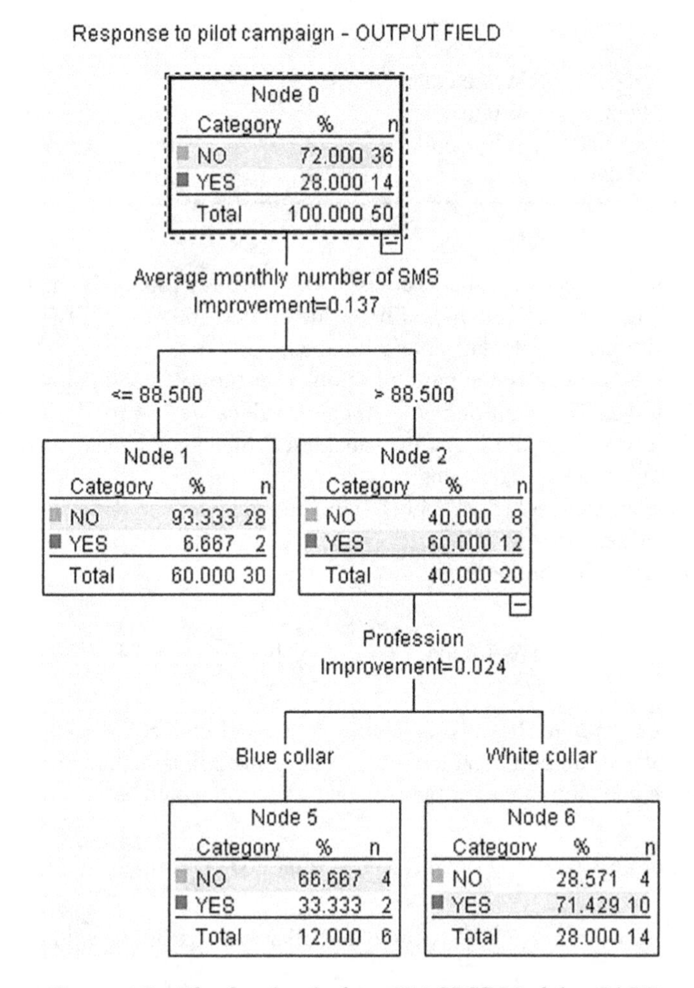

Figure 4.3 The first level of an IBM SPSS Modeler CART

The number of SMS field is selected for the first split as it has the minimum Gini index and returns the maximum purity improvement. The selected cutoff point of 88.5 SMS produces two child nodes which are then further split in a similar manner at subsequent levels of the CART.

The 2-level IBM SPSS Modeler CART and the respective first split on SMS usage are presented in Figure 4.3.

4.4.2 The Information Gain Ratio index used by C5.0/C4.5

The attribute selection measure used by C5.0/C4.5 is the Information Gain Ratio, an extension of the Information Gain used by their predecessor, the ID3 algorithm.

The Information Gain is based on information theory. The Information measure represents the bits needed to make out the outcome class of a particular node. It depends on the probabilities (proportions) of the outcome classes, and it can be expressed in bits which can be considered as the simple *yes/no* questions that are required to classify an instance. The formula for the Information or Entropy of the node N is as follows:

$$\text{Info } N = -\sum_{i=1}^{m} p_i \log_2(p_i)$$

where N is the node N and p_i is the nonzero probability of target class i in node N. p_i is calculated as the proportion of cases belonging to target class i in node N. The measure denotes the Information required to classify the records of node N.

The Information gets its maximum value in the case of equally distributed classes and a value of zero when the number of either positive or negative cases is zero.

The Information of a split denotes the Information still needed to classify a record after partitioning on a predictor. It is calculated as the weighted average of the Information of the resulting child nodes. Specifically,

$$\text{Info } S_{(P)} = \sum_{j=1}^{v} \text{Info } N_j \frac{n_j}{n}$$

where S is the split of N on predictor P, N_j are the partitions resulted from the split which corresponds to the j outcomes of P, n_j are the instances of partition j, and n are the instances of the parent node N.

The Information gained by a split S is measured with the Information Gain index as follows:

$$\text{Information Gain } S_{(P)} = \text{Info } N - \text{Info } S_{(P)}$$

This measure denotes the gain, the decrease in the required Information after branching the parent node N with predictor P. Before each split, all continuous predictors are optimally dichotomized as explained in 4.3.1 by identifying the ideal cutoff point for partitioning. Then, all predictors are evaluated in terms of Information reduction. The predictor which incurs the split with the maximum Information Gain is selected for the branch.

As an example, let's consider again the 50 cases of Table 4.1 as a root node with 50 cases, 14 belonging to the target class of responders and 36 to the class of nonresponders. The cutoff values of 134 and 67.5 are selected for the number of voice and SMS calls, respectively.

The Information of the root node is

$$\text{Info } N = -\left(\left(\frac{14}{50}\right)\log_2\left(\frac{14}{50}\right) + \left(\frac{36}{50}\right)\log_2\left(\frac{36}{50}\right)\right) = 0.86$$

The profession binary predictor has two outcomes, white and blue collar with 28 and 22 cases respectively. Twelve of the white-collar and only two of the blue-collar customers

attribute	weight
Gender	0.068
Profession	0.110
Average monthly number of Voice calls	0.118
Average monthly number of SMS	0.316

Figure 4.4 The Information gain measures for the four possible splits

responded to the campaign, returning an Information measure for the split on profession as follows:

$$\begin{aligned}\text{Information}(\text{split } S \text{ on profession}) &= \left(\frac{28}{50}\right)\text{Info}(\text{white-collar node partition, child node } N_1)\\&+\left(\frac{22}{50}\right)\text{Info}(\text{blue-collar node partition, child node } N_2)\\&=\left(\frac{28}{50}\right)\left\{-\left[\left(\frac{12}{28}\right)\log_2\left(\frac{12}{28}\right)+\left(\frac{16}{28}\right)\log_2\left(\frac{16}{28}\right)\right]\right\}\\&+\left(\frac{22}{50}\right)\left\{-\left[\left(\frac{2}{22}\right)\log_2\left(\frac{2}{22}\right)+\left(\frac{20}{22}\right)\log_2\left(\frac{20}{22}\right)\right]\right\}\\&=0.75\end{aligned}$$

The Information Gain by the split on profession is the simple difference of the above required Information values: $0.86-0.75=0.11$.

Figure 4.4 presents the Information Gain measures for all inputs and partitions as calculated by the Weight by Information Gain operator of RapidMiner.

The Information Gain measure is biased toward inputs with many outcomes. It is maximal when there is one case in each partition and tends to give highly branching splits. For example, a split on an input with unique values for each training instance such as the customer ID would yield pure, single instance partitions, hence a zero Information measure for the split and maximum Information Gain.

To compensate this bias, the C5.0 and C4.5 algorithms use a normalized format of the Information Gain, the Information Gain Ratio. The Information Gain Ratio takes into account a "complexity" factor, defined by the number and the size of the resulting child nodes, and "penalizes" splits with a large number of partitions.

The Information Gain Ratio of a split is calculated by dividing the respective Information Gain by the Split Information of the input. The Split Information increases with the outcomes of the split input. The formula of the index is presented below:

$$\text{Information Gain Ratio } S_{(P)} = \frac{\text{Information Gain } S_{(P)}}{\text{SplitInfo } S_{(P)}}$$

where SplitInfo $S_{(P)}$ is calculated as follows:

$$\text{SplitInfo } S_{(P)} = -\sum_{j=1}^{v} \frac{n_j}{n} \log_2\left(\frac{n_j}{n}\right)$$

where S is the split of node N on predictor P, j the outcomes of predictor P, n_j the instances of partition j, and n the instances of the parent node.

The input with the highest Information Gain Ratio, among those with at least average Information Gain, is selected for the partition.

A simple, 2-level C5.0 tree built by IBM SPSS Modeler using the Gain Ratio split criterion is presented in Figure 4.5. The SMS usage field is used for the first partition. In the second level of the tree, heavy SMS users are further partitioned according to their profession. Note that the C5.0 algorithm uses the Information Gain criterion for selecting the optimal cutoff for continuous inputs. This explains the use of the same threshold (67.5 SMS messages) used for the relative input. Once the threshold is chosen, the selection of the best input for the split is still made on the basis of the Gain Ratio.

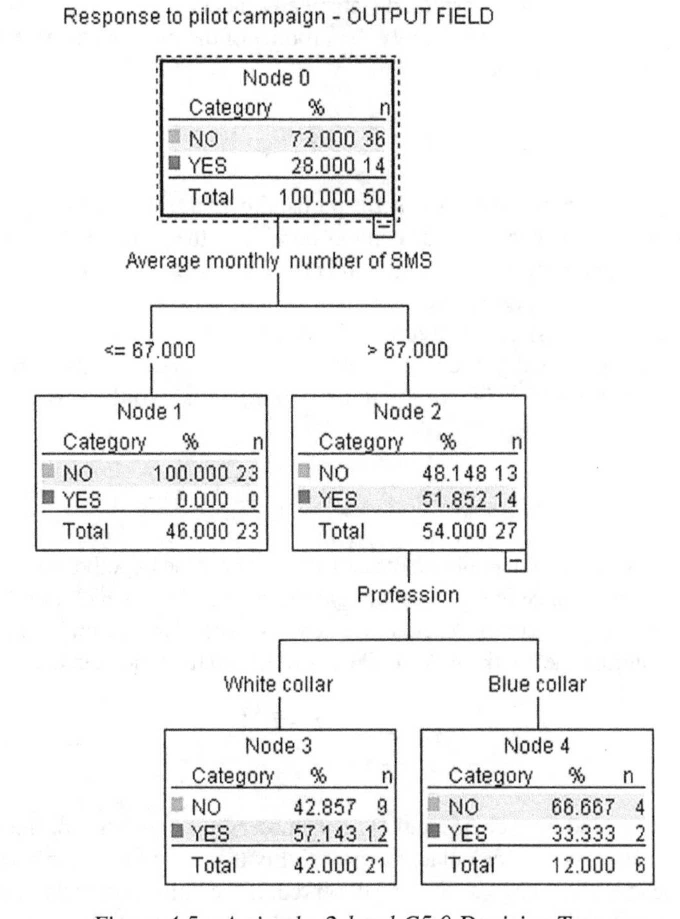

Figure 4.5 A simple, 2-level C5.0 Decision Tree

4.4.3 The chi-square test used by CHAID

The CHAID algorithm uses the statistical chi-square (χ^2) test of independence as the attribute selection measure.

The CHAID Decision Tree algorithm works in three main steps:

1. Before deciding the best split, it discretizes the continuous inputs into 10 or more bins of equal size.

2. Subsequently, as mentioned in 4.3.1, the categories of each input attribute, are further collapsed at an optimal number. Once again, the chi-square test of independence is applied for regrouping the input categories which are similar in respect to the target.

3. Finally, a series of chi-square tests are carried out to examine the null hypothesis of the independence of the binary target with each of the optimally regrouped input. The input with the strongest correlation with the target is selected for the split.

To present the basics of the chi-square test, let's consider a two-way contingency table, a simple cross-tab of the outcomes of two categorical variables, in our case the target and an input. These two variables are considered statistically independent if the probability that an instance belongs to a certain cell is simply the product of the marginal probabilities. Hence, under the independence hypothesis:

$$\hat{p}_{ij} = \hat{p}_i \hat{p}_j = \frac{r_i}{N}\frac{c_j}{N}$$

where $\hat{p}_{ij}$ is the probability of the cell i, j, $\hat{p}_i$ the probability of the outcome i of the row variable, $\hat{p}_j$ the probability of the outcome j of the column variable, N the grand total number of cases, r_i the ith row's subtotal number of cases, and c_j the jth column's subtotal number of cases.

Equivalently, by multiplying the two terms of the above equation with N the total number of cases, we can state our hypothesis in terms of the observed and the expected frequencies. More specifically, the two variables are independent if the expected number of cases that fall at the cell i, j of row i and column j equals to the observed number of cases in the cell:

$$\hat{p}_{ij} = \frac{r_i}{N}\frac{c_j}{N} \Leftrightarrow N\hat{p}_{ij} = N\frac{r_i}{N}\frac{c_j}{N} \Leftrightarrow \hat{E}_{ij} = \frac{r_i c_j}{N}$$

where $\hat{E}_{ij}$ denotes the expected number of cases in a cell under the hypothesis of independence.

The test of independence is based on comparing the residuals which are the differences between the observed and the expected frequencies. The Pearson chi-square statistic is calculated by summing the residuals over all cells of the contingency table as follows:

$$\chi^2 = \sum_i \sum_j \frac{\left(O_{ij} - \hat{E}_{ij}\right)^2}{\hat{E}_{ij}}$$

where $\hat{E}_{ij}$ is the expected frequencies and O_{ij} is the observed frequencies of the cell i, j.

By using statistics, we can calculate the probability that a random sample would return a chi-square value at least as large as the one observed, if the null hypothesis of independence holds true. This probability is called the p-value or the observed significance level, and it is

Table 4.2 A cross-tab of the outcome with the profession input

Profession–response to pilot campaign—OUTPUT FIELD cross-tabulation

			Response to pilot campaign—OUTPUT FIELD		Total
			No	Yes	
Profession	Blue collar	Count	20.00	2.00	22.00
		Expected count	15.84	6.16	22.00
		Residual	4.16	–4.16	
	White collar	Count	16.00	12.00	28.00
		Expected count	20.16	7.84	28.00
		Residual	–4.16	4.16	
Total		Count	36	36.00	14.00
		Expected count	36.0	36.00	14.00

tested against a predetermined threshold which is called the significance level of the statistical test or α. If the p-value is small enough, hence if the probability of observing such large residuals is small, typically lower than 0.05 (5%), the null hypothesis of independence is rejected.

The CHAID algorithm evaluates for the split all the statistically significant inputs for which the hypothesis of independence is rejected. The input with the lowest p-value is selected for the split.

As an example, let's consider once again the 50 cases of Table 4.1 as a root node and try to compare predictors in terms of the Pearson chi-square statistic. Table 4.2 presents the cross-tabulation of the target (response to pilot campaign) with the profession input and the relevant observed and expected frequencies.

The chi-square statistic for the test of independence is calculated as follows:

$$\chi^2 = \frac{(\text{Residual}_{11})^2}{\text{Expected frequency}_{11}} + \frac{(\text{Residual}_{12})^2}{\text{Expected frequency}_{12}} + \frac{(\text{Residual}_{21})^2}{\text{Expected frequency}_{21}} + \frac{(\text{Residual}_{22})^2}{\text{Expected frequency}_{22}}$$
$$= \frac{(4.16)^2}{15.84} + \frac{(-4.16)^2}{6.16} + \frac{(-4.16)^2}{20.16} + \frac{(4.16)^2}{7.84} = 6.97$$

The chi-square statistic of 6.97 corresponds to a p-value of 0.008. Since this value is lower than the significance level of 0.05, the independence hypothesis is rejected, and the profession input is considered a significant predictor. Hence, it'll compete with the rest of significant predictors for partitioning the root node.

Figure 4.6 shows a comparison of the inputs in terms of the chi-square statistical test, as presented by the IBM SPSS Modeler interactive CHAID tree algorithm. The last column designates the test's p-values for each input. The (binned) number of SMS input yields the lowest p-value value, and as shown in Figure 4.7, it is chosen for the first split.

Select Predictor

Predictor	N...	Statistic	DF	Adj. Prob.
Average monthly number of SMS	3	Chi-square=18.254	2	0.004
Profession	2	Chi-square=6.968	1	0.008
Gender	2	Chi-square=4.778	1	0.029
Average monthly number of Voice calls	2	Chi-square=4.837	1	0.251

OK Cancel Help

Figure 4.6 Selecting predictors for split using the CHAID algorithm and the chi-square statistic

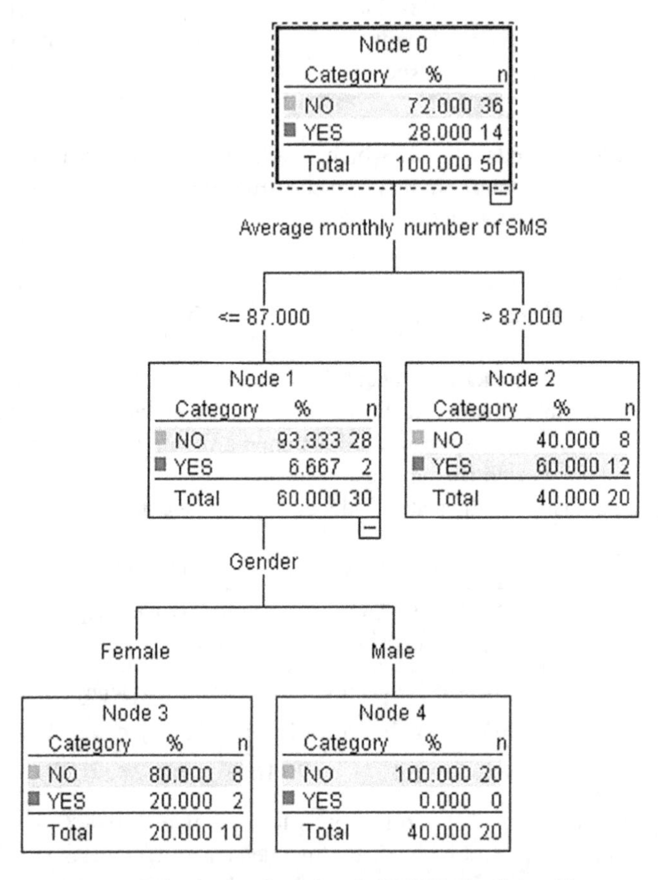

Figure 4.7 A simple, 2-level CHAID Decision Tree

@Modeling tech tips

The IBM SPSS Modeler CHAID Decision Trees

In Figures 4.8, 4.9, and 4.10 and in Table 4.3, the IBM SPSS Modeler CHAID parameters are presented and explained.

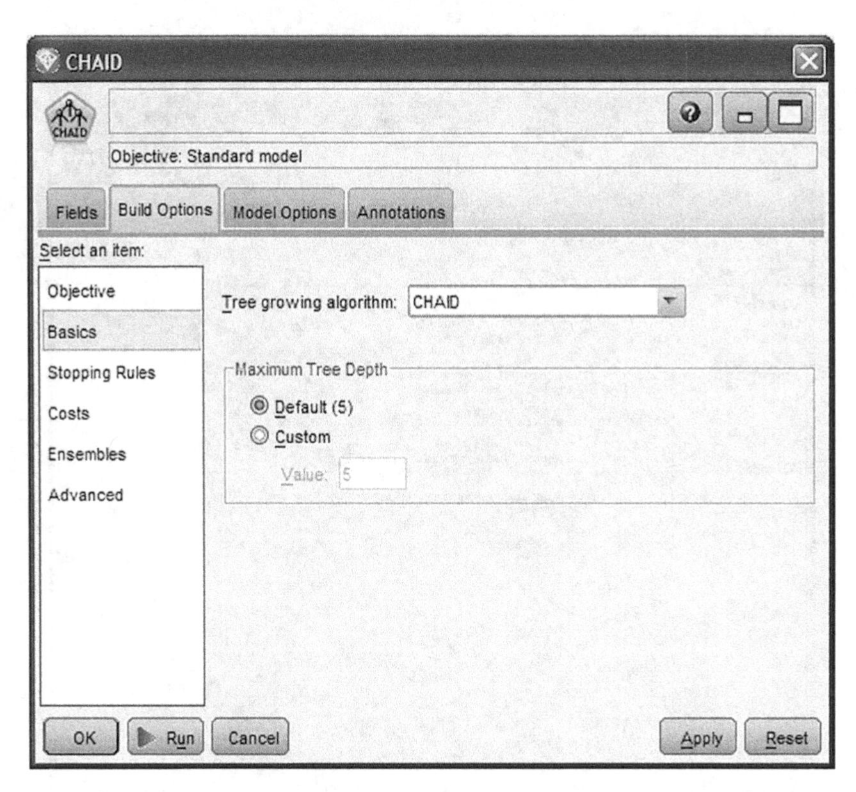

Figure 4.8 The IBM SPSS Modeler CHAID growing algorithm options

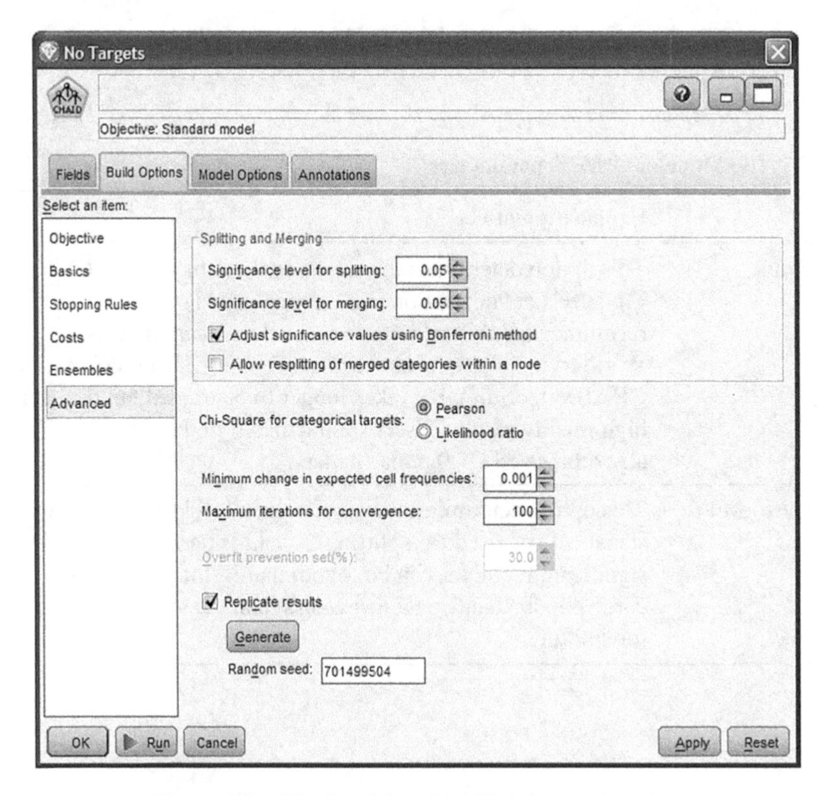

Figure 4.9 The Modeler CHAID Advanced options

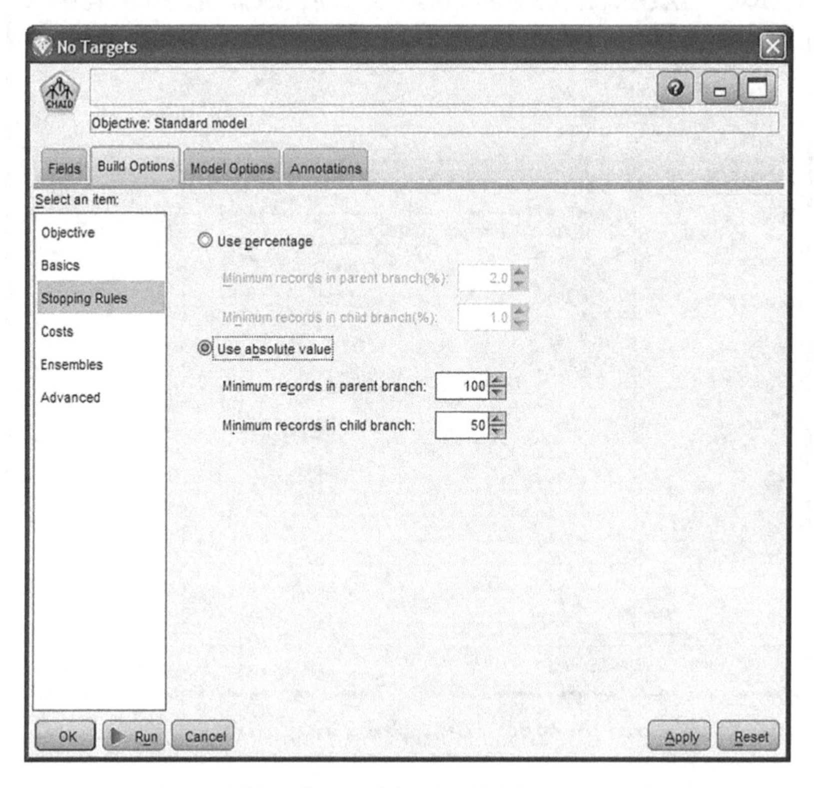

Figure 4.10 The Modeler CHAID Stopping Rules

Table 4.3 The Modeler CHAID parameters

Option	Explanation/effect
Tree growing algorithm	This option determines the tree growth method. CHAID is the Modeler's default option and the tree-growing method recommended to start with in most classification tasks Worth trying alternatives: Exhaustive CHAID, a modification of the CHAID algorithm that takes longer to be trained but often gives high-quality results. Users should also try other Decision Tree algorithms and C5.0 in particular
Significance level for splitting	This option determines the significance level for the chi-square statistical test used for splitting. A split is performed only if a significant predictor can be found, that is, the *p*-value of the corresponding statistical test is less than the specified alpha for splitting

Table 4.3 (**Continued**)

Option	Explanation/effect
	In plain language, this means that by increasing the alpha for splitting value (normally up to 0.10), the test for splitting becomes less strict, and splitting is made easier, resulting in potentially larger trees. A tree previously terminated due to nonsignificant predictors may be further grown because of loosening the test's criteria. Lower values (normally up to 0.01) tend to give smaller trees. The default value of 0.05 is adequate in most situations
Significance level for merging	This option determines the significance level for the merging of predictor categories. Higher values (normally up to 0.10) hinder the merging of predictor categories, and a value of 1.0 totally prevents merging. Lower values (normally up to 0.01) facilitate the collapsing of predictor categories. Once again, the default value of 0.05 is the most common choice
Minimum records in parent and child branch	These options specify the minimum allowable number of records in the parent and the child nodes of the tree. The tree growth stops if the size of the parent or of the resulting child nodes is less than the specified values. The size of the parent node should always be set higher (typically two times higher) than the corresponding size of the child nodes The requested values can be expressed in terms of percentage of overall training data or in terms of absolute number of records Although the respective settings also depend on the total number of records available, it is generally recommended to keep the number of records for the terminal (child) nodes at least above 50 and if possibly even higher. Large values on these settings provide robust rules "supported" by many cases/records which we expect to perform well when used in new datasets. Remember that the CHAID algorithm does not employ a postpruning method
Maximum tree depth/levels below root	This option determines the maximum allowable number of consecutive partitions of the data. Although this option is also related to the available number of records, users should try to achieve effective results without ending up with bushy and complicated trees For trees mainly constructed for profiling purposes, a depth of 3–4 levels is typically adequate For modeling purposes, a depth of 5–7 levels is sufficient, whereas larger trees, even when the available number of records allows them, would probably provide complicated rules, hard to examine and evaluate

The IBM SPSS Modeler C5.0 Decision Trees

The C5.0, the successor of the C4.5 and of the ID3 algorithm, is offered by the IBM SPSS Modeler with the parameters presented in Figure 4.11 and Table 4.4.

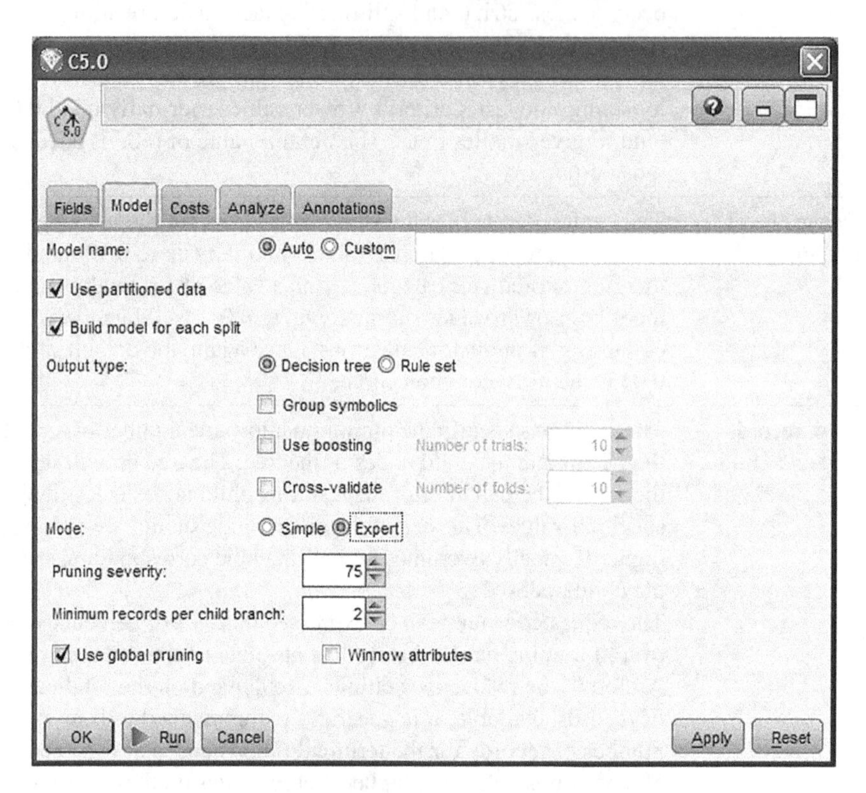

Figure 4.11 The Modeler C5.0 Model settings

Table 4.4 The Modeler C5.0 parameters

Option	Explanation/effect
Output type	The Decision Tree setting (recommended) is selected for building a Decision Tree with exhaustive and mutually exclusive rules which cover all the instances. The Rule set setting is used for developing decision rule sets
Group symbolics	Enables merging of predictors' outcomes for simpler and clearer trees. It can be applied to categorical inputs. Although it is not the default setting, it is recommended since it leads to less fragmented trees

Table 4.4 (**Continued**)

Option	Explanation/effect
Use Boosting	A worth-trying ensemble-modeling technique, although demanding in terms of modeling time and resources. It uses the Adaptive Boosting (AdaBoost) method to build and combine a set of individual models. The first model is built the usual way, the second one focuses on the "hard" cases which were misclassified by the first one, and so on. A (confidence-weighted) voting procedure is applied to determine the final prediction and confidences
Pruning severity (Expert option)	C5.0 builds Decision Trees in two phases. A large tree is first grown which is then "pruned" by removing parts with high error rates. At the first stage of "local" pruning, each subtree is tested to decide whether it should be replaced by a leaf node or a subbranch. Then, a second, global pruning stage assesses the performance of the tree as a whole for further pruning. The pruning severity option affects the first stage of pruning. It corresponds to (1—the confidence level) used for the pessimistic error calculation of pruning, and hence it affects the way that error rates are estimated and the severity of pruning The higher the value, the more severe the first stage pruning. The default value of 75% corresponds to a confidence value of 25%, and it is suitable for most purposes. A higher value could be tested for more drastic pruning
Winnow attributes (Expert option)	This option applies an integrated feature selection procedure for prepruning. Irrelevant predictors are identified and omitted from the model training. This option might be proved useful when there are a large number of inputs
Use global pruning (Expert option)	This option toggles on/off the second phase of global pruning. Global pruning is recommended for most Decision Tree models. Turning off global pruning can be tried especially when developing rule sets
Minimum records per child branch (Expert option)	A branch will be spit only if two or more of the resulting subbranches have at least this number of records. Values around 50 are a reasonable starting point, especially in cases with many noisy data. In any case, leaf nodes with a trivial number of instances should be avoided for stability reasons

The IBM SPSS Modeler CART

The main parameters of the IBM SPSS Modeler CART algorithm are summarized in Figures 4.12, 4.13, and 4.14 and Table 4.5.

The RapidMiner Decision Trees

The RapidMiner Decision Tree algorithm works similar with the C4.5 and CART algorithms. It applies recursive partitioning with successive splits based on the Gini,

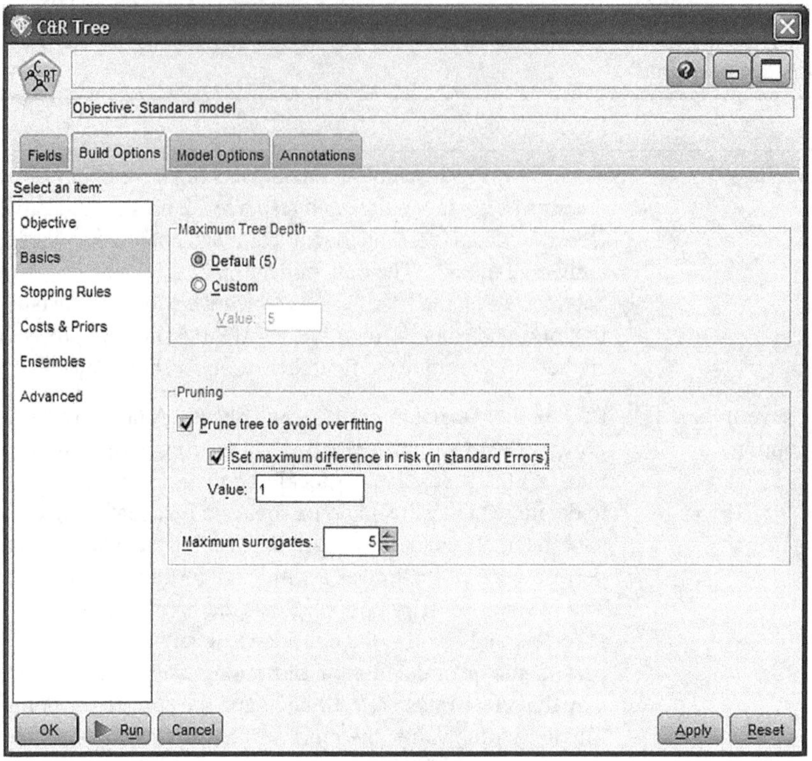

C&R Tree

Objective: Standard model

Fields | Build Options | Model Options | Annotations

Select an item:

Objective
Basics
Stopping Rules
Costs & Priors
Ensembles
Advanced

Maximum Tree Depth
Default (5)
Custom
Value: 5

Pruning
Prune tree to avoid overfitting
Set maximum difference in risk (in standard Errors)
Value: 1
Maximum surrogates: 5

OK | Run | Cancel | Apply | Reset

Figure 4.12 The Modeler CART Basics options

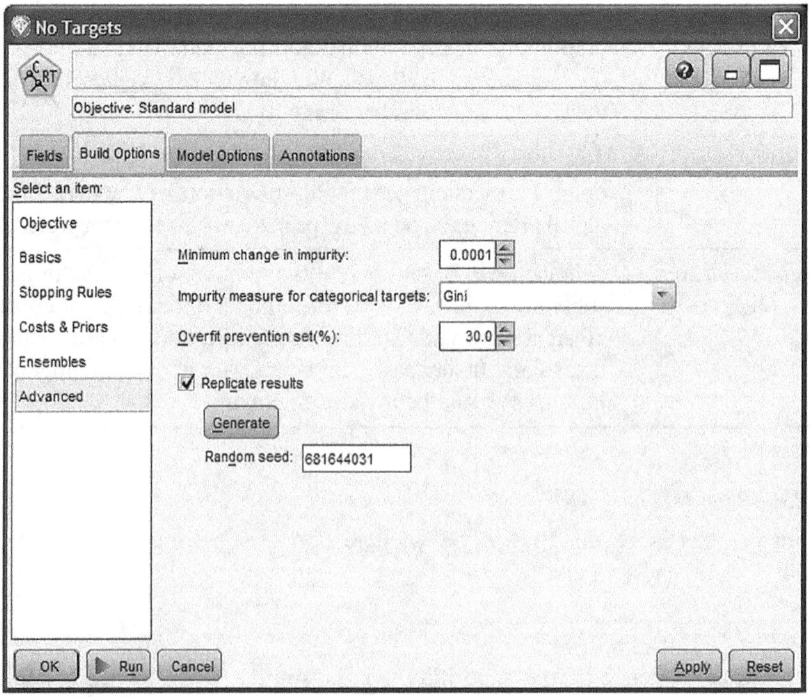

No Targets

Objective: Standard model

Fields | Build Options | Model Options | Annotations

Select an item:

Objective
Basics
Stopping Rules
Costs & Priors
Ensembles
Advanced

Minimum change in impurity: 0.0001
Impurity measure for categorical targets: Gini
Overfit prevention set(%): 30.0
Replicate results
Generate
Random seed: 681644031

OK | Run | Cancel | Apply | Reset

Figure 4.13 The Modeler CART Advanced options

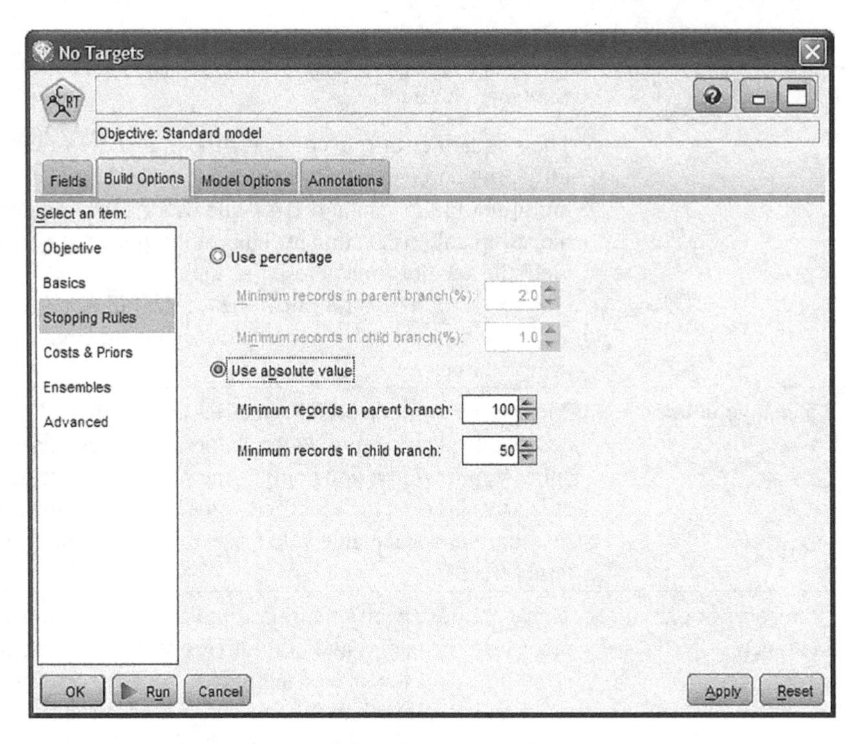

Figure 4.14 The Modeler CART Stopping criteria

Table 4.5 The Modeler CART parameters

Option	Explanation/effect
Impurity measure for categorical targets	This setting is available in the Expert tab, and users can select the impurity measure and hence the attribute selection method for the tree among Gini, Twoing, and Ordered. The Gini index presented in 4.4.1 is the recommended option
Prune tree (maximum difference in risk in standard errors)	The prune tree option employs cost-complexity pruning; it is turned on and should be normally kept on to produce trimmed, generalizable trees which are not memorizing the noise of the training data
	Initially, the tree is fully grown according to the specified stopping criteria before trimmed. The cost-complexity pruning method removes a tree branch if the cost associated with having a more complex tree exceeds the gain associated with having another branch. It uses an index that measures both the risk estimate (misclassification rate) and the complexity, the number of terminal nodes of the tree

(*Continued*)

Table 4.5 **(Continued)**

Option	Explanation/effect
	The standard error rule compares the pruned trees with the fully grown tree in terms of their risk estimates. The multiplier of the standard error rule defines the boundary of the acceptable risk estimate. Thus, a multiplier of one will yield the smallest tree whose risk estimate is up to (1 × standard error) larger than that of the full tree. Lower values of the multiplier would produce larger, slightly trimmed trees
Minimum change in impurity	A prepruning option which specifies the minimum acceptable purity improvement in order to create a new split. A split will be done only if the purity improvement is at least as large as the specified value. Hence, by lowering the minimum acceptable value, we allow more splits and larger trees
Minimum records in parent and child branch	These options specify the minimum allowable number of records in the parent and child nodes of the tree. The tree growth stops if the size of the parent or of the resulting child nodes is less than the specified values. The size of the parent node should always be set higher (typically two times higher) than the corresponding size of the child nodes The requested values can be expressed in terms of percentage of overall training data or in terms of absolute number of records Since CART employs a postpruning method, the number of records of the terminal nodes should not be set as high as in CHAID trees. Values around 50 are usually a good starting point for trial and error
Maximum tree depth/levels below root	As in CHAID, this option specifies the maximum allowable number of consecutive partitions of the data. Given that the fully grown tree will be finally trimmed, a larger number of levels than in CHAID are selected, typically 6+
Overfit prevention set (%)	Specifies the portion of the dataset used to track errors during model training in order to prevent overfitting. The default value is 30%

the Information Gain, or the Information Gain Ratio measures. The algorithm also employs prepruning as well as prostpruning techniques to avoid overspecific and overfitted models.

The relevant parameters are shown in Figure 4.15 and discussed in Table 4.6.

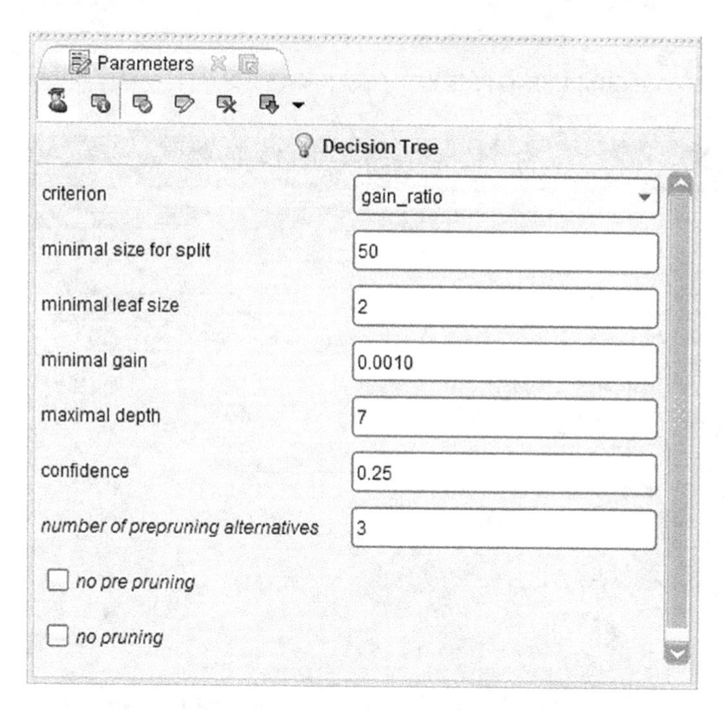

Figure 4.15 The RapidMiner Decision Tree options

Table 4.6 The RapidMiner Decision Tree parameters

Option	Explanation/effect
Criterion	This option determines the attribute selection method used for producing the tree. The available grow methods include the Information Gain and the Information Gain Ratio presented in 4.4.2 and the Gini index presented in 4.4.1. The Information Gain Ratio is the default option; the other two methods are certainly worth trying
Minimal size for split/minimal leaf size	The next four options—minimal size for split, minimal leaf size, minimal gain, and maximal depth—are related with the prepruning and define the terminating criteria of the model. The tree is prevented from growing if the sizes of the parent or terminal nodes are lower than the respective thresholds. Large enough threshold values should be set in order to identify rules with adequate support
Maximal depth	The maximum allowable number of consecutive splits. Although the produced tree is finally trimmed, you'll typically have to decrease the number of levels from the default (20) to values between 6 and 10
Minimal gain	The minimum acceptable purity improvement in order to branch a node. Lower values correspond to less stringent split criteria and lead to larger trees. The default threshold of 0.1 is quite strict in many situations, and you may have to try lower values
Confidence	The confidence level used for the pessimistic error rate calculation of pruning. Although the default value works well in most situations, it can be tweaked to control the severity of pruning. Smaller values than the default (25%) allow heavier pruning and produce smaller trees, while larger values result in less pruning
No prepruning/no pruning	By default, prepruning and postpruning are applied, and normally, this is the recommended approach for model training. However, analysts can disable these options by selecting the relevant checkboxes

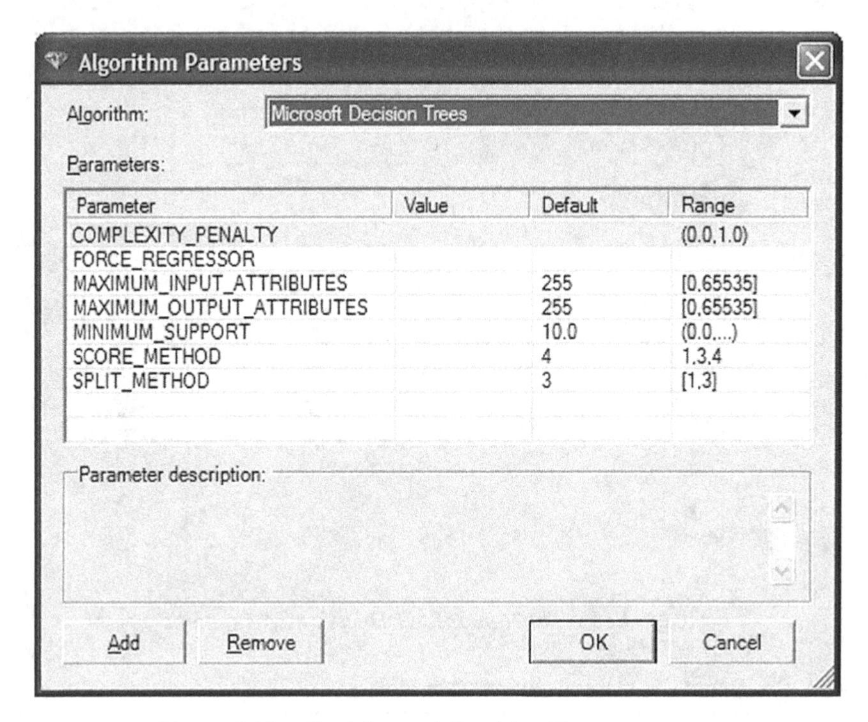

Figure 4.16 The Microsoft Decision Tree parameters

Microsoft Decision Trees

Decision Tree models are also available in Data Mining for Excel. The classification algorithm and the settings for fine-tuning are set through the Algorithm Parameters menu of the Classify Wizard as seen in Figure 4.16 and discussed in Table 4.7.

Microsoft Decision Trees is a recursive partitioning algorithm. The attribute selection methods used to determine the splits include measures based on Entropy (the Information measure presented in 4.4.2) and Bayesian networks.

The algorithm incorporates a preprocessing feature selection step to screen inputs and focus on the most useful predictors. In terms of inputs' handling, continuous predictors are optimally discretized for the needs of classification models. Initially, the continuous inputs are binned into chunks of equal range. Then, the bins are optimally regrouped based on the selected split measure in order to yield an optimized split score. Finally, the input with the best score is selected for branching.

The same approach of optimal regrouping is also the default approach for categorical inputs.

4.5 Bayesian networks

Naïve Bayesian networks are statistical models based on the Bayes theorem. They are probabilistic models as they estimate the probabilities of belonging to each target class. In fact, their classification is based on comparing the estimated target class probabilities, given the information provided by the inputs.

Table 4.7 The Microsoft Decision Trees parameters

Option	Explanation/effect
Complexity penalty	A parameter related with prepruning which controls the growth of the Decision Tree. Its values range from 0 to 1. Lower values increase the allowable number of splits, producing bushier trees. On the contrary, higher values, close to 1, penalize the tree growth, resulting smaller trees. The default value is based on the number of inputs as follows: For 1 through 9 predictors, the default value of 0.5 is applied For 10 through 99 predictors, the default value of 0.9 is applied For 100+ predictors, the default value of 0.99 is applied.
Maximum input attributes	A threshold parameter for the integrated feature selection. Feature selection is invoked if the number of inputs is greater than the specified value. A value of 0 turns off feature selection (the MAXIMUM OUTPUT ATTRIBUTES option is relevant only for regression tree models)
Minimum support	This parameter is a terminating criterion related with prepruning, and it defines the acceptable leaf size. More specifically, it determines the minimum support which is the minimum number of cases in a leaf node (as a percentage of the total cases or as absolute frequency) in order for a split to be allowed
Score method	The available attribute selection methods include the Information measure based on Entropy and two measures based on Bayesian networks: the Bayesian with K2 prior and the Bayesian Dirichlet Equivalent (BDE) with Uniform prior (the default), all of which are scalable and worth trying
Split method	Determines the handling of predictors for split. SPLIT METHOD = 1 produces binary splits. SPLIT METHOD = 2 is used for multiway splits with a branch for each outcome of the predictor. The default and recommended SPLIT METHOD = 3 is a hybrid method, which combines the previous two methods for optimal regrouping of predictors to yield the best split

In spite their simplicity and their naïve assumption of independence, Naïve Bayesian classifiers have shown remarkable performance in many situations comparable to more "sophisticated" classifiers. However, this is not always the case. Bayesian belief networks are an advancement of the Naïve Bayes algorithm. The dependence restrictions of Naïve Bayes are relaxed, achieving improved classification accuracy. Additionally, they are graphical models, providing a visual representation of the relationships between the variables.

4.6 Naïve Bayesian networks

The Bayes rule of conditional probability, named after the clergyman Thomas Bayes, states that if we have a hypothesis H and evidence E, then

$$P(H|E) = \frac{P(E|H)P(H)}{P(E)}$$

In classification problems, our hypothesis is the membership to a target class. We want to estimate the corresponding probabilities, given the evidence, the particular combination of input values. Hence, assuming n inputs X_i and m target classes Y_j, for an instance with input values $X=(x_1, x_2, \ldots, x_n)$, the probability of belonging to the target class Y_j is

$$P(Y_j|X) = \frac{P(X|Y_j)P(Y_j)}{P(X)}$$

or equivalently

$$P(Y_j|X_1 = x_1, X_2 = x_2, \ldots, X_n = x_n) = \frac{P(X_1 = x_1, X_2 = x_2, \ldots, X_n = x_n|Y_j)P(Y_j)}{P(X_1 = x_1, X_2 = x_2, \ldots, X_n = x_n)}$$

$P(H)$ or $P(Y_j)$ is the prior probability of the target class (of the hypothesis), and it is independent of the values of the inputs X. $P(Y_j|X)$ is the posterior probability of the hypothesis given the evidence. It is the probability of the target class conditioned on the observed values of the predictors.

An instance is classified to the target class with the maximum posterior probability. Hence, we need to estimate and compare the probabilities of $P(Y_j|X)$ for each target class Y_j. We need to maximize the $P(Y_j|X)$ and label the instance with the class which returns the maximum conditional probability. Therefore, we don't need to worry about the probability of evidence. As it is the common denominator for all target classes, it can be ignored. However, in a final step, the estimated probabilities must be normalized by dividing with their sum so that they add to 1.

The prior probability $P(Y_j)$ can be easily calculated as the percentage of the target class j in the training dataset.

So that leaves us with the calculation of $P(X|Y_j)$. To simplify calculations, in the Naïve Bayesian models, we "naïvely" assume class conditional independence. More specifically, we assume that the inputs are independent, given the target.

Having assumed independent events, we can multiply their probabilities, and the aforementioned equation of the conditional probability can be stated as follows:

$$P(Y_j|X_1 = x_1, X_2 = x_2, \ldots, X_n = x_n) = \frac{P(X_1 = x_1|Y_j)P(X_2 = x_2|Y_j)\ldots P(X_n = x_n|Y_j)P(Y_j)}{P(X_1 = x_1, X_2 = x_2, \ldots, X_n = x_n)}$$

The probabilities of the nominator $P(X_i = x_i|Y_j)$ can be easily computed for each training instance. If X_i is a categorical input, $P(X_i = x_i|Y_j)$ is simply the percentage of cases with the value x_i for the input X_i within the target class Y_j.

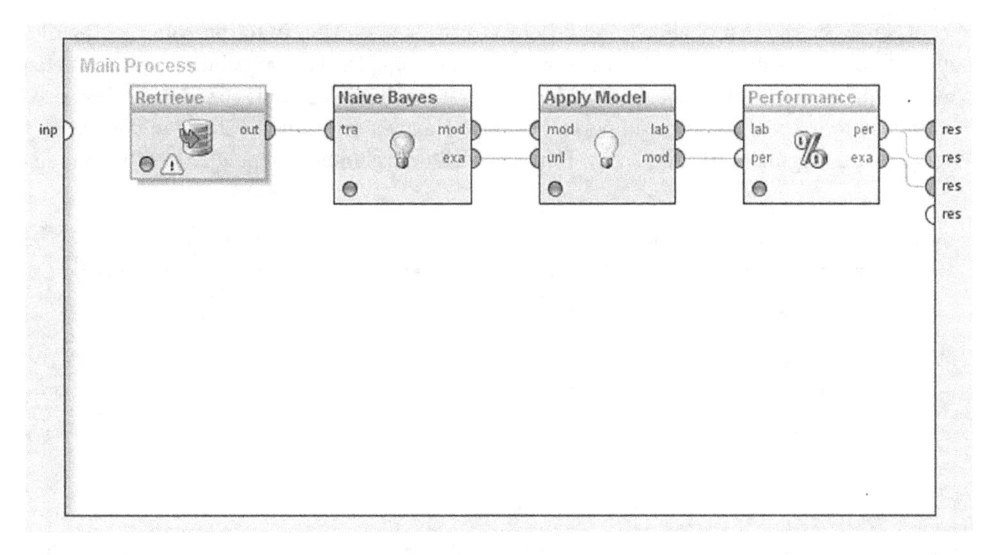

Figure 4.17 A RapidMiner process for training a Naïve Bayes classification model

But what if the input X_i is continuous? A common approach, which is also employed in IBM SPSS Modeler and Data Mining for Excel, is discretization. The continuous attribute is binned into an ordinal predictor, and the conditional probabilities are estimated for each bin based on the relevant distributions. Alternatively, a normal distribution can be assumed for the continuous input, and the probability density function can be used to estimate the respective probabilities.

The probability density function of a continuous random variable with the normal distribution, mean μ and standard deviation σ, is a function that describes the relative likelihood for this random variable to take on a given value x, and it is estimated as follows:

$$f(x)=\frac{1}{\sigma\sqrt{2\pi}}e^{-\frac{(x-\mu)^2}{2\sigma^2}}$$

So, under the normal distribution assumption, in order to estimate the conditional probabilities $P(X_i = x_i|Y_j)$, we only need to calculate the mean and the standard deviation of the input X_i for each target class Y_j. By plugging these two values (respectively, $\mu_{X_i|Y_j}, \sigma_{X_i|Y_i}$) in the above equation, the estimation of $P(X_i = x_i|Y_j)$ through $f(x_i|y_j)$ is straightforward for any observed value x_i of the input X_i for the cases of the target class Y_j.

Since the estimation is based on the multiplication of conditional probabilities, things get messy if we end up with a conditional probability of 0. This occurs when we have no cases for a particular combination of input–output values. In such situations, a single 0 $P(X_i = x_i|Y_j)$ leads to an overall $P(X|Y_j)$ equal to 0. A common approach to override this problem is to add 1 or a small constant to each count. This adjustment, known as the Laplace correction, ensures that possible zero counts would not distort our estimation.

In the final step of the algorithm, for each training (or scoring) instance, the conditional probabilities $P(Y_j|X)$, which as explained are analogous to $P(X|Y_j)P(Y_j)$, are assessed for each target class j. The instance is classified to the target class with the maximum probability.

Figure 4.17 shows a RapidMiner process with a Naïve Bayes operator used for developing a classification model.

As Naïve Bayes is a probabilistic algorithm, it provides estimates of the probabilities (prediction confidences) of the output classes as shown in Figure 4.18. The probabilities are estimated when the data pass through the Apply Model operator which is used for model scoring.

Data Mining for Excel also supports the Naïve Bayes algorithm. As stated in the Classify Wizard menu (Figure 4.19), the Naïve Bayes algorithm assumes independence of the inputs,

ExampleSet (50 examples, 5 special attributes, 4 regular attributes) View Filter (50 / 50): all

Row No.	Customer ID	Response t...	confidence(NO)	confidence(YES)	prediction(Respo...	Gender	Profession	Average mo...	A'
1	1	NO	0.989	0.011	NO	Male	White collar	45	1
2	2	NO	1.000	0.000	NO	Male	Blue collar	32	6
3	3	NO	0.998	0.002	NO	Male	Blue collar	57	3
4	4	YES	0.196	0.804	YES	Male	Blue collar	120	1
5	5	NO	0.426	0.574	YES	Female	White collar	87	8
6	6	NO	0.958	0.042	NO	Male	Blue collar	98	3
7	7	NO	0.181	0.819	YES	Female	White collar	110	7
8	8	YES	0.726	0.274	NO	Female	White collar	68	9
9	9	NO	0.922	0.078	NO	Male	White collar	67	9
10	10	NO	1.000	0.000	NO	Male	Blue collar	23	1
11	11	NO	0.993	0.007	NO	Female	Blue collar	57	3
12	12	YES	0.046	0.954	YES	Female	White collar	125	1
13	13	NO	0.426	0.574	YES	Female	White collar	87	

Figure 4.18 The prediction and probabilities estimated from a RapidMiner Naïve Bayes classification model

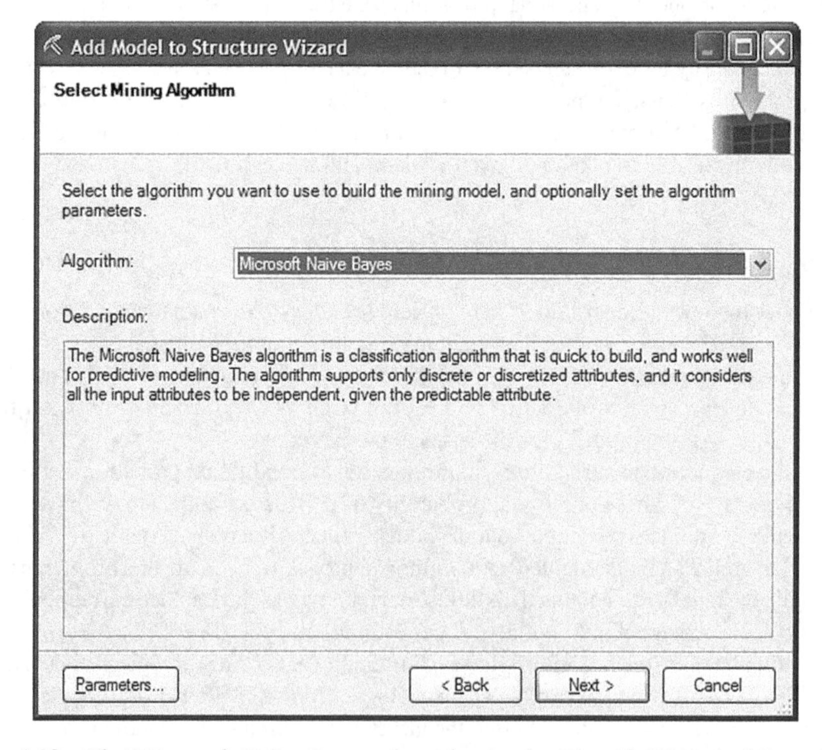

Figure 4.19 The Microsoft Naïve Bayes algorithm in the Classify Wizard of Data Mining for Excel

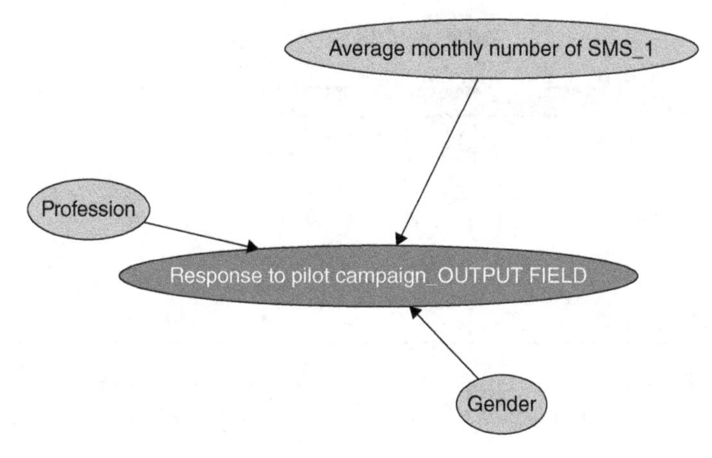

Figure 4.20 The Microsoft Naïve Bayes Dependency network in Data Mining for Excel

Attribute Profiles

Response to pilot campaign _OUTPUT FIELD

Attributes	States	Population (All)	NO	YES
Size		50	36	14
Average monthly number of SMS_1	45 - 70	16	42 %	7 %
Average monthly number of SMS_1	94 - 119	11	14 %	43 %
Average monthly number of SMS_1	21 - 45	8	22 %	0 %
Average monthly number of SMS_1	70 - 94	8	17 %	14 %
Average monthly number of SMS_1	119 - 143	7	6 %	36 %
Gender	Male	30	69 %	36 %
Gender	Female	20	31 %	64 %
Profession	White collar	28	44 %	86 %
Profession	Blue collar	22	56 %	14 %

Figure 4.21 The Data Mining for Excel Attribute Profiles report presents the conditional probabilities used by the Naïve Bayes algorithm

given the output. It also requires discretized inputs. Continuous inputs can be discretized into a specified number of equal-width buckets using the Explore wizard.

The Dependency network included in the model's results illustrates the inputs–output relationships as shown in Figure 4.20.

The Attribute Profiles output presented in Figure 4.21 depicts the percentages of the input categories, which are named as "states" in Data Mining for Excel, within each target class (in this example, response to pilot campaign: Yes/No). In other words, it illustrates the conditional probabilities $P(X_i = x_i | Y_j)$ discussed earlier.

The Attribute Characteristics shown in Figure 4.22 sorts the input categories in descending order of their conditional probabilities (given the selected target class). In other words, it depicts the conditional probabilities $P(X_i = x_i | Y_j)$ discussed earlier.

Like the Decision Tree models presented previously, the Naïve Bayes results indicate that the response to the pilot campaign is more likely among white-collar professionals, heavy SMS users, and women.

Attribute Characteristics		
Response to pilot campaign OUTPUT FIELD=YES		
Attributes	Values	Probability
Profession	White collar	86 %
Gender	Female	64 %
Average monthly number of SMS_1	94 - 119	43 %
Average monthly number of SMS_1	119 - 143	36 %
Gender	Male	36 %
Average monthly number of SMS_1	70 - 94	14 %
Profession	Blue collar	14 %
Average monthly number of SMS_1	45 - 70	7 %

Figure 4.22 The Data Mining for Excel Attribute Characteristics report sorts the input categories according to their conditional probabilities for the selected target class

4.7 Bayesian belief networks

Bayesian belief networks are an improvement over Naïve Bayes models. The class conditional independence, assumed by the Naïve Bayes algorithm, is relaxed. Instead of assuming independence between the inputs, we allow dependencies between subsets of variables. In Naïve Bayes, each input only depends on the target variable. In Bayesian belief networks, however, a subset of "parent" variables is identified for each variable. Our assumption now is less strict since each variable is assumed to be independent of other variables given its parents.

Bayesian belief networks are graphical, probabilistic models, represented by acyclic graphs. Each variable, including the output, is represented by a node in the graph. The nodes are connected by directed edges, in a way that they are no cycles (directed acyclic graph). The connecting arcs of the graph point from the parent to the child node, depicting conditional probabilities. Hence, each belief network consists of a graph and the conditional probability table (CPT) with the conditional probabilities for each node. The Naïve Bayes network is a simple belief network with an arc leading from the output node to each input attribute.

Similarly with Naïve Bayes models, the goal is to estimate the target class probabilities. As in 4.6, assuming n inputs X_i and m target classes Y_j, for an instance with input values $X=(x_1, x_2, \ldots, x_n)$, the posterior probabilities $P(Y_j|X)$, given the evidence, are calculated as follows:

$$P(Y_j|X) = \frac{P(X|Y_j)P(Y_j)}{P(X)}$$

or equivalently

$$P(Y_j|X_1 = x_1, X_2 = x_2, \ldots, X_n = x_n) = \frac{P(X_1 = x_1, X_2 = x_2, \ldots, X_n = x_n|Y_j)P(Y_j)}{P(X_1 = x_1, X_2 = x_2, \ldots X_n = x_n)}$$

Since each variable is conditionally independent of its nondescendants in the graph given its parents, the above formula becomes

$$P(Y_j|X_1 = x_1, X_2 = x_2, \ldots X_n = x_n) = \frac{\prod_{i=1}^{n} P(X_i = x_i|\pi_i, Y_j)P(Y_j)}{P(X_1 = x_1, X_2 = x_2, \ldots, X_n = x_n)}$$

where π_i is the parent set of the input X_i, besides Y.

Similarly with Naïve Bayes, an instance is classified to the target class which maximizes the posterior probability $P(Y_j|X)$. The probability of the denominator can be ignored, and the estimated probabilities are normalized to sum to 1.

In the Tree Augmented Naïve Bayes models (TAN Bayesian models), a class of belief networks, we allow each predictor to depend on another predictor, in addition to the output. The network has the structure shown in Figure 4.23.

A CPT is associated with each input node of the network denoting the conditional probabilities $(X_i = x_i | \pi_i, Y_j)$. Each column of the CPT corresponds to a category value x_i of the input X_i (Figure 4.26). Note that the continuous inputs are discretized. Each row of the table corresponds to a target class–parent input value combination. Hence, the CPT summarizes the conditional probabilities for each value of the input across all combinations of the values of its parents.

An instance is defined by its input values x_i. To classify the instance, we have to look up the CPT entries which correspond to its observed values. Then, we simply multiply the respective entries. This procedure is repeated for each target class, and the instance is classified to the class which yields the maximum conditional probability.

A TAN network trained in IBM SPSS Modeler is shown in Figure 4.24.

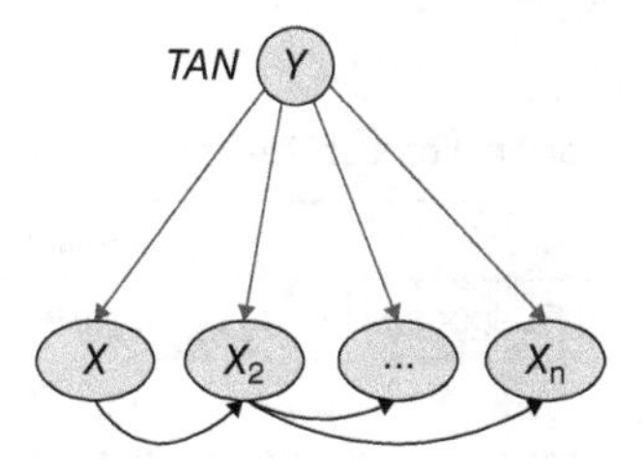

Figure 4.23 The structure of Tree Augmented Naïve Bayes models (TAN Bayesian models)

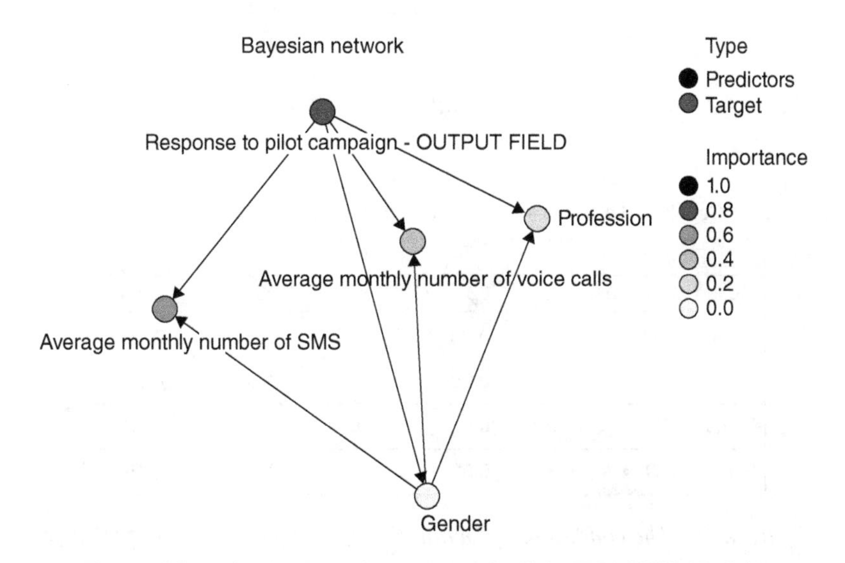

Figure 4.24 A TAN Bayesian network built in IBM SPSS Modeler

The network is consisted of five nodes, one for the output (response to pilot campaign field) and one for each input. Directed arcs connect each parent with its descendants. Each input node has the output as its parent. Furthermore, the Gender node is the second parent for the other inputs: Profession, Average monthly number of Voice calls, and Average monthly number of SMS fields.

Besides its topology, the network is defined by the CPT which explain each node. The CPT of the example are depicted in Figures 4.25, 4.26, 4.27, 4.28, and 4.29. Note that the continuous inputs have been automatically discretized in five bins of equal width.

Conditional probabilities of response

Probability	
YES	NO
0.29	0.71

Figure 4.25 The probabilities of the output (response to pilot campaign)

Conditional probabilities of profession

Parents		Probability	
Gender	Response	White collar	Blue collar
Male	YES	0.77	0.23
Male	NO	0.36	0.64
Female	YES	0.87	0.13
Female	NO	0.63	0.37

Figure 4.26 The conditional probabilities of the profession input attribute

Conditional probabilities of average monthly number of voice calls

Parents		Probability				
Gender	Response	<50.4	50.4-72.8	72.8-95.2	95.2-117.6	>117.6
Male	YES	0.02	0.38	0.20	0.20	0.20
Male	NO	0.40	0.28	0.12	0.20	0.00
Female	YES	0.22	0.01	0.33	0.22	0.22
Female	NO	0.36	0.01	0.44	0.01	0.18

Figure 4.27 The conditional probabilities of the voice calls input attribute

Conditional probabilities of average monthly number of SMS

Parents		Probability				
Gender	Response	<51	51 - 74	74 - 97	97 - 120	>120
Male	YES	0.02	0.02	0.02	0.75	0.20
Male	NO	0.40	0.32	0.12	0.12	0.04
Female	YES	0.01	0.12	0.22	0.33	0.33
Female	NO	0.01	0.44	0.27	0.18	0.10

Figure 4.28 The conditional probabilities of the SMS calls input attribute

Conditional probabilities of gender

Parents	Probability	
Response	Male	Female
YES	0.37	0.63
NO	0.69	0.31

Figure 4.29 The conditional probabilities of the gender input attribute

Now, let's consider a case with the following characteristics: Gender = Female, Profession = White collar, Average monthly number of SMS = 87, and Average monthly number of Voice calls = 81. This case is in row 5 of the training dataset shown in Figure 4.30.

In order to classify this instance, the relevant CPT entries have to be examined.

The conditional probability of Response = Yes given the evidence is calculated as follows:

P(Response = Yes | Gender = Female, Profession = White collar, Average monthly number of SMS = 87, Average monthly number of Voice calls = 81) ~ (is analogous to)

P(Gender = Female | Response = Yes)*

P(Profession = White collar | Response = Yes, Gender = Female)*

P(Average monthly number of SMS in [74, 97] | Response = Yes, Gender = Female)*

P(Average monthly number of Voice calls in [72.8, 95.2] | Response = Yes, Gender = Female)*

P(Response = Yes) = 0.63 * 0.87 * 0.22 * 0.33 * 0.29 = 0.011

Table (10 fields, 50 records) #6

File Edit Generate

	Customer ID	Gender	Profession	SMS	Voice calls	Response...	$B-Response...	$BP-Response ...	$BP-NO	$BP-YES
1	1.000	Male	White collar	45.0...	100.000	NO	NO	0.979	0.979	0.021
2	2.000	Male	Blue collar	32.0...	67.000	NO	NO	0.995	0.995	0.005
3	3.000	Male	Blue collar	57.0...	30.000	NO	NO	1.000	1.000	0.000
4	4.000	Male	Blue collar	120....	150.000	YES	YES	0.960	0.040	0.960
5	5.000	Female	White collar	87.0...	81.000	NO	NO	0.593	0.593	0.407
6	6.000	Male	Blue collar	98.0...	32.000	NO	NO	0.979	0.979	0.021
7	7.000	Female	White collar	110....	78.000	NO	YES	0.599	0.401	0.599
8	8.000	Female	White collar	68.0...	90.000	YES	NO	0.821	0.821	0.179
9	9.000	Male	White collar	67.0...	90.000	NO	NO	0.959	0.959	0.041
10	10.000	Male	Blue collar	23.0...	110.000	NO	NO	0.996	0.996	0.004
11	11.000	Female	Blue collar	57.0...	30.000	NO	NO	0.955	0.955	0.045
12	12.000	Female	White collar	125....	110.000	YES	YES	0.990	0.010	0.990
13	13.000	Female	White collar	87.0...	81.000	NO	NO	0.593	0.593	0.407
14	14.000	Male	Blue collar	56.0...	56.000	NO	NO	0.994	0.994	0.006
15	15.000	Female	White collar	56.0...	120.000	NO	NO	0.735	0.735	0.265
16	16.000	Female	White collar	90.0...	30.000	YES	NO	0.633	0.633	0.367
17	17.000	Male	White collar	78.0...	45.000	NO	NO	0.997	0.997	0.003
18	18.000	Male	Blue collar	32.0...	87.000	NO	NO	0.994	0.994	0.006
19	19.000	Male	Blue collar	57.0...	30.000	NO	NO	1.000	1.000	0.000
20	20.000	Male	White collar	110....	100.000	YES	YES	0.738	0.262	0.738

Table Annotations

OK

Figure 4.30 The training dataset of the TAN Bayesian network

The conditional probability of Response = No, for the same case, given the evidence is calculated as follows:

P(Response = No | Gender = Female, Profession = White collar, Average monthly number of SMS = 87, Average monthly number of Voice calls = 81) ~

P(Gender = Female | Response = No)*

P(Profession = White collar | Response = No, Gender = Female)*

P(Average monthly number of SMS in [74, 97] | Response = No, Gender = Female)*

P(Average monthly number of Voice calls in [72.8, 95.2] | Response = No, Gender = Female)*

P(Response = No) = 0.31 * 0.63 * 0.27 * 0.44 * 0.71 = 0.016

Finally, the case is classified as nonresponder since the *no* conditional probability is higher. The probability of *no* is normalized to 0.016/(0.011 + 0.016) = 0.59. The probability of *yes* is simply (1 − 0.59) or 0.41.

As shown in Figure 4.30, the predicted class for the scored instance is denoted in the field named $B-Response, along with its confidence, listed in the field named $BP-Response. The estimated probabilities for each target class are listed in the fields named as $BP-*no* and $BP-YES. The probability of *no* is equal to the prediction confidence.

@Modeling tech tips

The IBM SPSS Modeler Bayesian networks

Bayesian belief networks are available in IBM SPSS Modeler. The main parameters of the algorithm are presented in Figures 4.31 and 4.32 and in Table 4.8.

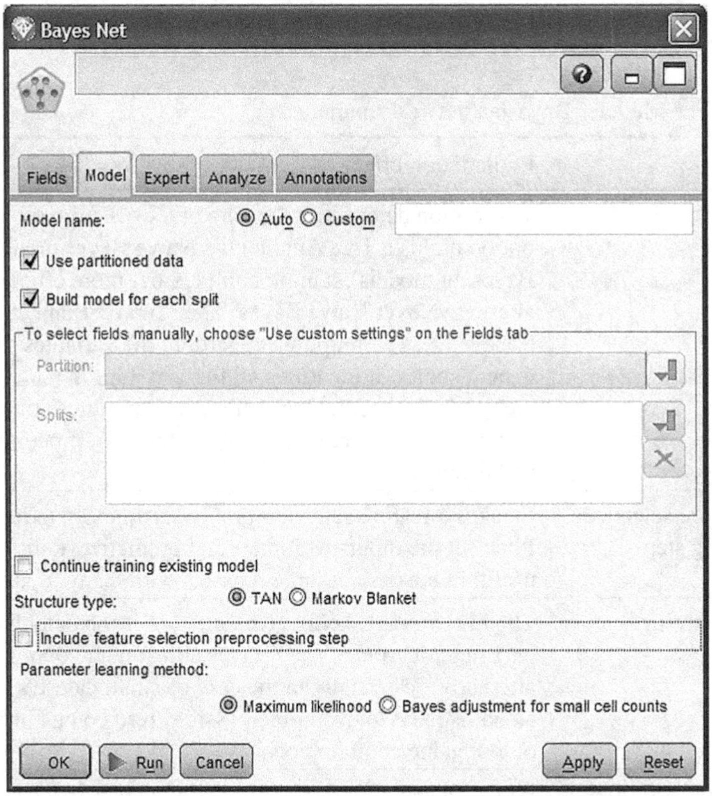

Figure 4.31 The IBM SPSS Modeler Bayesian networks model options

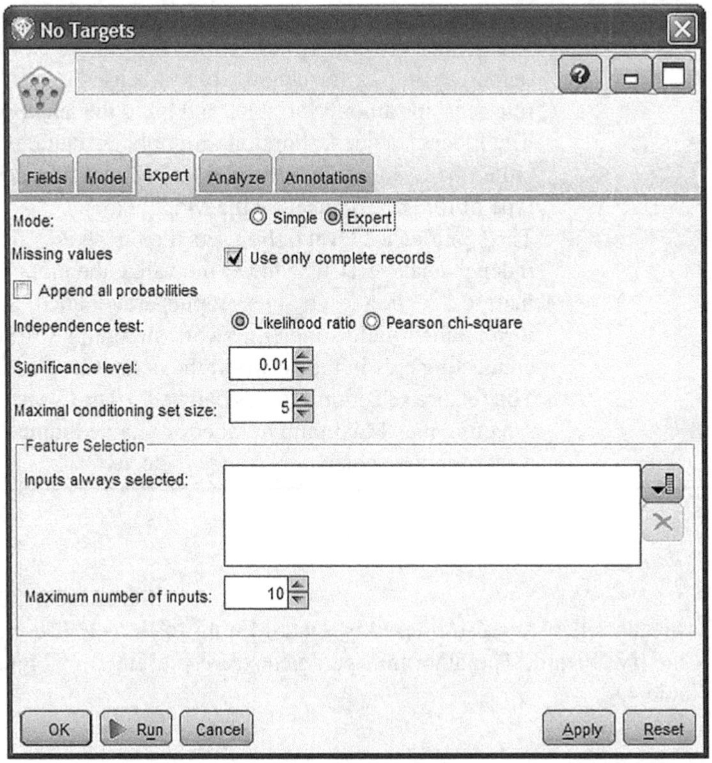

Figure 4.32 The IBM SPSS Modeler Bayesian networks model Expert options

Table 4.8 The Modeler Bayesian networks parameters

Option	Explanation/effect
Structure type	This option determines the network's structure type. The TAN option builds a Tree Augmented Naïve Bayes model (TAN Bayesian models), a more complex but more efficient alternative over Naïve Bayes. The Markov Blanket estimation has even less assumptions regarding the variables' dependencies. It identifies all the variables in the network that are needed to predict the output. As it produces more complex networks, it is more demanding in terms of processing times and resources
Include feature selection preprocessing step	Invokes a feature selection preprocessing step to identify and filter out the inputs with marginal predictive ability. Especially useful in the case of large datasets with many inputs
Paremeter learning method	The Maximum Likelihood estimation is the default option for parameter learning. The "Bayes adjustment" option is a useful alternative, especially in the case of small datasets. It applies an adjustment to overcome possible zero counts in the input–output value combinations
Append all probabilities	Bayesian networks are probabilistic models that can estimate the probabilities of belonging to each target class. This option, available in the Expert options tab, calculates the relevant fields
Independence test	Determines the test to be applied to identify the independencies among variables. The Likelihood ratio (the default option) or the chi-square test is used to find the relationships among variables and build the network The Independence test options are only available if you select either the feature selection preprocessing step or a Structure type of Markov Blanket on the Model tab The significance level option sets the threshold value for the independence tests. The lower the value, the more strict the feature selection test and less inputs are retained. Similarly, lower values yield simpler network structures with fewer connecting arcs in the network; the default level is 0.01 The feature selection process can be further tweaked by selecting the "Maximum number of retained inputs" as well as a set of inputs which are "Always selected"

Microsoft Naïve Bayes

The Naïve Bayes classifiers are also offered by Data Mining for Excel. The model is built through the Classify Wizard. The algorithm parameters are presented in Figure 4.33 and discussed in Table 4.9.

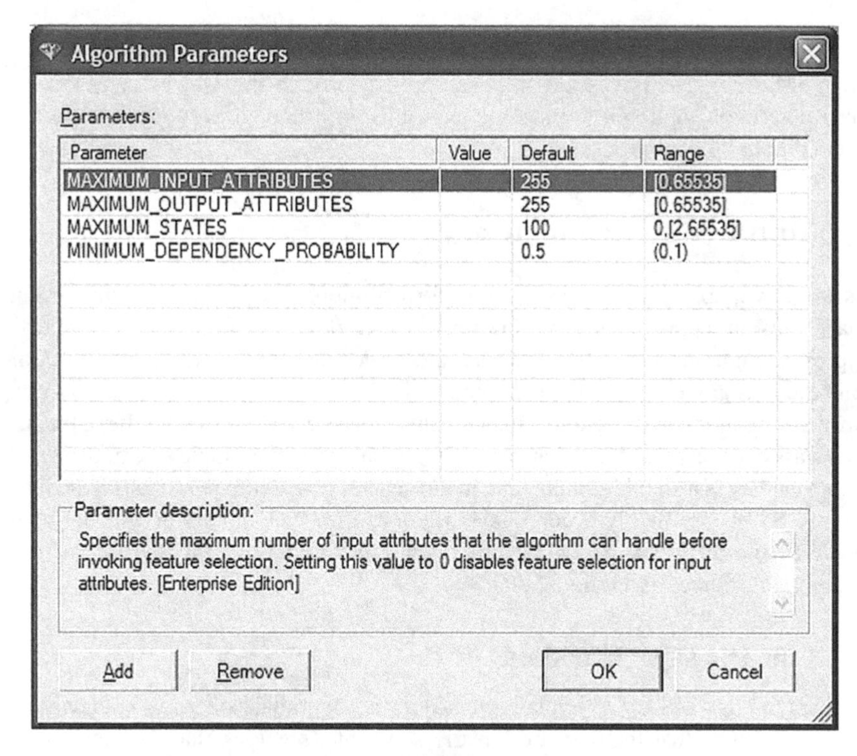

Figure 4.33 The Microsoft Naïve Bayes parameters

Table 4.9 The Microsoft Naïve Bayes parameters

Option	Explanation/effect
Maximum input attributes	A threshold parameter for the integrated feature selection preprocessing step. Feature selection is invoked if the number of inputs is greater than the specified value A value of 0 turns off feature selection The default value is 255
Maximum states	This option determines the maximum number of input categories ("states") that the algorithm will analyze. If an attribute has more categories than the specified threshold, then only the most frequent categories will be used, and the remaining and rare categories are treated as missing The default value is 100
Minimum dependency probability	It sets a minimum threshold probability value for the independence tests between the inputs and the outputs. This parameter does not affect the model training or prediction, but it limits the size of the output content returned by the algorithm. Increasing the value reduces the number of inputs in the network It ranges from 0 to 1; the default level is 0.1

The RapidMiner Naïve Bayes

The RapidMiner Naïve Bayes classifier is applied through the Naïve Bayes operator. A "Laplace correction" option is available to reduce the impact of zero counts in the estimation of the conditional probabilities.

4.8 Support vector machines

SVMs work by mapping the inputs to a transformed input space of a higher dimension. The original inputs are appropriately transformed through nonlinear functions. The selection of the appropriate nonlinear mapping is a trial-and-error procedure in which different transformation functions are tested in turn and compared.

Once the inputs are mapped into a richer feature space, SVMs search for the optimal, linear decision class boundary that best separates the target classes. In the simple case of two inputs, the decision boundary is a simple straight line. In the general case of n inputs, it is a hyperplane.

Overall SVMs are highly accurate classifiers as they can capture and model nonlinear data patterns and relationships. On the other hand, they are also highly demanding in terms of processing resources and training time.

4.8.1 Linearly separable data

In this paragraph, we'll try to explain what does an "optimal" decision boundary means. Initially, we assume that the data are linearly separable, meaning that there exists a linear decision boundary that separates the positive from the negative instances.

A linear classifier is based on a linear discriminant function of the form

$$f(\boldsymbol{X}) = \boldsymbol{w} \cdot \boldsymbol{X} + b$$

where $\boldsymbol{X}$ denotes the vector of the inputs x_i, $\boldsymbol{w}$ denotes the vector of weights, and b can be considered an additional weight (w0). Hence, the discriminant function is the dot product of the inputs' vector with the vector of weights w, which, along with b, are estimated by the algorithm.[1]

A separating hyperplane can be written as

$$f(\boldsymbol{X}) = \boldsymbol{w} \cdot \boldsymbol{X} + b = 0$$

The hyperplane separates the space into two regions. The decision boundary is defined by the hyperplane. The sign of the discriminant function $f(\boldsymbol{X})$ denotes on which side of the hyperplane a case falls and hence on which class is classified.

For illustrative purposes, let's consider a classification problem with two inputs x_1 and x_2. In such cases, the decision boundary is a simple straight line. Because we have assumed that the data are linearly separable, no mapping to higher input dimensions is required. In the next paragraph, we'll generalize the SVM approach to situations where the data are linearly inseparable.

The separating hyperplane for the simple case of two inputs is a straight line:

$$w_1 x_1 + w_2 x_2 + b = 0$$

[1] A dot or inner product of two vectors $\boldsymbol{w} = (w_1, w_2)$ and $\boldsymbol{x} = (x_1, x_2)$ is calculated as $w_1 x_1 + w_2 x_2$.

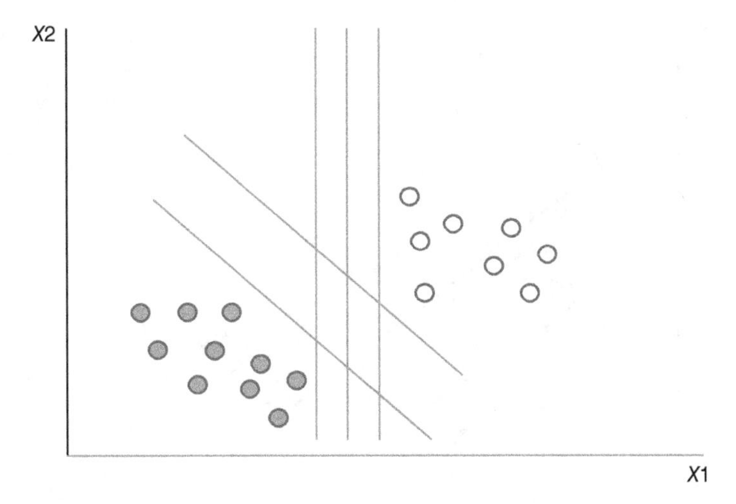

Figure 4.34 A set of separating lines in the input feature space, in the case of linearly separable data

Cases for which $w_1x_1 + w_2x_2 + b > 0$ lie above the separating line, while cases for which $w_1x_1 + w_2x_2 + b < 0$ lie below. The instances are classified based on which side of the hyperplane they land.

As shown in Figure 4.34, there are an infinite number of lines that can provide complete separation in the feature space of the inputs.

The SVM approach is to identify the maximum marginal hyperplane (MMH). It seeks for the optimal separating hyperplane between the two classes by maximizing the margin between the classes' closest points. The points that lie on the margins are known as the support vectors, and the middle of the margin is the optimal separating hyperplane.

In Figure 4.35, a separating line is shown. It completely separates the positive cases, depicted as transparent dots, from the negative ones which are depicted as solid dots. The marked data points correspond to the support vectors since they are the instances which are closest to the decision boundary. They define the margin with which the two classes are separated. The MMH is the straight line in the middle of the margin. The support vectors are the hardest cases to be classified as they are equally close to the MMH.

In Figure 4.36, a different separating line is shown. This line also provides complete separation between the target classes. So which one is the optimal and should be selected for classifying new cases?

As explained earlier, SVMs seek for large margin classifiers. Hence, the MMH depicted in Figure 4.35 is preferred to the one shown in Figure 4.36. Due to its larger margin, we expect it to be more accurate in classifying unseen cases. A wider margin indicates a generalizable classifier, not inclined to overfitting.

In practice, a greater margin is achieved on the cost of sacrificing accuracy. A small amount of misclassification can be accepted in order to widen the margin and favor general classifiers. On the other hand, the margin can also be narrowed a little to increase the accuracy. The SVM algorithms incorporate a regularization parameter C which sets the relative importance of maximizing the margin. It balances the trade-off between the margin and the

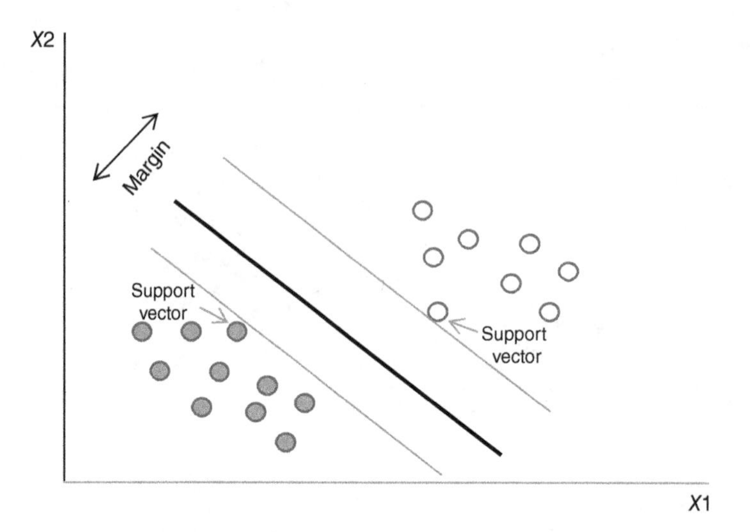

Figure 4.35 A separating hyperplane with the support vectors and the margin in a classification problem with two inputs and linear data

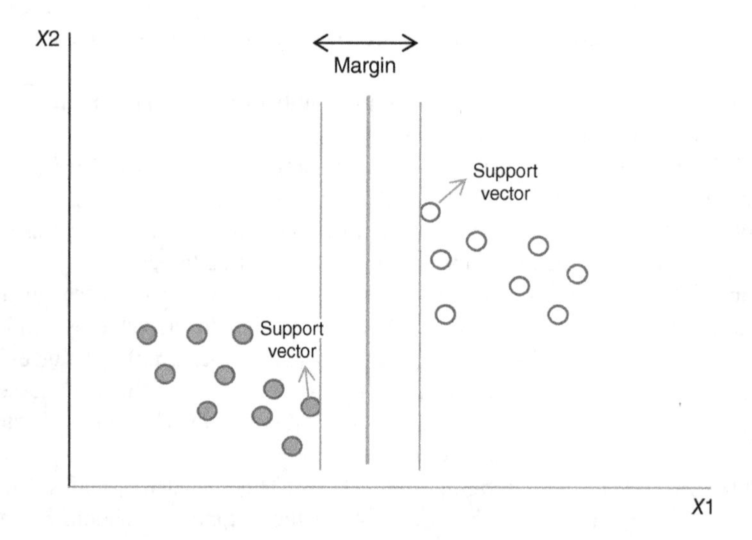

Figure 4.36 An alternative separating hyperplane with its support vectors and its margin for the same linear data

accuracy. Typically, experimentation is needed with different values of this parameter to yield the optimal classifier.

The training of an SVM model and the search for the MMH and its support vectors are a quadratic optimization problem. Once the classifier is trained, it can be used to classify new cases. Cases with null values at one or more inputs are not scored.

4.8.2 Linearly inseparable data

Let's consider again a classification problem with two predictors, x_1, x_2. Only in this case let's consider linearly inseparable data. Therefore, no straight line can be drawn in the plane defined by x_1, x_2 that could adequately separate the cases of the target classes. In such cases, a nonlinear mapping of the original inputs can be applied. This mapping shifts the dimensions of the input space in such a way that an optimal, linear separating hyperplane can then be found. The cases are projected into a (usually) higher-dimensional space where they become linearly separable.

For example, in the simple case of two inputs, a Polynomial transformation φ: $R^2 \rightarrow R^3$ is

$$\varphi(X) = \varphi(x_1, x_2) \overset{\text{yields}}{\rightarrow} \left(x_1^2, \sqrt{2}x_1x_2, x_2^2\right)$$

It maps the initial two inputs into three. Once the data are transformed, the algorithm searches for the optimal linear decision class boundary in the new feature space.

The decision class boundary in the new space is defined by

$$f(X) = \boldsymbol{w} \cdot \varphi(\boldsymbol{x}) + b = 0$$

where $\varphi(x)$ is the mapping applied to the original inputs.

In the example of the simple Polynomial transformation, it can be written as

$$f(X) = w_1x_1^2 + w_2\sqrt{2}x_1x_2 + w_3x_2^2 + b = 0$$

Hence, the decision boundary is linear in the new space but corresponds to an ellipse in the original space.

However, the above approach does not scale well with wide datasets and a large number of inputs. This problem is solved with the application of kernel functions. Kernel functions are used to make (implicit) nonlinear feature mapping. The training of SVM models involves the calculation of dot products between the vectors of instances. In the new feature space, this would be translated as dot products of the form

$$\varphi(X) \cdot \varphi(X')$$

where φ denotes the selected mapping function.

But these calculations are computationally expensive and time consuming. But if there was a function K such that $K(\boldsymbol{X}, \boldsymbol{X}') = \varphi(\boldsymbol{X}) \cdot \varphi(\boldsymbol{X}')$, then we wouldn't need to make the φ mapping at all. Fortunately, certain functions named the kernel functions can achieve this. So, instead of mapping the inputs with φ and computing their dot products in the new space, we can apply a kernel function to the original inputs.

The dot product is replaced by a nonlinear kernel function. In this way, all calculations are made in the original input space, which is desirable since it is typically of much lower dimensionality. The decision boundary in the high-dimensional new feature set is identified. However, due to kernel functions, the φ mapping is never explicitly applied.

Two commonly used Kernel functions are:

Polynomial (of degree d): $K(\boldsymbol{X}, \boldsymbol{X}') = (\boldsymbol{X} \cdot \boldsymbol{X}' + 1)^d$ where d is a tunable parameter

(Gaussian) Radial Basis Function (RBF): $K(\boldsymbol{X}, \boldsymbol{X}') = e^{-\frac{\|X - X'\|^2}{2\sigma^2}} = e^{-\gamma\|X - X'\|^2}, \gamma = 2\sigma^2$

Sigmoid (or neural): $K(\boldsymbol{X}, \boldsymbol{X}') = \tanh(\gamma \boldsymbol{X} \cdot \boldsymbol{X}' + b)$

For example, the Polynomial, degree 2 kernel (quadratic kernel) for the case of two inputs would be:

$$K(\boldsymbol{X},\boldsymbol{X}')=(\boldsymbol{X}\cdot\boldsymbol{X}'+1)^2=((x_1,x_2)\cdot(x_1',x_2')+1)^2$$

$$=(x_1x_1'+x_2x_2'+1)^2=x_1^2x_1'^2+x_2^2x_2'^2+1+2x_1x_1'x_2x_2'+2x_1x_1'+2x_2x_2'$$

The above is equal to the dot product of two instances in the feature space defined by the following transformation φ: $R^2 \rightarrow R^6$:

$$\varphi(\boldsymbol{X})\cdot\varphi(\boldsymbol{X}')=\varphi(x_1,x_2)\cdot\varphi(x_1',x_2')=(x_1^2,\sqrt{2}x_1x_2,x_2^2,\sqrt{2}x_1,\sqrt{2}x_2,1)\cdot$$
$$(x_1'^2,\sqrt{2}x_1'x_2',x_2'^2,\sqrt{2}x_1',\sqrt{2}x_2',1)=x_1^2x_1'^2+2x_1x_1'x_2x_2'+x_2^2x_2'^2+2x_1x_1'+2x_2x_2'+1$$

@Modeling tech tips

The IBM SPSS Modeler SVM model

Figure 4.37 and Table 4.10 present the parameters of the IBM SPSS Modeler SVM classifier.

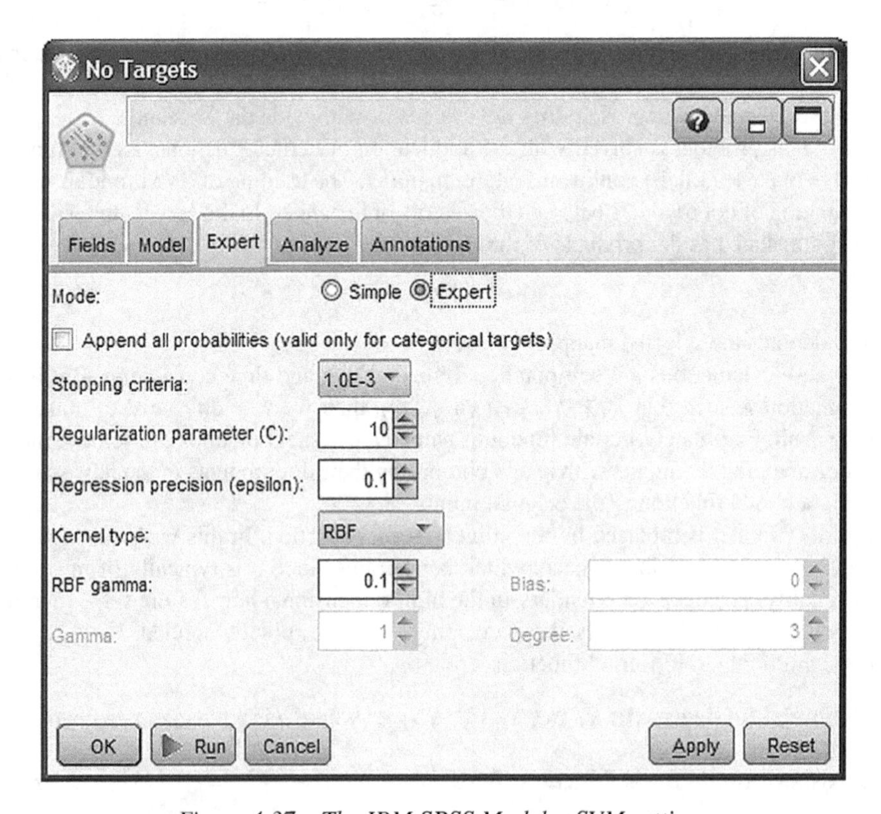

Figure 4.37 The IBM SPSS Modeler SVM settings

Table 4.10 The IBM SPSS Modeler SVM parameters

Option	Explanation/effect
Stopping criteria	Determines when the optimization algorithm converges and stops. Values range from 1.0E–1 to 1.0E–6. The default value of 1.0E–3 is appropriate for most classification tasks; smaller values increase the accuracy of the model but at a cost of increased training time
Regularization parameter (C)	The regularization parameter C which balances the trade-off between generalization (wider margin) and accuracy. Default value is 10. Smaller values correspond to "wider" margins, and larger values produce "broader" ones. Thus, increasing the value improves the model accuracy but with a risk of overfitting
Kernel type	It determines the kernel function to be applied. Available kernels include Radial Basis Function (RBF), Polynomial, Sigmoid, and Linear. The default function is RBF which is a reasonable first choice
RBF gamma	Available only when RBF kernel function is selected. It corresponds to the $\gamma = 2\sigma^2$ parameter of the kernel function. As a rule of thumb, values between $3/k$ and $6/k$, where k is the number of predictors in the model, should be set as initial values. Increased values lead to improved accuracy but may also lead to overfitting
Degree	The kernel degree parameter is enabled when the Polynomial kernel function is selected. It sets the Polynomial degree, the d parameter of the kernel function. It determines the complexity of the mapping and hence of the model training time. The selected values should rarely exceed 10
Gamma	The γ adjustable parameter for the Polynomial and the Sigmoid kernels. The default value is 1. Increased values improve the mode accuracy with a risk of overfitting
Bias	Enabled only when the Polynomial or the Sigmoid kernel is selected. It corresponds to the b parameter. (Note that in IBM SPSS Modeler, the Polynomial kernel function is defined as $K(\boldsymbol{X},\boldsymbol{X}') = (\gamma \boldsymbol{X} \cdot \boldsymbol{X}' + b)^d$) The default value of 0 is appropriate in most situations

The RapidMiner SVM models

RapidMiner includes several SVM models. The SVM classifier, available through the modeling operator of the same name, supports the RBF, the Polynomial, and the Sigmoid kernel functions. Its parameters are presented in Figure 4.38 (for the RBF kernel function). Since the role and the significance of SVM parameters are already explained in Table 4.10, Table 4.11 briefly presents the RapidMiner parameters.

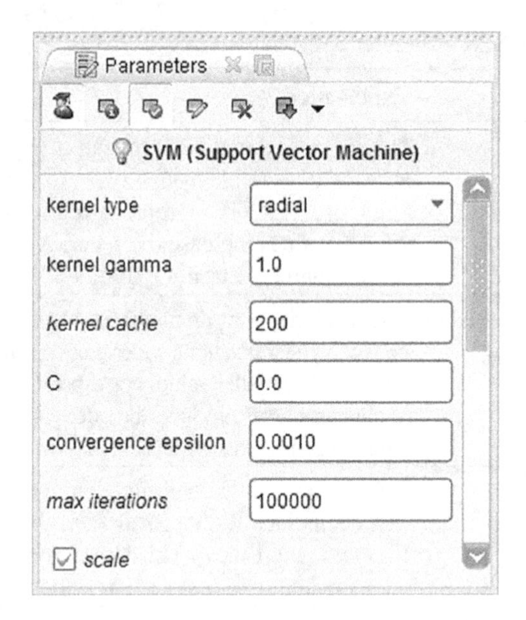

Figure 4.38 The RapidMiner RBF Support Vector Machine model options

Table 4.11 The RapidMiner SVM parameters

Option	Explanation/effect
Convergence epsilon	Determines when the optimization algorithm converges and stops. Default value is 1.0E−1. Smaller values increase the accuracy of the model but at a cost of increased training time
C	The regularization parameter *C*
Kernel type	It determines the kernel function to be applied. Available kernels include radial (Radial Basis Function), Polynomial, and neural (Sigmoid)
Kernel gamma	Available when the radial (RBF) kernel function is selected. It corresponds to the $\gamma = 2\sigma^2$ parameter of the kernel function. The default value is 2
Kernel degree	The kernel degree parameter is enabled when the Polynomial kernel function is selected. It sets the Polynomial degree, the *d* parameter of the kernel function
Kernel *a*	The γ adjustable parameter for the Sigmoid (neural) kernel
Kernel *b*	The *b* (bias) parameter of the neural kernel

Overall, the fine-tuning of the SVM parameters is a trial-and-error process. The RBF kernel function is a usual initial kernel choice. However, different values of the C and γ should be tested with validation or grid search in order to select the optimal parameter values.

4.9 Summary

In this chapter, we presented three of the most accurate and powerful classification algorithms: Decision Trees, SVMs, and Bayesian networks.

SVMs are highly accurate classifiers. However, they suffer in terms of speed and transparency. They are demanding in terms of processing resources and time. This hinders their application in real business problems, especially in the case of "long" datasets with a large number of training instances. Apart from their speed problem, SVMs also lack transparency. They belong to the class of opaque classifiers whose results do not provide insight on the target event.

This is not the case with the Bayesian networks. Especially the class of Bayesian belief networks, with their graphical results, provides a comprehensive visual representation of the results and of the underlying relationships among attributes. Although faster than SVMs, the Bayesian networks in most cases are not as accurate.

Decision Trees on the other hand seem to efficiently combine accuracy, speed, and transparency. They generate simple, straightforward, and understandable rules which provide insight in the way that predictors are associated with the output. They are fast and scalable, and they can efficiently handle a large number of records and predictors.

5

Segmentation algorithms

5.1 Segmenting customers with data mining algorithms

In this chapter, we'll focus on the data mining modeling techniques used for segmentation. Although clustering algorithms can be directly applied to input data, a recommended preprocessing step is the application of a data reduction technique that can simplify and enhance the segmentation process by removing redundant information. This approach, although optional, is highly recommended, as it can lead to rich and unbiased segmentation solutions that account for all the underlying data dimensions without being affected and biased by possible intercorrelations of the inputs. Therefore, this chapter also focuses in Principal components analysis (PCA), an established data reduction technique that can construct meaningful and uncorrelated compound measures which can then be used in clustering.

5.2 Principal components analysis

PCA is a statistical technique used for the data reduction of the original inputs. It analyzes the correlations between input fields and derives a core set of component measures that can efficiently reduce the data dimensionality without sacrificing much of the information of the original inputs.

PCA examines the correlations among the original inputs and then appropriately constructs composite measures that take these correlations into account. A brief note on linear correlations: if two or more continuous fields tend to covary, then they are correlated. If their relationship is expressed adequately by a straight line, then they have a strong linear correlation. The scatter plot in Figure 5.1 depicts the relationship between the monthly average sms and mms usage for a group of mobile telephony customers.

Effective CRM using Predictive Analytics, First Edition. Antonios Chorianopoulos.

Companion website: www.wiley.com/go/chorianopoulos/effective_crm

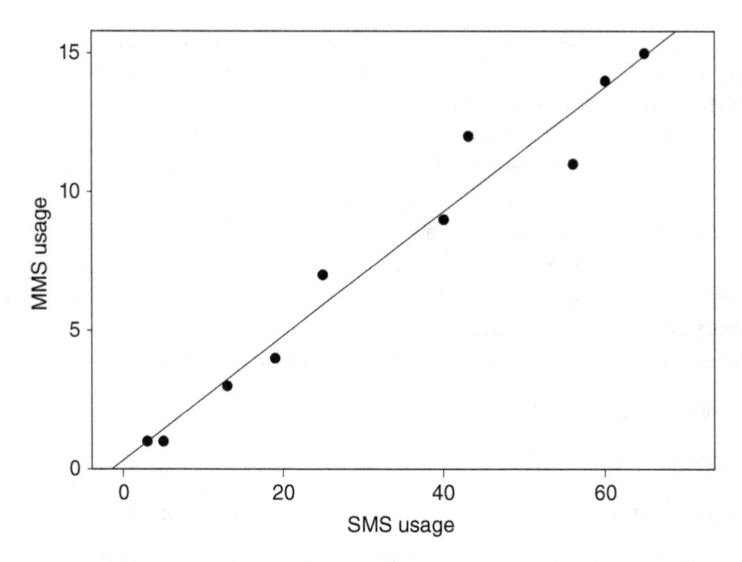

Figure 5.1 Linear correlation between two continuous measures

As seen in the above plot, most customer dots cluster around a straight line with a positive slope that slants upward to the right. Customers with increased sms usage also tend to be mms users as well. These two services are related in a linear manner, and they present a strong, positive linear correlation, since high values of one field tend to correspond with high values of the other. On the other hand, in negative linear correlations, the direction of the relationship is reversed. These relationships are summarized by straight lines with a negative slope that slant downward. In such cases, high values of one field tend to correspond with low values of the other. The strength of linear correlation is quantified by a measure named Pearson correlation coefficient. It ranges from −1 to +1. The sign of the coefficient reveals the direction of the relationship. Values close to +1 denote strong positive correlation while values close to −1 negative correlation. Values around 0 denote no discernable linear correlation, yet this does not exclude the possibility of nonlinear correlation.

So PCA is identifying strong linear relationships between the original inputs, and its goal is to replace them with the smallest number of components that account for as much as possible of their information. Moreover, a typical PCA analysis derives uncorrelated components, appropriate for input in many other modeling techniques, including clustering. The derived components are produced by linear transformations of the inputs, as shown by the following formula:

$$C_k = \sum_{i=1}^{n} a_{ki} F_i$$

where C_k denotes the component k produced by the inputs F_i. The n inputs can produce a maximum of n components; however, in practice, the number of components retained is much smaller since the main goal is the reduction of dimensionality.

The coefficients a_{ki} are calculated by the algorithm so that the loss of information is minimal. The components are extracted in decreasing order of importance with the first one being the most significant as it accounts for the largest amount of input data information. Specifically, the first component is the linear combination that explains as much as possible of the total variability of the input fields. Thus, it retains most of their information.

The second component accounts for the largest amount of the unexplained variability, and it is also uncorrelated with the first component. Subsequent components are constructed to account for the remaining information.

Since *n* components are required to fully account for the original information of the *n* fields, the question is where to stop and how many components to extract? Although there are specific technical criteria that can be applied to guide analysts on the procedure, the final decision should also take into account business criteria such as the interpretability and the business meaning of the components. The final solution should balance between simplicity and effectiveness, consisting of a reduced and interpretable set of compound attributes that can sufficiently represent the original fields.

Key issues that a data miner has to face in a PCA analysis include:

- How many components to extract?
- Is the derived solution efficient and useful?
- Which original fields are mostly related with each component?
- What does each component represent? In other words, what is the meaning of each component?

The next paragraphs will try to clarify these issues.

5.2.1 How many components to extract?

In the next paragraphs, we'll present PCA by examining the fixed telephony usage fields of Table 5.1.

Table 5.2 lists the pairwise Pearson correlation coefficients among the original fields.

As shown in the table, there are significant correlations among specific usage fields. Statistically significant correlations (at the 0.01 level) are denoted by the symbol "**" The number of outgoing calls to fixed lines, for instance, is positively correlated with the minutes

Table 5.1 Behavioral fields used in the PCA example

Field name	Description
FIXED_OUT_CALLS	Monthly average of outgoing calls to fixed telephone lines
FIXED_OUT_MINS	Monthly average number of minutes of outgoing calls to fixed telephone lines
PRC_FIXED_OUT_CALLS	Percentage of outgoing calls to fixed telephone lines: outgoing calls to fixed lines as a percentage of the total outgoing calls
MOBILE_OUT_CALLS	Monthly average of outgoing calls to mobile telephone lines
PRC_MOBILE_OUT_CALLS	Percentage of outgoing calls to mobile telephone lines
BROADBAND_DAYS	Days with broadband traffic
BROADBAND_TRAFFIC	Monthly average broadband traffic
INTERNATIONAL_OUT_CALLS	Monthly average of outgoing calls to international telephone lines
PRC_INTERNATIONAL_OUT_CALLS	Percentage of outgoing international calls

Table 5.2 Pairwise correlation coefficients among inputs

	FIXED_OUT_CALLS	FIXED_OUT_MINS	MOBILE_OUT_CALLS	INTERNATIONAL_OUT_CALLS	BROADBAND_TRAFFIC	BROADBAND_DAYS	PRC_FIXED_OUT_CALLS	PRC_MOBILE_OUT_CALLS	PRC_INTERNATIONAL_OUT_CALLS
FIXED_OUT_CALLS	1	0.816**	0.225**	0.305**	0.037	−0.213**	0.199**	−0.099**	−0.232**
FIXED_OUT_MINS	0.816**	1	0.195**	0.179**	0.122**	−0.147**	0.154**	−0.064**	−0.206**
MOBILE_OUT_CALLS	0.225**	0.195**	1	0.260**	0.070**	−0.106**	-0.644**	0.720**	−0.134**
INTERNATIONAL_OUT_CALLS	0.305**	0.179**	0.260**	1	−0.104**	−0.142**	−0.285**	0.060**	0.514**
BROADBAND_TRAFFIC	0.037	0.122**	0.070**	−0.104**	1	0.389**	−0.001	0.095**	−0.207**
BROADBAND_DAYS	−0.213**	−0.147**	−0.106**	−0.142**	0.389**	1	0.084**	0.011	−0.216**
PRC_FIXED_OUT_CALLS	0.199**	0.154**	−0.644**	−0.285**	−0.001	0.084**	1	−0.901**	−0.272**
PRC_MOBILE_OUT_CALLS	−0.099**	−0.064**	0.720**	0.060**	0.095**	0.011	−0.901**	1	−0.173**
PRC_INTERNATIONAL_OUT_CALLS	−0.232**	−0.206**	−0.134**	0.514**	−0.207**	−0.216**	−0.272**	−0.173**	1

of such calls, denoting that customers that make a large number of voice calls also tend to talk a lot. Other fields are negatively correlated such as the percentage of calls to fixed and mobile lines. This signifies a contrast between calls to fixed and mobile phone numbers, not in terms of usage volume but in terms of the total usage ratio that each service accounts for. Studying correlation tables in order to come up to conclusions is a cumbersome job. That's where PCA comes in. It analyzes such tables and identifies the groups of related fields.

In order to reach a conclusion on how many components to extract, the following criteria are commonly used:

5.2.1.1 The eigenvalue (or latent root) criterion

The eigenvalue is a measure of the variance that each component accounts for. The eigenvalue criterion is perhaps the most widely used criterion for selecting which components to keep. It is based on the idea that a component should be considered insignificant if it does not perform better than a single input. Each single input contains 1 unit of standardized variance; thus, components with eigenvalues below 1 are not extracted.

Figure 5.2 presents the "Total Variance Explained" part of the Modeler PCA output for the fixed telephony data.

This table presents the eigenvalues and the percentage of variance/information attributable to each component. The components are listed in the rows of the table. The first 4 first rows of the table correspond to the extracted components. A total of nine components is needed to

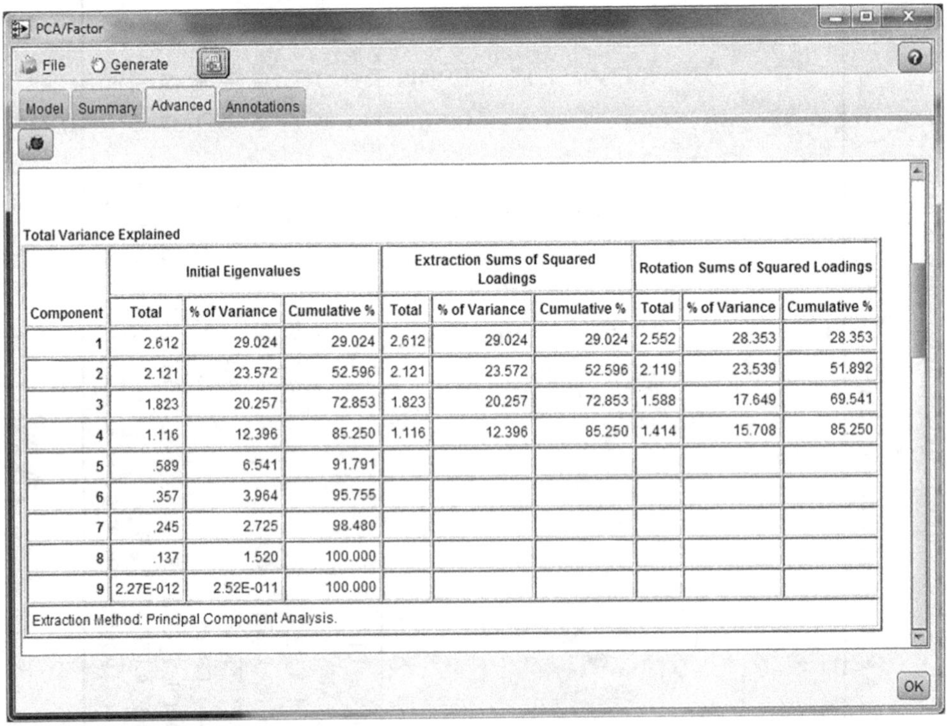

PCA/Factor

File Generate

Model Summary Advanced Annotations

Total Variance Explained

Component	Initial Eigenvalues			Extraction Sums of Squared Loadings			Rotation Sums of Squared Loadings		
	Total	% of Variance	Cumulative %	Total	% of Variance	Cumulative %	Total	% of Variance	Cumulative %
1	2.612	29.024	29.024	2.612	29.024	29.024	2.552	28.353	28.353
2	2.121	23.572	52.596	2.121	23.572	52.596	2.119	23.539	51.892
3	1.823	20.257	72.853	1.823	20.257	72.853	1.588	17.649	69.541
4	1.116	12.396	85.250	1.116	12.396	85.250	1.414	15.708	85.250
5	.589	6.541	91.791						
6	.357	3.964	95.755						
7	.245	2.725	98.480						
8	.137	1.520	100.000						
9	2.27E-012	2.52E-011	100.000						

Extraction Method: Principal Component Analysis.

OK

Figure 5.2 The "Total Variance Explained" part of the Modeler PCA output for the fixed telephony data

fully account for the information of the nine original fields. That's why the table contains nine rows. However, not all these components are retained. The algorithm extracted four of them, based on the eigenvalue criterion that we have specified when we set up the model.

The second column of the table denotes the eigenvalue of each component. Components are extracted in descending order of importance so the first one carries over the largest part of the variance of the original fields. Extraction stops at component 4 since the component 5 has an eigenvalue below the requested threshold of 1.

Eigenvalues can also be expressed in terms of the percentage of the total variance of the original fields. The third column of the table, titled "% of Variance," denotes the variance proportion attributable to each component, while the next column titled "Cumulative %" denotes the total proportion of variance explained by all the components up to that point. The percentage of the initial variance attributable to the four extracted components is about 85%. Not bad at all, if you consider that by only keeping four of the nine original fields, we lose just a small part of their initial information.

Tech tips on the eigenvalue criterion

Variance is a measure of the variability of a field. It summarizes the dispersion of the field values around the mean. It is calculated by dividing the sum of the squared deviations from the mean with the total number of records (minus 1). Standard deviation is another measure of variability, and it is the square root of the variance. A standardized field with the *z*-score method is created as:

$$\frac{\text{Value} - \text{Field mean}}{\text{Standard deviation}}$$

The variance can be considered as a measure of the information of a field. A standardized field has a standard deviation and a variance value of 1 hence it carries 1 unit of information.

In PCA, each component is related to the original inputs, and these correlations are referred to as loadings. The proportion of variance of a field that can be interpreted by another field is represented by the square of their correlation. The eigenvalue of each component is the sum of squared loadings (correlations) across all inputs. Thus, each eigenvalue denotes the total variance or total information interpreted by the respective component.

Since a single standardized field contains one unit of information, the total information of the original inputs is equal to their number. The ratio of the eigenvalue to the total units of information (9 in our example) denotes the percentage of variance that each component represents.

By comparing the eigenvalue to the value of 1, we examine if a component is more useful and informative than a single input.

5.2.1.2 The percentage of variance criterion

According to that criterion, the number of components to be extracted is determined by the total explained percentage of variance. The threshold value for extraction depends on the specific situation, but in general, a solution should not fall below 70%.

Component	Standard Deviation	Proportion of Variance	Cumulative Variance
PC 1	1.616	0.290	0.290
PC 2	1.457	0.236	0.526
PC 3	1.350	0.203	0.729
PC 4	1.056	0.124	0.852
PC 5	0.767	0.065	0.918
PC 6	0.597	0.040	0.958
PC 7	0.495	0.027	0.985
PC 8	0.370	0.015	1.000
PC 9	0.000	0.000	1.000

Figure 5.3 The eigenvalues of the RapidMiner PCA model results

Figure 5.3 presents the eigenvalues part of the RapidMiner PCA output for the fixed telephony example. Note that the inputs have been normalized with a z-transformation before the model training.

The rows of the output table correspond to the derived components. The proportion of variance attributable to each component is displayed in the third column. The cumulative variance explained is listed in the fourth column. Note that these values are equal to the ones produced by the Modeler PCA model. A variance threshold of 85% has been set for the dimensionality reduction. This model setting leads to the retaining of the first four components.

5.2.1.3 The scree test criterion

Eigenvalues decrease in a descending order along with the extracted components. According to that criterion, we should look for a sudden large drop in the eigenvalues which indicate transition from large to small values. At that point, the unique variance (variance attributable to a single field) that a component carries over starts to dominate the common variance. This criterion is graphically illustrated by the scree plot which plots the eigenvalues against the number of extracted components. What we should look for is a steep downward slope of the eigenvalues' curve followed by a straight line. In other words, the first "bend" or "elbow" that would resemble the scree at the bottom of a mountain. The starting point of the "bend" indicates the maximum number of components to extract while the point immediately before the "bend" could be selected for a more "compact" solution.

5.2.1.4 The interpretability and business meaning of the components

Like any other model, data miners should experiment and try different extraction solutions before reaching a decision. The final solution should include directly interpretable, understandable, and useful components. Since the retained components will be used for subsequent modeling and reporting purposes, analysts should be able to recognize the information that they convey. A component should have a clear business meaning otherwise it would be of little value for further usage. In the next paragraph, we shall present the way to interpret components and to recognize their meaning.

5.2.2 What is the meaning of each component?

The next task is to determine the meaning of the derived components, in respect to the original fields. The goal is to understand the information that they convey and name them accordingly. This interpretation is based on the correlations among the derived components and the original inputs. These correlations are referred to as loadings. In RapidMiner, they are presented in the "Eigenvectors" view of the PCA results.

The rotation is a recommended technique to apply in order to facilitate the components' interpretation process. The rotation minimizes the number of fields that are strongly correlated with many components and attempts to associate each input with one component. There are numerous rotation techniques, with varimax being the most popular for data reduction purposes since it yields transparent components which are also uncorrelated, a characteristic usually wanted for next tasks. Thus, instead of looking at the original loadings, we typically examine the loadings after the application of a rotation method.

Tech tips on rotation methods

Rotation is a method used in order to simplify the interpretation of components. It attempts to clarify in-between situations in which fields seem associated with more than one component. It tries to produce a solution in which each component has large (+1, −1) correlations with a specific set of original inputs and negligible correlations (close to 0) with the rest of the fields.

As mentioned before, components are extracted in order of significance, with the first one accounting for as much of the input variance as possible and subsequent components accounting for residual variance. Subsequently, the first component is usually a general component with most of the inputs associated with it. Rotation tries to fix this and redistributes the explained variance in order to produce a more efficient and meaningful solution.

It does so by rotating the reference axes of the components, as shown in Figure 5.4, so that the correlation of each component with the original fields is either minimized or maximized.

Varimax rotation moves the axes so that the angle between them remains perpendicular, resulting in uncorrelated components (orthogonal rotation). Other rotation methods (like promax and oblimin) are not constrained to produce uncorrelated components (oblique rotations) and are mostly used when the main objective is data interpretation instead of reduction.

Rotation reattributes the percentage of variance explained by each component in favor of the components extracted last, while the total variance jointly explained by the derived components remains unchanged.

Figure 5.5 presents the rotated component matrix of the Modeler PCA model with varimax rotation. It lists the loadings of the inputs and the components of our example. Note that loadings with absolute values below 0.4 have been suppressed for easier interpretation. Moreover, the original inputs have been sorted according to their loadings so that fields associated with the same component appear together as a set.

To understand what each component represents, we should identify the original fields with which it is associated and the extent and the direction of this association. Hence, the

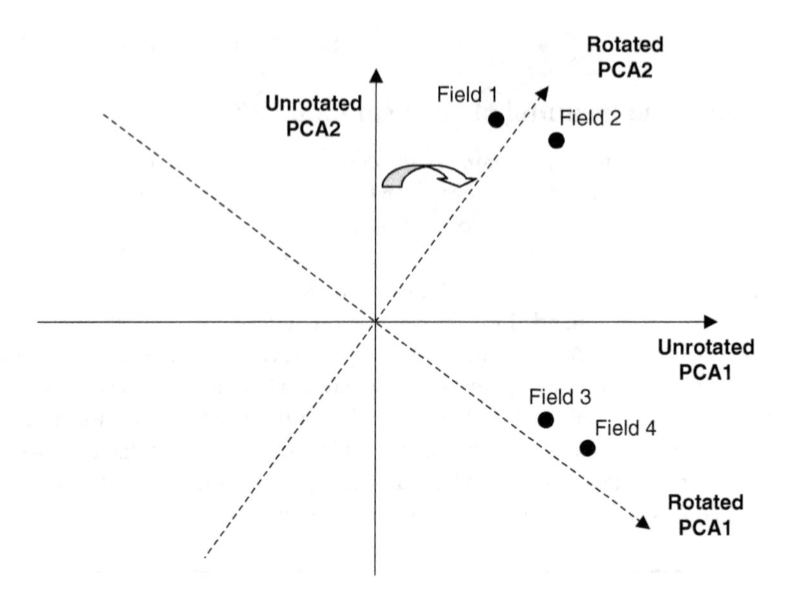

Figure 5.4 An orthogonal rotation of the derived components

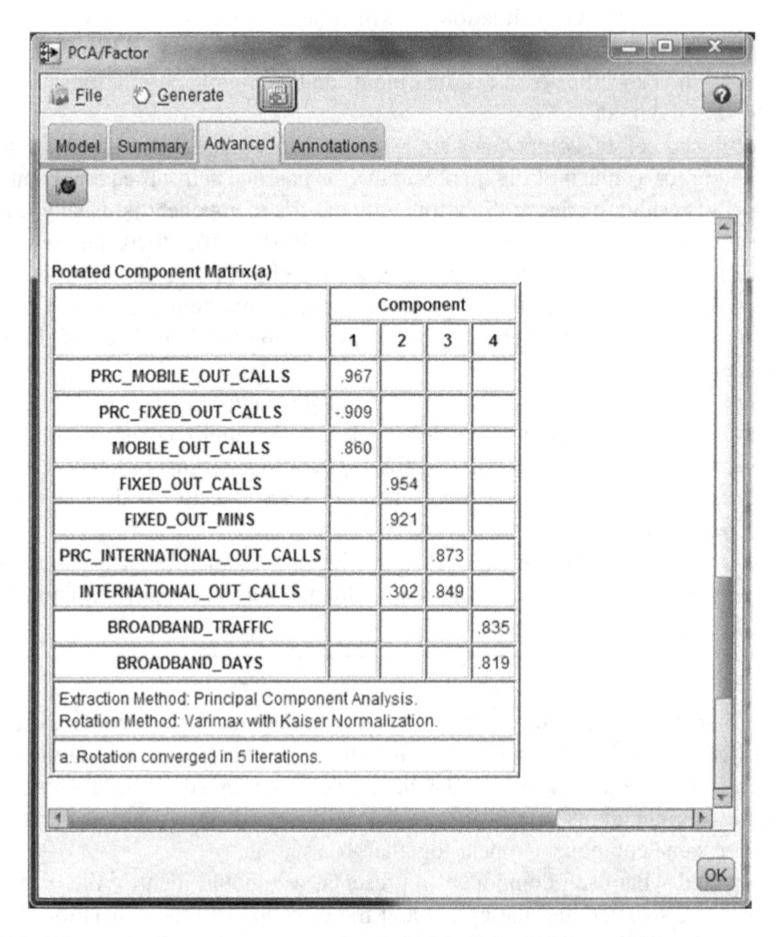

Rotated Component Matrix(a)

	Component			
	1	2	3	4
PRC_MOBILE_OUT_CALLS	.967			
PRC_FIXED_OUT_CALLS	-.909			
MOBILE_OUT_CALLS	.860			
FIXED_OUT_CALLS		.954		
FIXED_OUT_MINS		.921		
PRC_INTERNATIONAL_OUT_CALLS			.873	
INTERNATIONAL_OUT_CALLS		.302	.849	
BROADBAND_TRAFFIC				.835
BROADBAND_DAYS				.819

Extraction Method: Principal Component Analysis.
Rotation Method: Varimax with Kaiser Normalization.

a. Rotation converged in 5 iterations.

Figure 5.5 The rotated component matrix of the Modeler PCA model for the interpretation of the components

interpretation process involves the examination of the loading values and their signs and the identification of significant correlations. Typically, correlations above 0.4 in absolute value are considered of practical significance, and they denote the original fields which are representative for each component. The interpretation process ends up with the naming of the derived components with names that appropriately summarize their meaning.

In the fixed telephony example, component 1 seems to be strongly associated with calls to mobile numbers. Both the number (MOBILE_OUT_CALLS) and the ratio of calls to mobile lines (PRC_ MOBILE_OUT_CALLS) load heavily on this component. The negative sign in the loading of the fixed-calls ratio (PRC_FIXED_OUT_CALLS) indicates a strong negative correlation with component 1. It suggests a contrast between calls to fixed and mobile numbers, in terms of usage ratio. We can label this component as "CALLS_TO_MOBILE" as it seems to measure this usage aspect.

The number and minutes of outgoing calls to fixed lines (FIXED_OUT_CALLS and FIXED_OUT_MINS) seem to covary, and they are combined to form the second component which can be named as "CALLS_TO_FIXED." By examining the loading in a similar way, we can conclude that the component 3 represents "INTERNATIONAL_CALLS" and the component 4 "BROADBAND" usage.

Ideally, each field would heavily load on a single component and the original inputs would be clearly separated into distinct groups. This is not always the case though. Fields that do not load on any of the extracted components or fields not clearly separated are indications of a solution that fails to explain all the original fields. In these cases, the situation could be improved by requesting a richer solution with more extracted components.

5.2.3 Moving along with the component scores

The identified components can be used in the subsequent data mining models, provided of course they comprise a conceptually clear and meaningful representation of the original inputs. The PCA algorithm derives new composite fields, named component scores, that denote the values of each instance in the revealed components. Component scores are produced through linear transformations of the original fields, by using coefficients that correspond to the loading values. They can be used as any other inputs in subsequent tasks.

The derived component scores are continuous numeric fields with standardized values; hence, they have a mean of 0 and a standard deviation of 1, and they designate the deviation from the average behavior. As noted before, the derived component scores, apart from being fewer in number than the original inputs, are standardized and uncorrelated and equally represent all the input dimensions. These characteristics make them perfect candidates for inclusion in subsequent clustering algorithms for segmentation.

@Modeling tech tips

The IBM SPSS Modeler PCA model

Figure 5.6 and Table 5.3 present the parameter settings of the PCA algorithm in IBM SPSS Modeler.

The RapidMiner PCA model

Figure 5.7 and Table 5.4 present the parameter settings of the PCA model in RapidMiner.

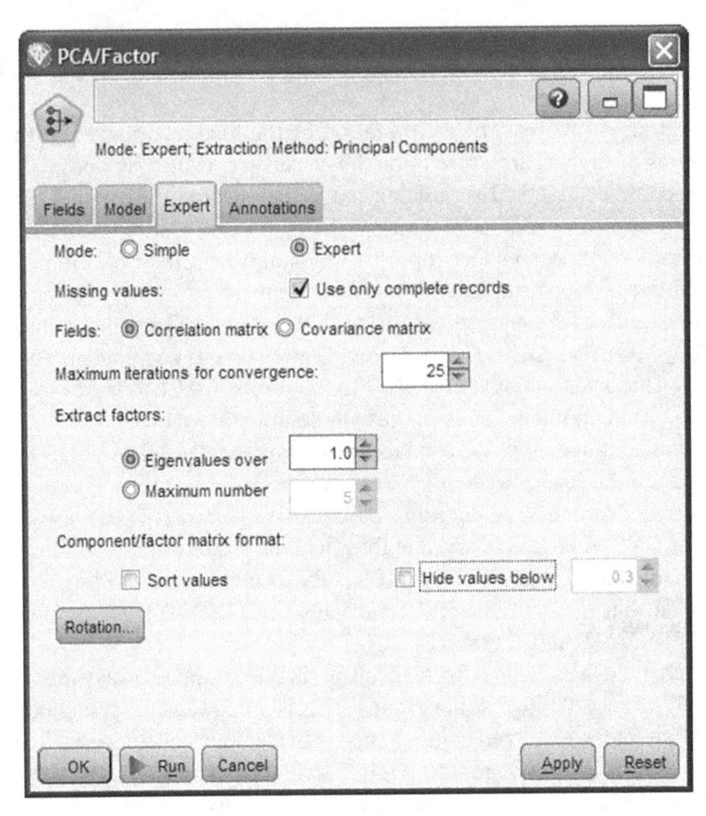

Figure 5.6 The PCA model settings in IBM SPSS Modeler

Table 5.3 The PCA model parameters in IBM SPSS Modeler

Option	Explanation/effect
Extraction method	PCA is the recommended extraction method in cases where data reduction is the first priority Worth trying alternatives: other data reduction algorithms like Principal Axis Factoring and Maximum Likelihood which are the most commonly used factor analysis methods
Extract factors	Sets the criterion for the number of components to be extracted. The most common criterion is based on the eigenvalues: only components with eigenvalues above 1 are retained Worth trying alternatives: analysts can experiment by setting a different threshold value. Alternatively, based on the examined results, they can ask for a specific number of components to be retained through the "maximum number (of factors)" option
Rotation	Applies a rotation method. Varimax rotation can facilitate the interpretation of components by producing clearer loading patterns
Sort values	This option aids the interpretation of the components by improving the readability of the rotated component matrix. When selected, it places together the original fields associated with the same component
Hide values below	This option aids the component interpretation by improving the readability of the rotated component matrix. When selected, it suppresses loadings with absolute values lower than 0.4 and allows users to focus on significant loadings

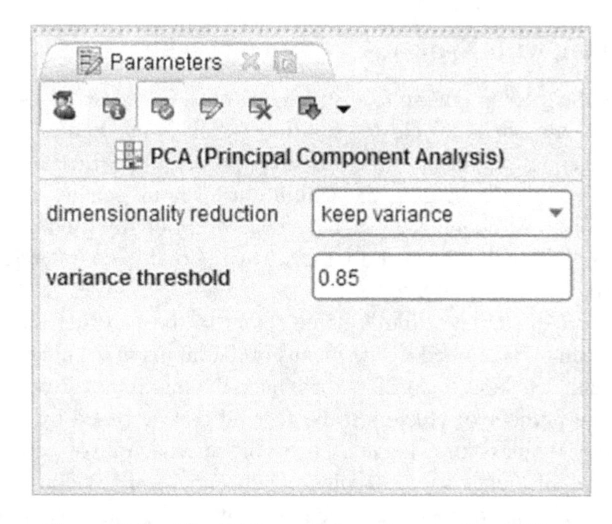

Figure 5.7 The PCA model settings in RapidMiner

Table 5.4 The PCA model parameters in RapidMiner

Option	Explanation/effect
Dimensionality reduction	Sets the criterion for the number of components to be retained. The 'keep variance" option removes the components with a cumulative variance explained greater than the specified threshold. The "fixed number" option keeps the specified number of components
Variance threshold (or fixed number)	The "variance threshold" option is invoked with the 'keep variance" dimensionality reduction criterion. The default threshold is 0.85, and in most cases, it is a good starting point for experimenting. The "fixed number" option allows users to request a specific number of components to be extracted

5.3 Clustering algorithms

Clustering models are unsupervised techniques, appropriate when there is no target output. They analyze the input data patterns and reveal the natural groupings of records. The groups identified are not known in advance, as is the case in classification modeling. Instead, they are recognized by the algorithms. The instances are assigned to the detected clusters based on their similarities in respect to the inputs.

According to the way they work and their outputs, the clustering algorithms can be categorized in two classes, the hard and the soft clustering algorithms. The hard clustering algorithms, such as K-means, assess the distances (dissimilarities) of the instances. The revealed clusters do not overlap and each instance is assigned to one cluster. The soft clustering techniques on the other hand, such as Expectation Maximization (EM) clustering, use probabilistic measures to assign the instances to clusters with certain probabilities. The clusters can overlap and the instances can belong to more than one cluster with certain, estimated probabilities.

5.3.1 Clustering with K-means

K-means is one of the most popular clustering algorithms. It is a centroid-based partitioning technique. It starts with an initial cluster solution which is updated and adjusted until no further refinement is possible (or until the iterations exceed a specified number). Each iteration refines the solution by increasing the within cluster homogeneity and by reducing the within cluster variation. Although K-means has been traditionally used to cluster numerical inputs, with the use of special distance measures or internal encodings, it can also handle mixed numeric and categorical inputs.

The "means" part of the algorithm's name refers to the fact that the centroid of each cluster, the cluster center, is defined by the means of the inputs in the cluster. The "K" comes from the fact that analysts have to specify in advance the number of *k* clusters to be formed. The selection of the number of clusters to be formed can be based on preliminary model training and/or prior business knowledge. A heuristic approach involves building candidate models for a range of different *k* values (typically between 4 and 15), comparing the obtained solutions in respect to technical measures (such as the within cluster variation or the Bayesian Information Criterion (BIC)) and conclude the best *k*. Such approaches are followed by Excel's K-means and RapidMiner's X-means models. The goal is to balance the trade-off between quality and complexity and produce actionable and homogenous clusters of a manageable number.

K-means uses a distance measure, for instance, the Euclidean distance, and iteratively assigns each record in the derived clusters. It is very quick since it does not need to calculate distances between all the pairs of the records. Clusters are refined through an iterative procedure during which records move between the clusters until the procedure becomes stable. The procedure starts by selecting *k* well-spaced initial records as cluster centers (initial seeds) and assigns each record to its "nearest" cluster. As new records are added to the clusters, the cluster centers are recalculated to reflect their new members. Then, cases are reassigned to the adjusted clusters. This iterative procedure is repeated until it reaches convergence and the migration of records between the clusters no longer refines the solution.

The K-means procedure is graphically illustrated in Figure 5.8. For easier interpretation, we assume two inputs and the use of the Euclidean distance measure.

A drawback of the algorithm is that, at an extent, it depends on the initial selected cluster centers. A recommended approach is to perform multiple runs and train different models, each with a different initialization, then pick the best model. K-means is also sensitive to outliers since cases with extreme values can have a big influence on the cluster centroids. K-medoids, a K-means variant, addresses this drawback by using centroids which are actual data points. Each cluster is represented by an actual data point instead of using the hypothetical instance defined by the cluster means. Each remaining case is then assigned to the cluster with the most similar representative data point, according to the selected distance measure.

On the other hand, K-means advantages include its speed and scalability: it is perhaps the fastest clustering model, and it can efficiently handle long and wide datasets with many records and input clustering fields.

Figure 5.9 presents the distribution of the five clusters revealed by Modeler's K-means algorithm on the fixed telephony data.

Figure 5.10 presents the centroids of the identified clusters.

The table of centroids presents the means of each input across clusters. Since in this example the clustering inputs are normalized PCA scores, a mean above or below 0 designates

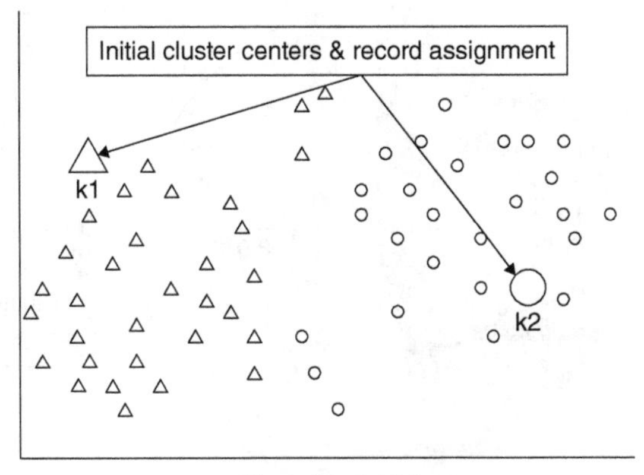

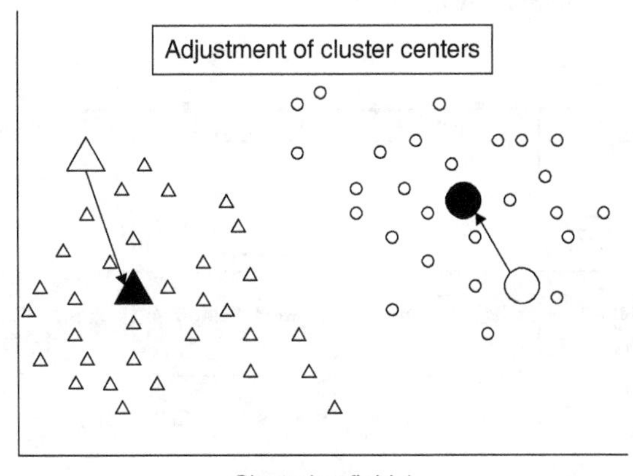

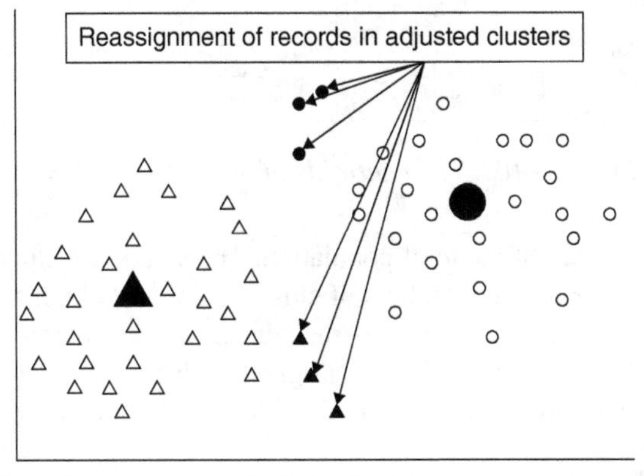

Figure 5.8 An illustration of the K-means clustering process

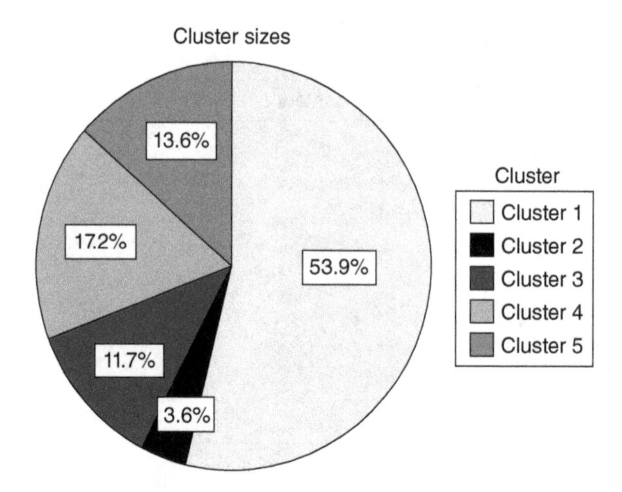

Figure 5.9 The distribution of the five clusters revealed by Modeler's K-means algorithm on the fixed telephony data

Cluster	cluster-1	cluster-2	cluster-3	cluster-4	cluster-5
Label					
Size	53.9% (1109)	3.6% (75)	11.7% (240)	17.2% (353)	13.6% (280)
Features	BROADBAND -0.19	BROADBAND 4.03	BROADBAND -0.08	BROADBAND -0.14	BROADBAND -0.10
	CALLS_TO_FIXED -0.26	CALLS_TO_FIXED -0.73	CALLS_TO_FIXED -0.29	CALLS_TO_FIXED -0.18	CALLS_TO_FIXED 1.70
	CALLS_TO_MOBILE -0.44	CALLS_TO_MOBILE -0.26	CALLS_TO_MOBILE 0.14	CALLS_TO_MOBILE 1.65	CALLS_TO_MOBILE -0.37
	INTERNATIONAL_CALLS -0.22	INTERNATIONAL_CALLS -0.37	INTERNATIONAL_CALLS 2.05	INTERNATIONAL_CALLS -0.41	INTERNATIONAL_CALLS -0.28

Figure 5.10 The centroids of the fixed telephony clusters

deviation from the mean of the total population. The solution includes a dominant cluster of basic users, the cluster 1. The members of cluster 2 present a high average on the broadband component. Hence, cluster 2 seems to include the heavy internet users. Customers with increased usage of international calls are assigned to cluster 3, while cluster 4 is characterized by increased usage of mobile calls. Finally, cluster 5 is dominated by fixed-calls customers.

The IBM SPSS Modeler K-means cluster

The parameters of the IBM SPSS Modeler K-means algorithm are presented in Figure 5.11 and in Table 5.5.

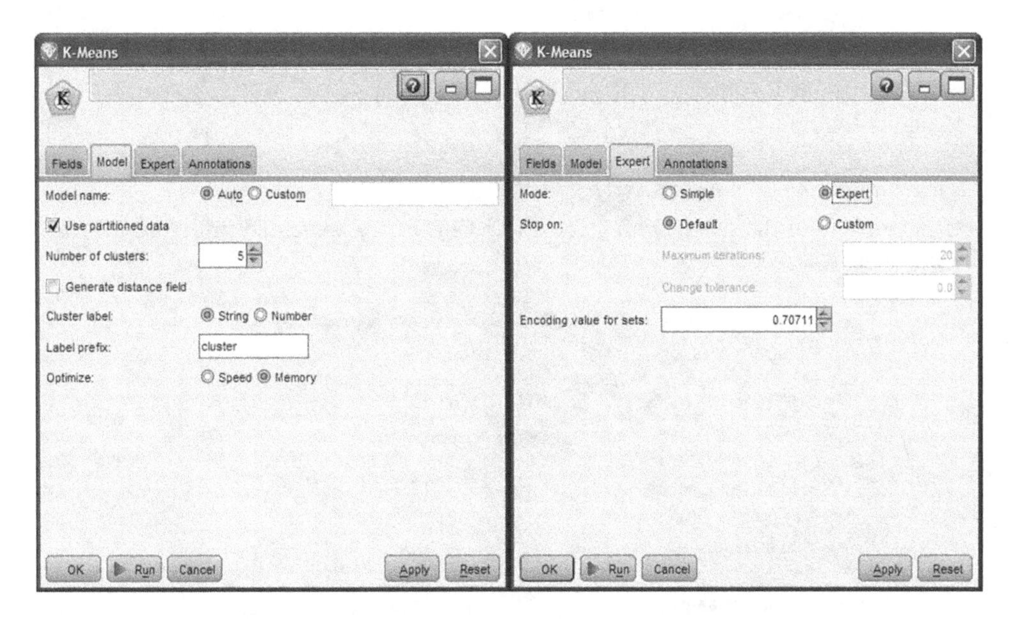

Figure 5.11 IBM SPSS Modeler K-means options

Table 5.5 IBM SPSS Modeler K-means parameters

Option	Explanation/effect
Number of clusters	Users should experiment with different number of clusters to find the one that best fits their specific business goals Worth trying alternatives: users can get an indication on the underlying number of clusters by trying other clustering techniques that incorporate specific criteria to automatically detect the number of clusters
Generate distance field	It generates an additional field denoting the distance of each record to the center of the cluster assigned. It can be used to assess whether a case is a typical representative of its cluster or lies far apart from the rest of the other members
Maximum iterations and change tolerance	The algorithm stops if after a data pass (iteration) the cluster centers do not change (0 change tolerance) or after the specified number of iterations have been completed, regardless of the change of the cluster centers. Higher tolerance values lead to faster but most likely worse models Users can examine the solution steps and the change of cluster centers at each iteration. In the case of nonconvergence after the specified number of iterations, they should rerun the algorithm with the number of iterations increased
Encoding value for sets	In K-means, each categorical field is transformed to a series of numeric flags. The default encoding value is the square root of 0.5 (approximately 0.7071 07). Higher values closer to 1.0 will weight categorical fields more heavily

The RapidMiner K-means and K-medoids cluster

The parameters of RapidMiner's K-means and K-medoid models are presented in Figure 5.12 and in Table 5.6.

Clustering in Data Mining for Excel

The parameters of the Data Mining for Excel Cluster are presented in Figure 5.13 and discussed in Table 5.7.

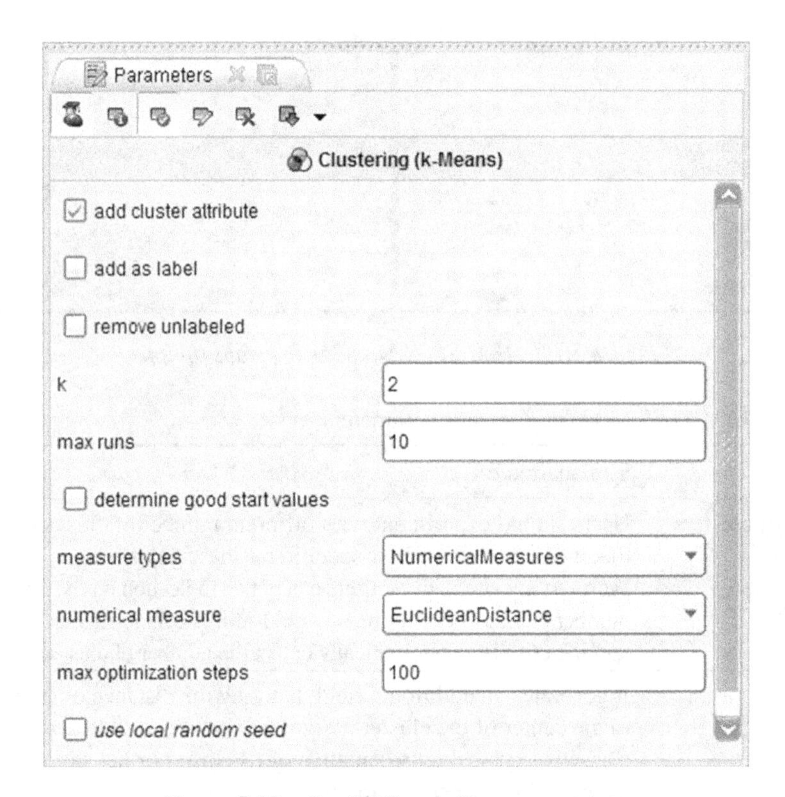

Figure 5.12 RapidMiner's K-means options

Table 5.6 RapidMiner's K-means and K-medoid parameters

Option	Explanation/effect
Add cluster attribute	Adds a cluster membership field. By default is selected
Add as label	When checked, the cluster attribute field is assigned a label (target) role instead of a cluster role
Remove unlabeled	When on, records not assigned to a cluster are deleted
k	Specifies the number of clusters to be formed
Determine good start values (for K-means)	A recommended option for K-means cluster which ensures that *k* well-spaced records are chosen as the start centroids for the algorithm

Table 5.6 (**Continued**)

Option	Explanation/effect
Max runs/max optimization steps	The "max optimization steps" option specifies the maximum number of optimization iterations for each run of the algorithm. The "max runs" option sets the maximum number of different times the algorithm will run. At each run, different random start centroids are used in order to decrease the dependency of the solution to the starting points
Measure types and measure	The combination of these two options sets the measure type and the measure which will be used to assess the similarities of records. The selection depends on the type and the measurement level of the inputs but as always, experimentation with different distance measures and number of clusters should be done. The available numerical measures include the classic Euclidean distance and other measures such as the Chebyshev and the Manhattan distances. The class of nominal measures includes the Dice similarity and others. The mixed Euclidean distance is included in the class of mixed measures, appropriate in the case of mixed inputs. Finally, the squared Euclidean distance as well as other measures such as the KL divergence and Mahalanobis distance are available in the class of Bregman divergences (distances)
(Use) local random seed	By specifying a local random seed, we "stabilize" the random selection of the initial cluster centroids, ensuring that the same clustering results can be reproduced

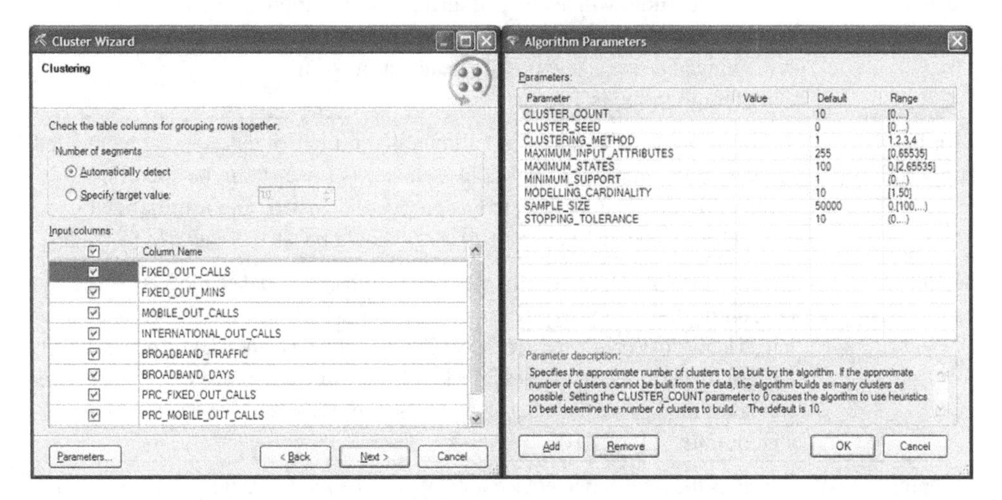

Figure 5.13 The cluster options of Data Mining for Excel

Table 5.7 The cluster parameters of Data Mining for Excel

Option	Explanation/effect
Number of segments	In the clustering step of the Cluster wizard, apart from the clustering inputs, analysts specify the number of clusters to be formed. They can specify a specific target value based on prior experimentation or they can let the algorithm use heuristics to autodetect the optimal number of clusters. In the heuristic procedure, many models are trained on a small sample of data and compared based on fit measures to find the optimal number of clusters
Cluster count	The "number of segments choice" made in the Clustering wizard is inherited in the "Cluster count" parameter. If the specified number of clusters cannot be built from the data, the algorithm builds as many clusters as possible. A value of 0 in the "Cluster count" parameter invokes autodetection of clusters
Cluster seed	By specifying a random seed, we "stabilize" the random sampling done in the initial steps of the algorithm, ensuring that the same clustering results can be reproduced
Clustering method	Available methods include scalable Expectation Maximization (EM), nonscalable EM, scalable K-means, and nonscalable K-means Scalable methods are based on sampling to address performance issues in the case of large datasets. Nonscalable methods (K-means and EM) load and analyze the entire dataset in one clustering pass. On the other hand, scalable methods use a sample of cases (the default sample size is 50 000 cases) and load more cases only if the model has not converged and it is necessary to improve its fit. Nonscalable methods might produce slightly more accurate solutions but are demanding in terms of process time and resources. In the case of large datasets, the scalable methods are recommended
Sample size	Sets the sample size of each data pass for the scalable clustering methods The default value is 50 000
Maximum input attributes	A threshold parameter for the integrated feature selection preprocessing step. Feature selection is invoked if the number of inputs is greater than the specified value A value of 0 turns off feature selection. The default value is 255
Maximum states	This option determines the maximum number of input categories ("states") that the algorithm will analyze. If an attribute has more categories than the specified threshold, then only the most frequent categories will be used and the remaining and rare categories are treated as missing The default value is 100
Minimum support	This option sets the minimum allowable number of cases in each detected cluster. If the number of cases in a cluster is lower than the specified threshold, the cluster is treated as empty, discarded, and reinitialized The default value of 1 in almost all cases should be altered with a higher value
Modeling cardinality	Controls the number of runs, the number of different candidate models that are built during the clustering process. Each model is built with a different random initialization to decrease the dependency of the solution to the starting points. Finally, the best model of the candidates is picked. Lower values improve speed, since fewer candidate models are built, but also increase the risk of a lower-quality solution The default value is 100
Stopping tolerance	Specifies the accepted tolerance for stopping the model training. The model stops and convergence is reached when the overall change in cluster centers or the overall change in cluster probabilities is less than the ratio of the specified "Stopping Tolerance" parameter divided by the number of the clusters The default value is 10. Higher tolerance values lead to faster but riskier models

5.3.2 Clustering with TwoStep

TwoStep is a scalable clustering model, based on the BIRCH algorithm, available in IBM SPSS Modeler that can efficiently handle large datasets. It uses a log-likelihood distance measure to accommodate mixed, continuous, and categorical inputs.

Unlike classic hierarchical techniques, it performs well with big data due to the initial resizing of the input records to subclusters. As the name implies, the clustering process is carried out in two steps. In the first step of preclustering, the entire dataset is scanned and a clustering feature tree (CF tree) is built, comprised by a large number of small preclusters ("microclusters"). Instead of using the detailed information of its member instances to characterize each precluster, summary statistics are used such as the centroid (mean of inputs), radius (average distance of member cases to the centroid), and diameter (average pairwise distance within the precluster) of the preclusters. Each case is recursively guided to its closest precluster. If the respective distance is below an accepted threshold, it is assigned to the precluster. Otherwise, it starts a precluster of its own. In the second phase of the process, an hierarchical (agglomerative) clustering algorithm is applied to the CF tree preclusters which are recursively merged until the final solution.

TwoStep uses heuristics to suggest the optimal number of clusters by taking into account:

- The goodness of fit of the solution. The Bayesian Information Criterion (BIC) or the Akaike Information Criterion (AIC) measure is used to assess how well the specific data are fit by the current number of clusters.

- The distance measure for merging the preclusters. The algorithm examines the final steps of the hierarchical procedure and tries to spot a sudden increase in the merging distances. This point indicates that the agglomerative procedure has started to join dissimilar clusters, and hence, it indicates the correct number of clusters to fit.

Another TwoStep advantage is that it integrates an outlier handling option that minimizes the effects of noisy records which could otherwise distort the segmentation solution. Outlier identification is taking place in the preclustering phase. Sparse preclusters with few members compared to other preclusters (less than 25% of the largest precluster) are considered as potential outliers. These outlier cases are set aside and the preclustering procedure is rerun without them. Outliers that still cannot fit the revised precluster solution are filtered out from the next step of hierarchical clustering, and they do not participate in the formation of the final clusters. Instead, they are assigned to a "noise" cluster.

The algorithm is fast and scalable; however, it seems to work best when the clusters are spherical in shape.

@Modeling tech tips

The IBM SPSS Modeler TwoStep cluster

The IBM SPSS Modeler TwoStep cluster options are listed in Figure 5.14 and Table 5.8.

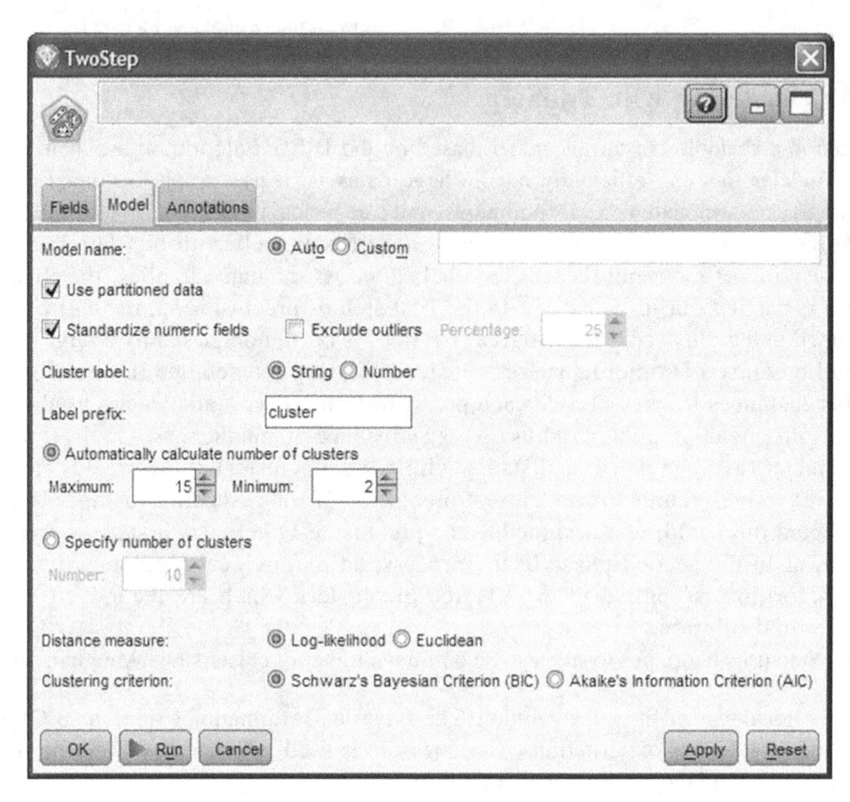

Figure 5.14 The IBM SPSS Modeler TwoStep cluster options

Table 5.8 The IBM SPSS Modeler TwoStep cluster parameters

Option	Explanation/effect
Standardize numeric fields	By default, the numeric inputs are standardized in the same scale (mean and standard deviation of 1)
Automatically calculate number of clusters	This option invokes autoclustering and lets the algorithm suggest the number of clusters to fit. The "Maximum" and "Minimum" text box values define the range of solutions that will be evaluated. The default setting limits the algorithm to evaluate the last 15 steps of the hierarchical clustering procedure and to propose a solution comprising from a minimum of 2 up to a maximum of 15 clusters The autoclustering option is the typical approach for starting the model training. Alternatively, analysts can set a specific number of clusters to be created through the Specify number of clusters option
Exclude outliers	When selected, handling of outliers is applied, an option generally recommended as it tends to yield richer clustering solutions
Distance measure	Sets the distance measure to be used for assessing the similarities of cases. Available measures include the log-likelihood (default) and the Euclidean distance. The latter can be used only when the inputs are continuous
Clustering criterion	Sets the criterion to be used in the autoclustering process to identify the optimal number of clusters. The criteria include the Bayesian Information Criterion (BIC) and the Akaike Information Criterion (AIC)

5.4 Summary

Clustering techniques are applied for unsupervised segmentation. They find natural groupings of records/customers with similar characteristics. The revealed clusters are directed by the data and not by subjective and predefined business opinions. There are not right or wrong clustering solutions. The value of each solution depends on its ability to represent transparent, meaningful, and actionable customer typologies.

In this chapter, we've presented only some of the available clustering techniques: K-means and K-medoids and the TwoStep cluster model based on the BIRCH algorithm. We've also dedicated a part of the chapter to PCA analysis. Although not required, PCA is a recommended data preparation step as it reduces the dimensionality of the inputs and identifies the distinct behavioral aspects to be used for unbiased segmentation.

Part III

The Case Studies

6

A voluntary churn propensity model for credit card holders

6.1 The business objective

A bank was losing credit card customers to its competitors, and the marketers of the organization decided to use analytics in order to deal with this issue. Their goal was to use data mining models for identifying customers with increased propensity to churn so they could fine-tune their retention campaigns.

Each credit card customer is unique. They all have different demographic profiles and usage characteristics. They vary in terms of spending frequency and intensity, payment behavior, and limit usage. Certain customers use their cards very often, while for others, the usage is infrequent and low, maybe because their main card is with a competitor. Some customers may have increased their usage compared to the past, while others may show signs of usage decrease. The so-called transactors usually pay their full balances every month as opposed to other customers who tend to carry past balances and use their entire credit limit. "Old" customers have a long relationship with the bank, whereas others are new customers. So what makes a customer loyal and how all these characteristics are related to customer attrition. Fortunately, all this information which defines each customer's unique "signature" could be retrieved from the bank's mining datamart. After proper transformations, they could be used as inputs in the churn model to be trained.

Since not all customers are equally important, the bank's marketers also intended to combine churn propensities with value information in order to prevent the attrition of high value at risk customers. Their final goal was more ambitious than using analytics just for a single campaign. They wanted, after validating the effectiveness of the models, to build a process and integrate analytics in the marketing operations for better targeting their retention campaigns.

Effective CRM using Predictive Analytics, First Edition. Antonios Chorianopoulos.

Companion website: www.wiley.com/go/chorianopoulos/effective_crm

6.2 The mining approach

The mining approach decided was to build a voluntary churn propensity model by analyzing customer behavior prior to churn. Therefore, the model had to be trained on customers that were active at the end of the observation period in order to discern the patterns that differentiate those that finally churned (target population) from those that stayed loyal in the following period.

6.2.1 Designing the churn propensity model process

The decisions made in the design process concluded on the following.

6.2.1.1 Selecting the data sources and the predictors

The organization had already in place a mining datamart with detailed information on credit cards' usage. This simplified the mining procedure since the tedious data preparation tasks needn't start from raw transactional data. The mining datamart contained historical information for more than 12 months which is typically necessary for building reliable churn models. The usage information that was considered as important and was decided to participate in the model included credit card spending (amount and transactions), balances, limits, and payments. This information was supplemented with card characteristics such as opening date, expiration date, and card type along with customer demographics.

6.2.1.2 Modeling population and level of data

The modeling population included "active" credit card holders, meaning customers with at least one open credit card at the end of the historical observation period.

Churn of high value customers

In churn models, a worth trying approach is to focus on high value customers. Since their churn really matters, instead of analyzing all customers, it is recommended to narrow the modeling population to the high value segment in order to build a more refined model that will perform better in identifying high value at risk customers. After all, the issue is not to prevent the attrition of customers of trivial importance but to succeed in retaining the customers that really contribute to the company's profit.

The campaign's level (customers) also imposed the level of the modeling population and dataset. Therefore, an additional data preparation phase was necessary for the aggregation of card-level data at the customer level. This procedure will be briefly described later in this chapter.

6.2.1.3 Target population and churn definition

There was a debate among the marketers on what to define as churn. Some argued about defining inactivity as churn. According to this approach, a card was to be flagged as churned after some months with no spending usage. The approach finally adopted was to go with actual card closing and consider a card as churned if voluntarily closed. Cards closed involuntarily, due to bad payment behavior, were excluded and left over for a credit risk model.

But how is this translated at the customer level, and what makes a customer with more than one cards a churner? The decision was to target full churn instead of partial churn. Therefore, a customer was considered as a churner if he/she closed all of his/her cards in the period examined.

6.2.1.4 Time periods and historical information required

For the needs of model training, the retrieved data covered three distinct time periods: the observation period, the latency period, and the event outcome period.

Twelve months of historical data was decided to be analyzed in order to identify data patterns relevant to attrition. Since the modeling procedure started at mid-2013 (start of July 2013), the observation window covered the whole 2012. A latency period of 2 months was reserved for preparing the relevant campaign. A period of 4 months was examined for churn. Thus, an active customer was considered as a churner if he/she had closed all his/she cards within the 4 months after the latency period, from March to July 2013. The approach is described in Figure 6.1 which is also a guide for the data setup for the model training.

6.3 The data dictionary

The fields retrieved from the mining datamart for the scope of the project are listed and briefly explained in Table 6.1. They include card characteristics and status information, customer demographics, and usage attributes.

Usage attributes cover an observation period of 12 months, from January to December 2012. These fields, after proper transformations, were used as predictors in the churn propensity model; therefore, they summarize customer behavior prior to the time period reserved for observing possible churn. Their names have a suffix of _n where *n* corresponds to the month they summarize. One stands for the earlier month (1/2012) and 12 for the most recent month (12/2012) of the observation period.

Status fields such as the closing date and reason were used for capturing churn and for the definitions of the target (output) field of the model.

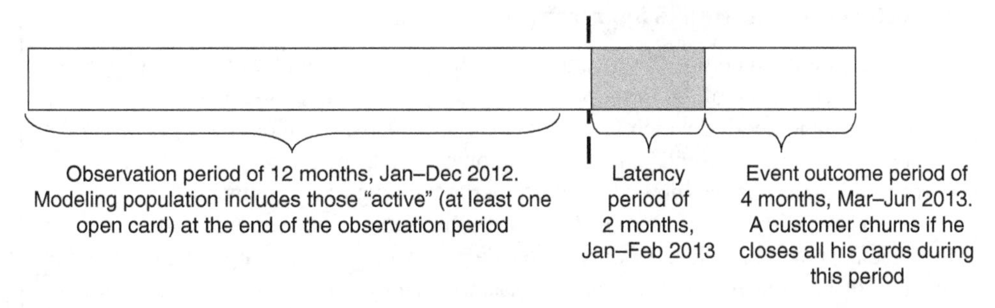

Figure 6.1 The three different time periods in the model training phase

Table 6.1 Data dictionary (card level) for the voluntary churn model

Data file: CC_CHURN_DP_DATA.TXT	
Field name	**Description**
Usage fields	
SPENDING_AMOUNT_(n)	Total card spending (purchases and cash advance) for month *n*: 1 stands for the oldest and 12 for the most recent month of the observation period
TRX_(n)	Total number of spending transactions for the month
BALANCE_(n)	Balance (daily average) for the month
LAST_TRX_DATE	Most recent date of spending transaction. To be used for calculating recency of spending
Card and customer characteristics fields	
CARD_ID	The (primary) card identification number
NUM_CARDS	The number of add-on cards for each primary card
LIMIT	The credit limit at the last month of the observation period
CARD_TYPE	The card type (Classic/Gold/Platinum)
CUSTOMER_ID	The customer identification number
GENDER	Gender of the customer
EDUCATION	Educational level of the customer
MARITAL_STATUS	Marital status of the customer
AGE	Age of the customer
Data file: CC_CHURN_ACCOUNTS.TXT	
Status fields	
OPENING_DATE	The opening date of the card
EXPIRY_DATE	The expiration date of the card
CLOSING_DATE	The closing date of the product account. A null value for open cards
CLOSING_REASON	The recorded closing reason for closed cards (voluntary churn/involuntary churn and null for open cards)

Predictors and time frames in propensity modeling

As we have outlined before, in propensity modeling, when we analyze customer behavior before the occurrence of the event of interest, candidate predictors should only be based in the observation period which should not overlap with the event outcome period. In the deployment phase, there will be no event outcome period, and the observation period will correspond to the current view of the customer at the time of deployment.

In our case study, the time frame of the observation period is the entire 2012. Information from the next months will serve only for the definition of the churn target field. When the model will be used for deployment, for instance in July 2013, it will again require 12 months of summarized usage data, July 2012 to June 2013, to score new cases.

Although, normally customer demographics, card characteristics, and usage attributes reside in different database tables, for simplicity in this example, we assume that they are stored at the same table.

6.4 The data preparation procedure

Section 6.4 describes in detail the data management procedures that had been carried out in the fields of Table 6.1 before being used as inputs in the churn model. Readers not interested in the data preparation phase might skip the next sections and go directly to Section 6.5 for modeling.

6.4.1 From cards to customers: aggregating card-level data

Since the scope was to analyze and target customers, individual card usage had to be combined to convey the behavior of each customer. Therefore, the initial card usage data, listed in Table 6.1, had to be aggregated at the customer level. This step included the grouping of individual card records to summarized customer records with the use of standard summary statistics and the construction of customer attributes such as the total spending amount and the number of transactions of all cards, the last transaction date, etc. This step is illustrated in Figure 6.2.

Even after aggregation, the data preparation phase was far from over since summarized data had to be enriched with informative KPIs.

But before we get into that, let's spend some more time examining the aggregation procedure. Before the actual grouping and the replacement of the card records with the customer records, cards already closed before the observation period (prior 2012-1-1) were filtered out. Obviously, it makes no sense to analyze cards that had already been terminated before the period under examination. At first, cards open at some point within the observation period were flagged as OPEN WITHIN. These cards comprise the population to be examined for the model.

Specifically, by examining the cards' closing dates, a card was flagged as OPEN WITHIN if it hadn't been closed (it had a null closing date) at all or it had been closed after the beginning of the observation period (it had a closing date after 2012-1-1). Additionally, only cards opened before the end of the observation period (the so-called REFERENCE DATE of

Card ID	Customer ID	Spending amount	Number of transactions	Most recent transaction date
CC1	CID1	$ 70	3	2012-12-20
CC2	CID1	$ 130	4	2012-6-2
CC3	CID2	$ 110	5	2012-11-26
CC3	CID2	$ 40	2	2012-10-15
.....				

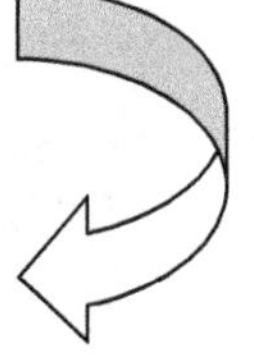

Customer ID	Spending amount	Number of transactions	Most recent transaction date
CID1	$ 200	7	2012-12-20
CID2	$ 150	7	2012-11-26
...			

Figure 6.2 Aggregating usage data per customer

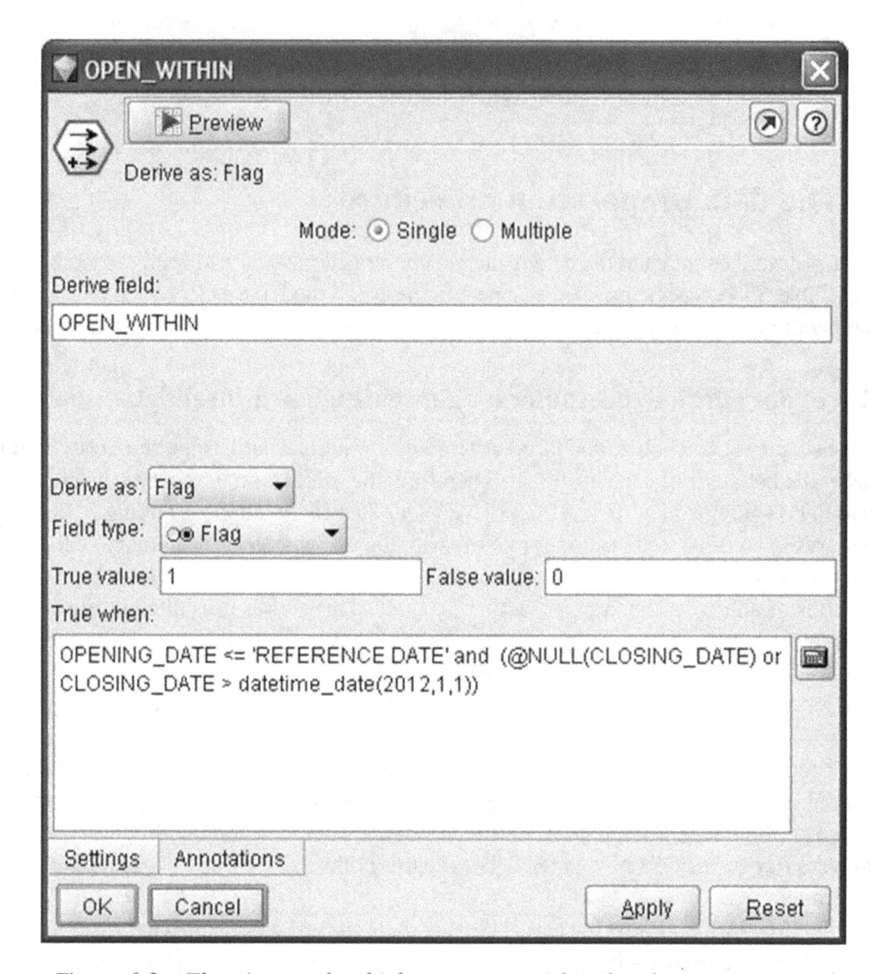

Figure 6.3 Flagging cards which were open within the observation period

2012-12-31) were retained. Initially, cards open within the observation period were flagged using a Modeler Derive node as shown in Figure 6.3.

Then, only cards flagged as OPEN WITHIN were included in the modeling file. Cards involuntarily closed at any time were also discarded before the aggregation through a Modeler Select node (Figure 6.4).

Analysts also wanted to monitor the number of open cards per customer at different time points. They achieved this with a series of flags and more specifically:

- Open cards at the beginning of the observation period (2012-1-1, OPEN_AT_START field)
- Open cards at the end of the observation period (2012-12-31, OPEN_AT_END field)
- Open cards at the end of the latency period (2013-3-1, OPEN_AT_LATENCY)
- Open cards at the end of the event outcome period (2013-7-1, OPEN_AT_EVENT_OUTCOME_PERIOD field)

The flags were created with Modeler Derive nodes as sown in Figure 6.5.

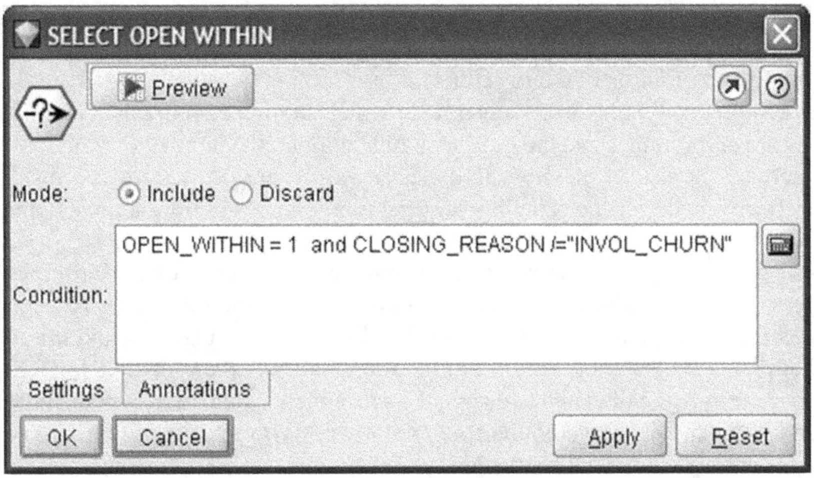

SELECT OPEN WITHIN
Preview
Mode:
Include
Discard
OPEN_WITHIN = 1 and CLOSING_REASON /="INVOL_CHURN"
Condition:
Settings
Annotations
OK
Cancel
Apply
Reset

Figure 6.4 Filtering out cards closed before the observation period

OPEN_AT_END
Preview
Derive as: Flag
Mode:
Single
Multiple
Derive field:
OPEN_AT_END
Derive as:
Flag
Field type:
Flag
True value:
1
False value:
0
True when:
OPENING_DATE <= 'REFERENCE DATE' and (@NULL(CLOSING_DATE) or
CLOSING_DATE > 'REFERENCE DATE')
Settings
Annotations
OK
Cancel
Apply
Reset

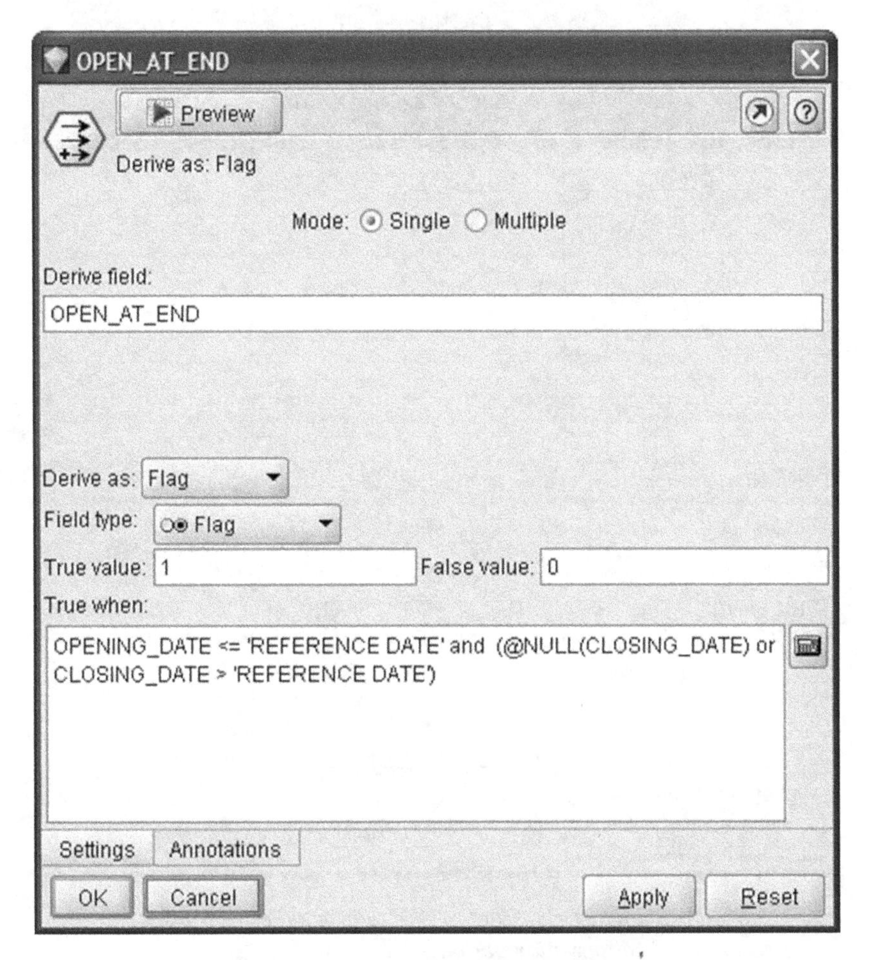

Figure 6.5 Flagging cards open at the end of the observation period. The REFERENCE DATE is the last day of the observation period (2012-12-31)

These series of flags enabled the analysts not only to capture the trend of card ownership for each customer but also, as we'll see in the next paragraphs, facilitated the identification of churners and the definition of the target field.

Furthermore, new fields were constructed which summated individual monthly fields for each card and summarized the total spending amount, the number of transactions, and the balances over the 12 months of the observation period. Figure 6.6 displays the Modeler Derive node for the calculation of the total number of transactions in the observation period.

These preparatory tasks were followed by the aggregation of the information per customer. Sums of spending amount, number of transactions, and balances were computed for each customer, and the card-level data were replaced with customer data form that point on. The most recent transaction date (maximum of LAST_TRX_DATE field) of all cards was calculated to enable the subsequent calculation of spending recency for each customer. Additionally, the first card opening (minimum of OPENING_DATE field) was derived to enable the calculation of the tenure for each customer. The Modeler Aggregate node was applied for aggregating the information at customer level as displayed in Figure 6.7.

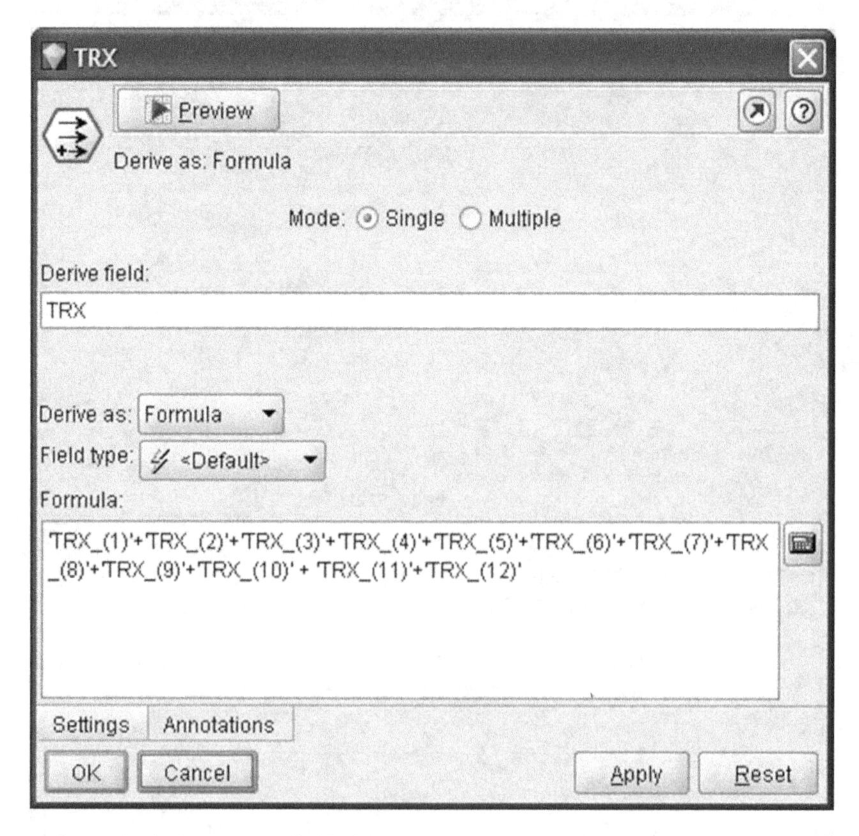

Figure 6.6 Calculating a new field denoting the total number of transactions over the 12 months of the observation period

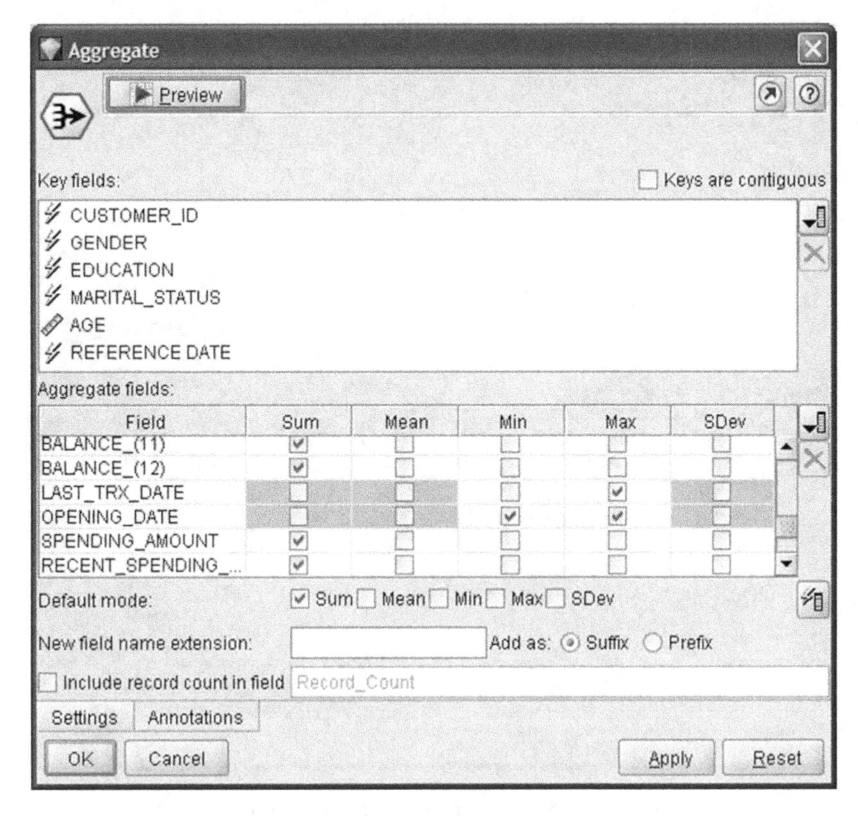

Figure 6.7 Using an IBM SPSS Modeler node for aggregating card data at the customer level

6.4.2 Enriching customer data

The data preparation procedure was continued with the enrichment of customer data with usage KPIs.

The derived fields included:

Customer tenure

The tenure (months on books) of each customer (TENURE_CUSTOMER field). A Modeler Derive node and a date_months_difference function was used to calculate the months of relationship with each customer, based on the opening date of each customer's first card (Figure 6.8).

Monthly average spending amount

The monthly average spending amount (AVG_SPENDING_AMOUNT field) of each customer over the 12 months of observation period. A Modeler Conditional Derive node was used to divide the total spending amount with the customer tenure, for new customers, or 12 months, for old customers as shown in Figure 6.9.

Spending recency and frequency

The spending recency and frequency were considered as churn predictors with potential predictive efficiency and were also derived. The spending recency (SPENDING_RECENCY

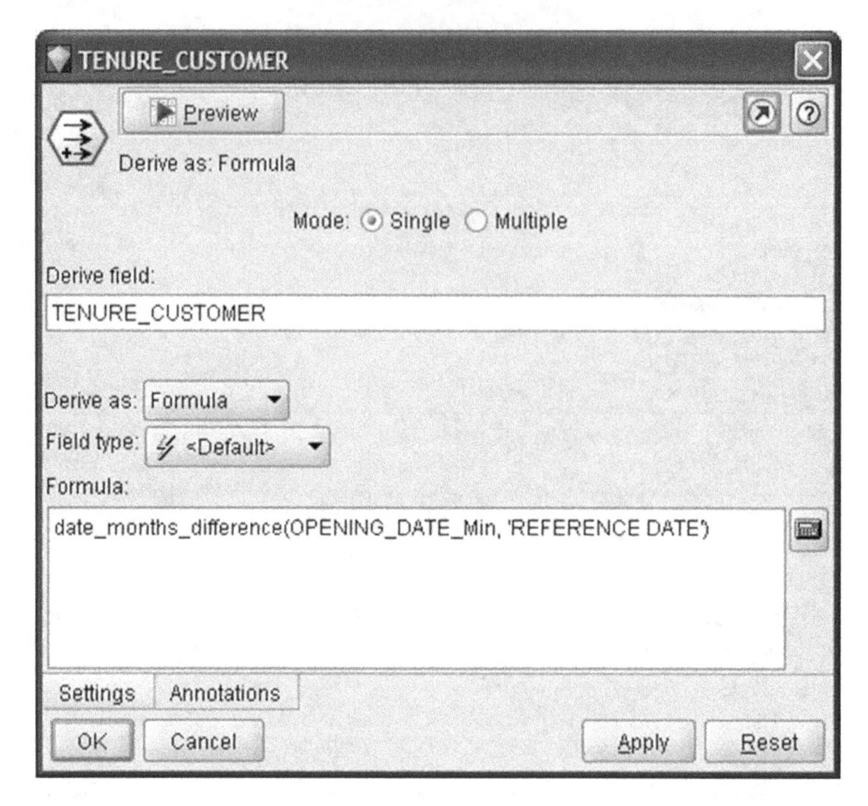

Figure 6.8 Deriving customer tenure

field) was computed as the time from the last spending transaction (LAST_TRX_DATE_Max) to the last day of the observation period (REFERENCE DATE field) as shown in Figure 6.10.

The spending frequency (SPENDING_FRECENCY field) was calculated as the percentage of the 12 months of the observation period with spending.

Spending Deltas

The average spending amount and spending frequency were also calculated for the last 4 months of the observation period, the 4 months nearest to potential churn. Deltas were then derived to denote changes during these 4 months. The idea was to capture a spending decline, something like a negative slope in spending, which might signify a churn prospective. The delta for spending amount (DELTA_SPENDING_AMOUNT field) was calculated as the signed ratio (±):

$$\frac{\text{Recent spending} - \text{Total spending}}{\text{Total spending}}.$$

It denotes the relative percentage increase or decrease of spending during the most recent months of the observation period. A Modeler Conditional Derive node was used for its computation as sown in Figure 6.11.

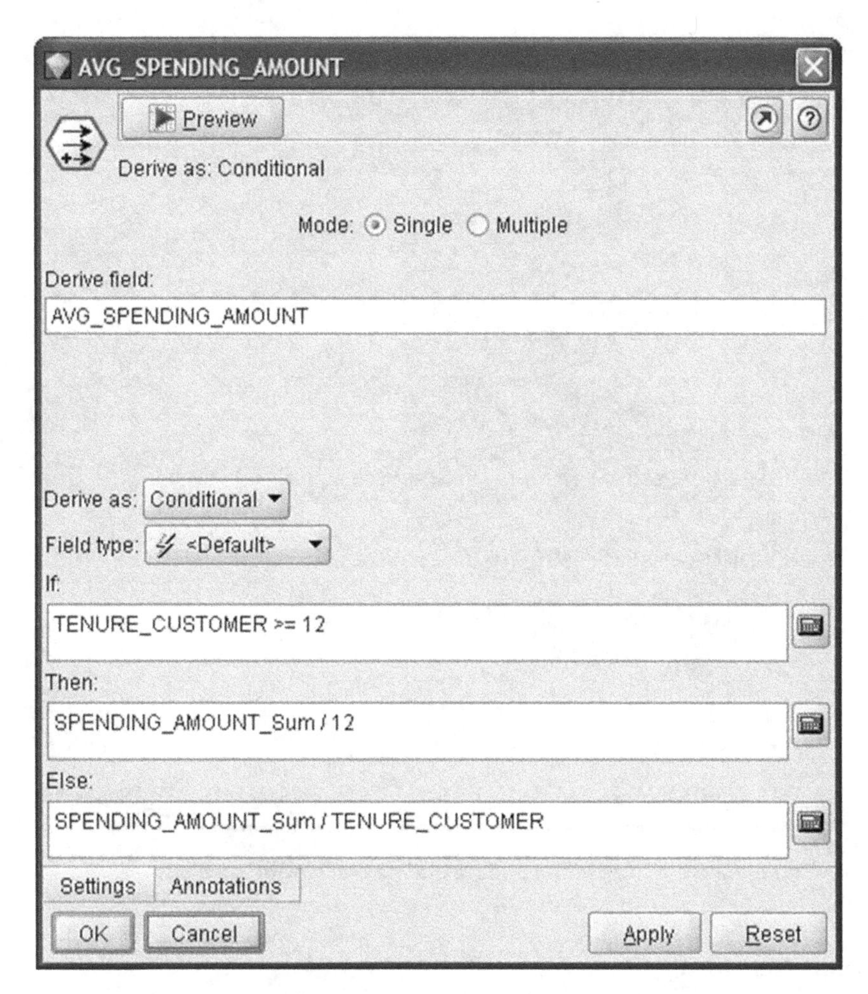

Figure 6.9 Deriving monthly average spending amount for each customer

Monthly average number of transactions and average transaction amount

In a way analogous to the one for spending amount, the monthly average number of transactions (AVG_TRX field) and the respective delta (DELTA_TRX field), the change in the most recent observation period, were also derived. An additional KPI calculated denoted the average amount per spending transaction (AVG_TRX_AMOUNT field).

Monthly average balances and balances frequency

Concerning balances, monthly average balances (AVG_BALANCES field), and the ratio of months with balances (BALANCES_FREQUENCY field) were also derived.

Limit ratios

Also, simple ratios of (monthly average) spending to credit limit (SPENDING_LIMIT_RATIO field) and (monthly average) balances to limit (LIMIT_USAGE) were constructed. The last KPI denotes limit usage and is a good indicator for identifying customers that seem to have problems with their credit limit.

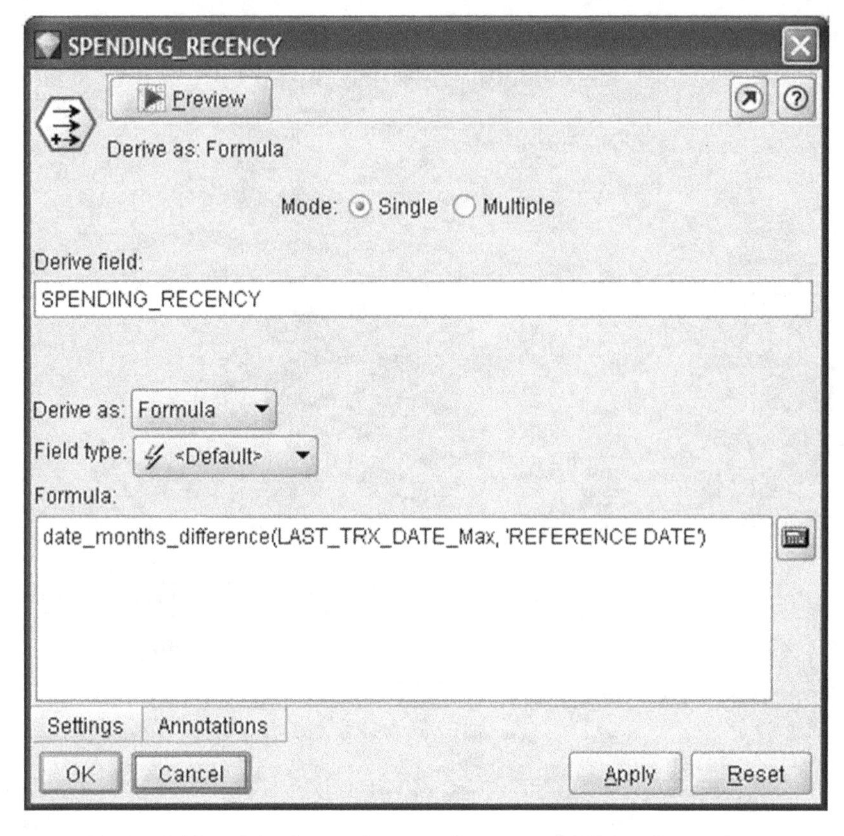

Figure 6.10 Calculating the spending recency for each customer

Card ownership trends

The ownership flags computed before the actual aggregation of data enabled the comparison of the number of open cards each customer had at different time points and the capturing of ownership trends. Specifically, the indicator END_START_CARDS_DELTAS denotes the signed (±) difference in the number of open cards a customer had at the end versus the beginning of the observation period. The second trend indicator END_WITHIN_CARDS_DELTAS captures the change in the number of open cards a customer had at the end of the observation period compared to all the cards he/she had during the whole observation period (Figure 6.12).

6.4.3 Defining the modeling population and the target field

After enriching predictors, it was time for selecting the modeling population and defining the target field. The approach followed was to include all customers with at least one open card at the end of the observation period (the so-called REFERENCE DATE, 2012-12-31). Remember that in deployment the REFERENCE DATE will correspond to the present. Therefore, in the scoring phase, the generated model will calculate

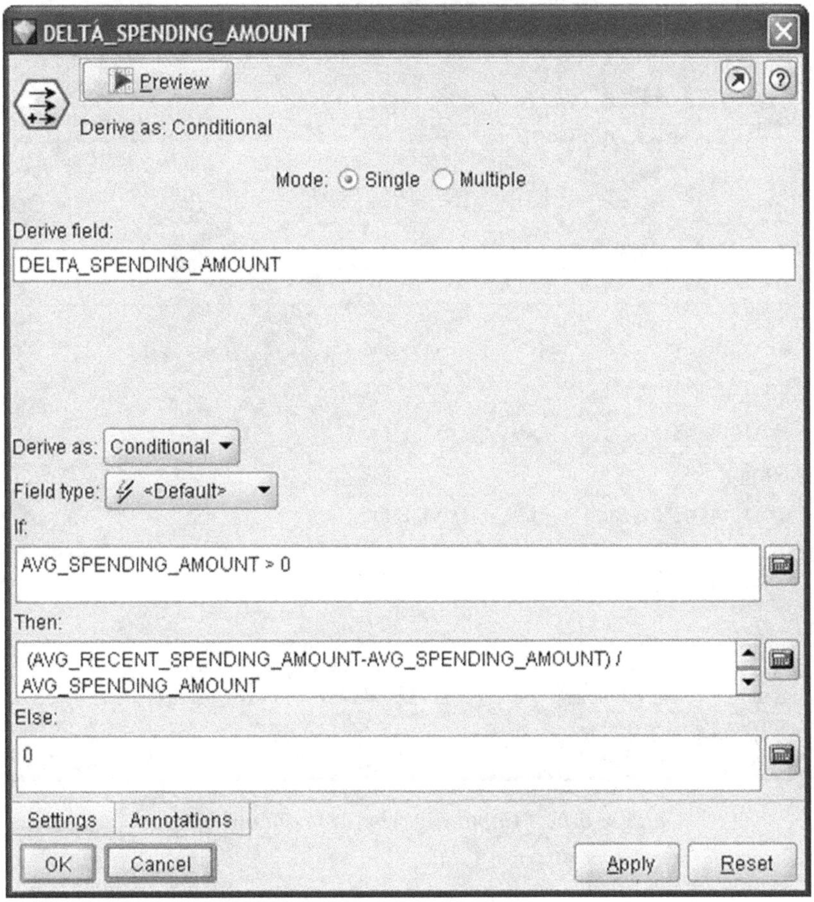

DELTA_SPENDING_AMOUNT
Preview
Derive as: Conditional
Mode: Single Multiple
Derive field:
DELTA_SPENDING_AMOUNT
Derive as: Conditional
Field type: <Default>
If:
AVG_SPENDING_AMOUNT > 0
Then:
(AVG_RECENT_SPENDING_AMOUNT-AVG_SPENDING_AMOUNT) / AVG_SPENDING_AMOUNT
Else:
0
Settings Annotations
OK Cancel Apply Reset

Figure 6.11 Capturing changes of spending with spending amount delta

propensities for all active customers at present. The modeling population was selected based on the number of open cards at the end of the observation period through a Modeler Select node (Figure 6.13).

Moreover, since latency period has been reserved, customers that have churned within 2 months after the observation period have been discarded. We want the model to be trained on churn cases for which there is a chance of retention. Immediate churners will probably leave during the campaign preparation. Even if the bank manages to contact them before their formal attrition, it'd be quite unlikely to change their decision. As shown in Figure 6.14, short-term churners have been discarded based on the number of open cards at the end of the latency period (OPEN_AT_LATENCY_Sum field).

Finally, the target field and population have been defined. Full churners, that is, customers without open cards at the end of the event outcome period (OPEN_AT_EVENT_OUTCOME_PERIOD_Sum field), have been flagged as churners (Figure 6.15, Modeler Flag Derive node).

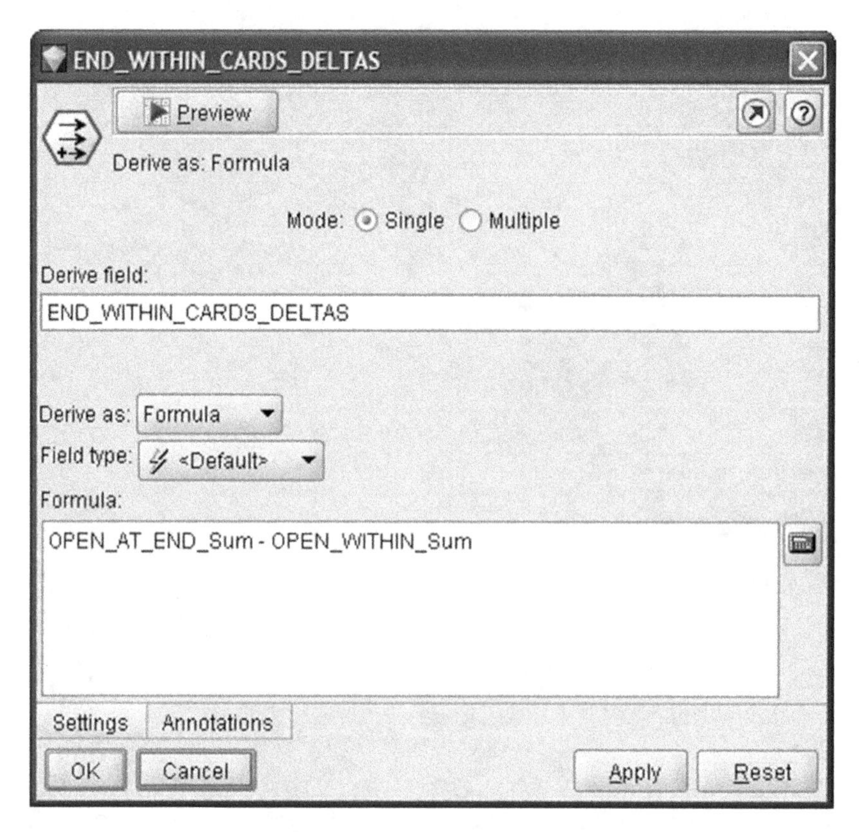

Figure 6.12 Capturing trends in card ownership

Figure 6.13 Selecting the modeling population

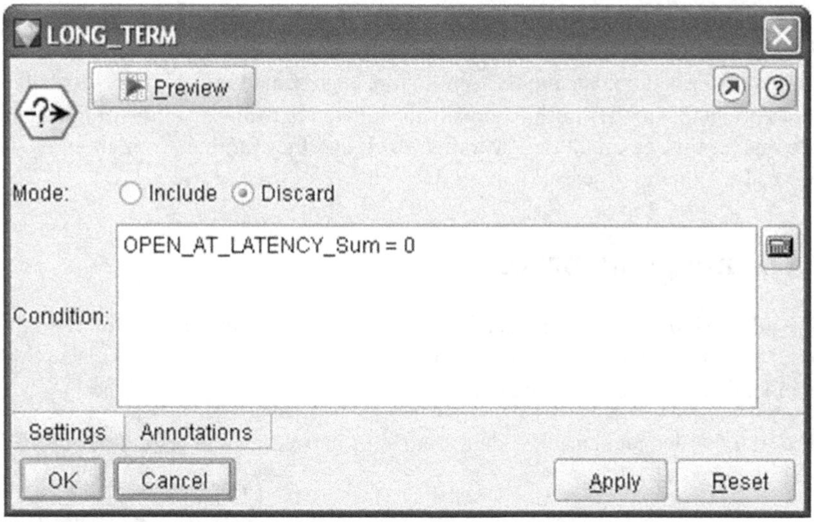

LONG_TERM

Preview

Mode: Include Discard

Condition: OPEN_AT_LATENCY_Sum = 0

Settings Annotations

OK Cancel Apply Reset

Figure 6.14 Discarding short-term churners from the model

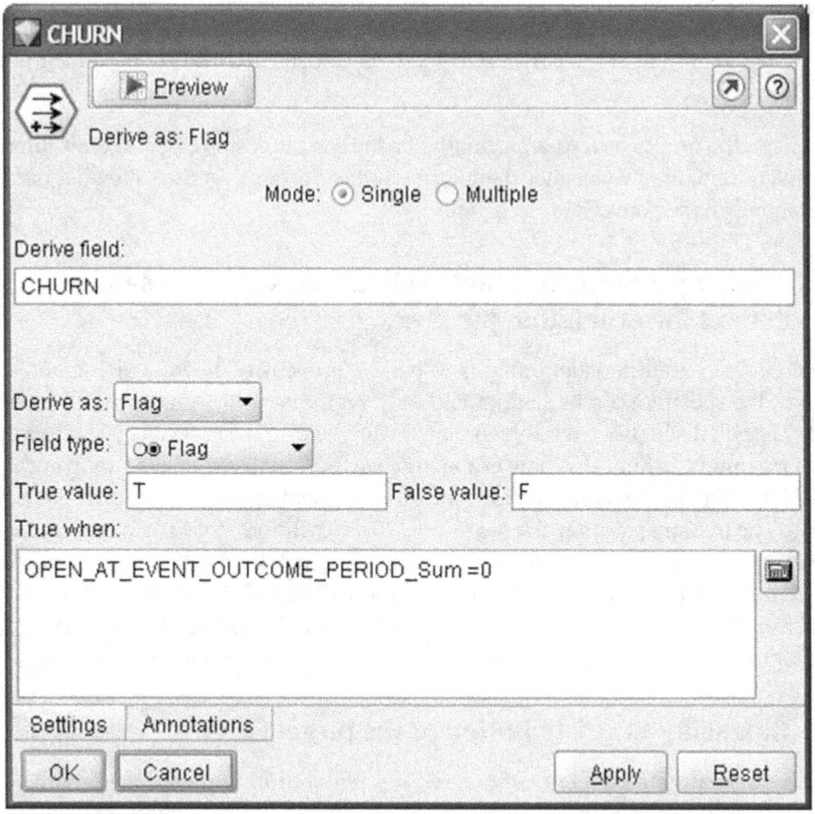

CHURN

Preview

Derive as: Flag

Mode: Single Multiple

Derive field:

CHURN

Derive as: Flag

Field type: Flag

True value: T False value: F

True when:

OPEN_AT_EVENT_OUTCOME_PERIOD_Sum =0

Settings Annotations

OK Cancel Apply Reset

Figure 6.15 Defining the target filed and population

6.5 Derived fields: the final data dictionary

The final list of derived fields/candidate predictors is presented in Table 6.2. Attributes finally included as predictors in the churn model are designated in the last column of the table. Fields summarizing card usage are at the customer level, and they are based on all cards that each customer had within the observation period.

6.6 The modeling procedure

The IBM SPSS Modeler stream (procedure) for churn modeling is displayed in Figure 6.16. The model is trained on the fields of Table 6.2.

The modeling steps followed included:

- The transformed and enriched data were split in training and testing partitions for validation purposes.
- Then, due to the relatively small number of the observed churned cases, the distribution of the churn field was balanced through stratified sampling.
- The role of each field was determined, and the predictors to be included in the model were selected.
- Finally, a series of propensity models were developed and evaluated for their predictive efficiency.

Although the first two steps are actually data preparation steps, due to their direct effect to the model training, we consider them as modeling presteps. In the following paragraphs, we'll examine all the above steps in detail.

6.6.1 Applying a Split (Holdout) validation: splitting the modeling dataset for evaluation purposes

A very common pitfall in data mining is to build a propensity model which memorizes the patterns of the specific training dataset and only performs well for the specific records. To avoid this pitfall, it's highly advisable to train the model in one dataset and test it in a different one. In our example, a Split (Holdout) validation was applied through a Partition node (Figure 6.17) which split the modeling dataset into training and testing datasets through random sampling. The training partition size was set to 70%. Therefore, the 70% of the initial 22.693 records were used for the training of the model. The remaining 30% of the records were allocated at the testing partition and were used for assessing the model accuracy.

Additionally, a random seed was specified in order to "stabilize" the underlying random sampling procedure and ensure the same partitioning on every model run.

6.6.2 Balancing the distribution of the target field

Since only a small percentage of the modeling population (almost 7%) belonged to the target group (class of churners), the analysts balanced the distribution of the outcome field by applying a Modeler Balance node. The balance technique is often applied in

Table 6.2 Data dictionary (customer level) for the voluntary churn model

Data file: CC_CHURN_MODELING_DATA.TXT		
Card and customer characteristics fields		
Field name	Description	Role in the model
CUSTOMER_ID	The customer identification number	
GENDER	Gender of the customer	PREDICTOR
EDUCATION	Educational level of the customer	
MARITAL_STATUS	Marital status of the customer	PREDICTOR
AGE	Age of the customer	PREDICTOR
AGE_BANDS	Age bins (18–24, 25–34, 35–44, 45–54, 55+)	
NUM_CARDS_Sum	The total number of add-on cards based on all primary cards a customer had within the observation period	PREDICTOR
LIMIT_Sum	The sum of credit limit of all cards at the end of the observation period	
Usage fields		
SPENDING_AMOUNT_(n)_Sum	Total card spending (purchases and cash advance) for the month *n*: 1 for the oldest and 12 for the most recent month of the observation period. Based on all cards that each customer had within the observation period	
TRX_(n)_Sum	Total number of spending transactions for the respective month	
BALANCE_(n)_Sum	Total balance (daily average) for the month	
LAST_TRX_DATE_Max	Most recent date of spending transaction	
OPENING_DATE_Min	The opening date of the first card	
OPENING_DATE_Max	The opening date of the last card	
REFERENCE DATE	The last day of the observation period (2012-12-31)	
OPEN_WITHIN_Sum	Number of cards open at some point within the observation period (2012)	
OPEN_AT_START_Sum	Number of cards open at the beginning of the observation period (2012-1-1)	

(*Continued*)

Table 6.2 (**Continued**)

Field name	Description	Role in the model
OPEN_AT_END_Sum	Number of cards open at the REFERENCE DATE, that is, at the end of the observation period (2012-12-31)	
OPEN_AT_LATENCY_Sum	Number of cards open at the end of the latency period (2012-3-1)	
OPEN_AT_EVENT_OUTCOME_PERIOD_Sum	Number of cards open 6 months after the observation period, that is, at the end of the event outcome period (2012-7-1)	
SPENDING_AMOUNT_Sum	Total card spending (purchases and cash advance) over the 12 months of the observation period. Based on all cards that each customer had within the observation period	
RECENT_SPENDING_AMOUNT_Sum	Total card spending over the last 4 months of the observation period	
TRX_Sum	Total number of spending transactions over the 12 months of the observation period	
RECENT_TRX_Sum	Total number of spending transactions over the last 4 months of the observation period	
BALANCES_Sum	Total balance (daily average) over the 12 months of the observation period	
MONTHS_TO_FEE_ANNIVERSARY_Min	Months till next fee of a card	
MONTHS_TO_EXPIRATION_Min	Months till next expiration of a card	
Classic_OPEN_WITHIN_Sum	Number of Classic cards open at some point within the observation period (2012)	
Gold_OPEN_WITHIN_Sum	Number of Gold cards open at some point within the observation period (2012)	
Platinum_OPEN_WITHIN_Sum	Number of Platinum cards open at some point within the observation period (2012)	
TENURE_CUSTOMER	Months since first registration of the customer	PREDICTOR

Table 6.2 (**Continued**)

Field name	Description	Role in the model
AVG_SPENDING_AMOUNT	Monthly average spending (purchases and cash advance) for the 12 months of the observation period. Based on all cards that each customer had within the observation period	PREDICTOR
SPENDING_RECENCY	Months since the last spending transaction	PREDICTOR
SPENDING_FREQUENCY	Percentage of the months with spending transactions	PREDICTOR
RECENT_SPENDING_FREQUENCY	Spending frequency for the last 4 months of the observation period	
DELTA_SPENDING_FREQUENCY	Relative change (percentage increase or decrease) in spending frequency during the last 4 months of the observation period	PREDICTOR
AVG_RECENT_SPENDING_AMOUNT	Monthly average spending (purchases and cash advance) for the last 4 months of the observation period	
DELTA_SPENDING_AMOUNT	Relative change (percentage increase or decrease) in monthly average spending during the last 4 months of the observation period	PREDICTOR
AVG_TRX	Monthly average number of spending transactions for the 12 months of the observation period	PREDICTOR
AVG_RECENT_TRX	Monthly average number of spending transactions for the last 4 months of the observation period	
DELTA_TRX	Relative change (percentage increase or decrease) in monthly average number of transactions during the last 4 months of the observation period	PREDICTOR
AVG_TRX_AMOUNT	Average spending amount per transaction	PREDICTOR
AVG_BALANCES	Monthly average balance for the 12 months of the observation period	PREDICTOR
BALANCES_FREQUENCY	Percentage of the months with balances	PREDICTOR

(Continued)

Table 6.2 (**Continued**)

Field name	Description	Role in the model
SPENDING_LIMIT_ RATIO	Ratio of spending amount to credit limit	PREDICTOR
LIMIT_USAGE	Ratio of balances to credit limit	PREDICTOR
END_START_CARDS_ DELTAS	Difference in the number of open cards between the end and the start of the observation period	PREDICTOR
END_WITHIN_ CARDS_DELTAS	Change in the number of open cards a customer had at the end of the observation period compared to all the cards that had possessed at some point within the observation period	PREDICTOR
CHURN	A flag indicating customers that closed all their cards within the event outcome period (Mar - Jun 2013)	TARGET

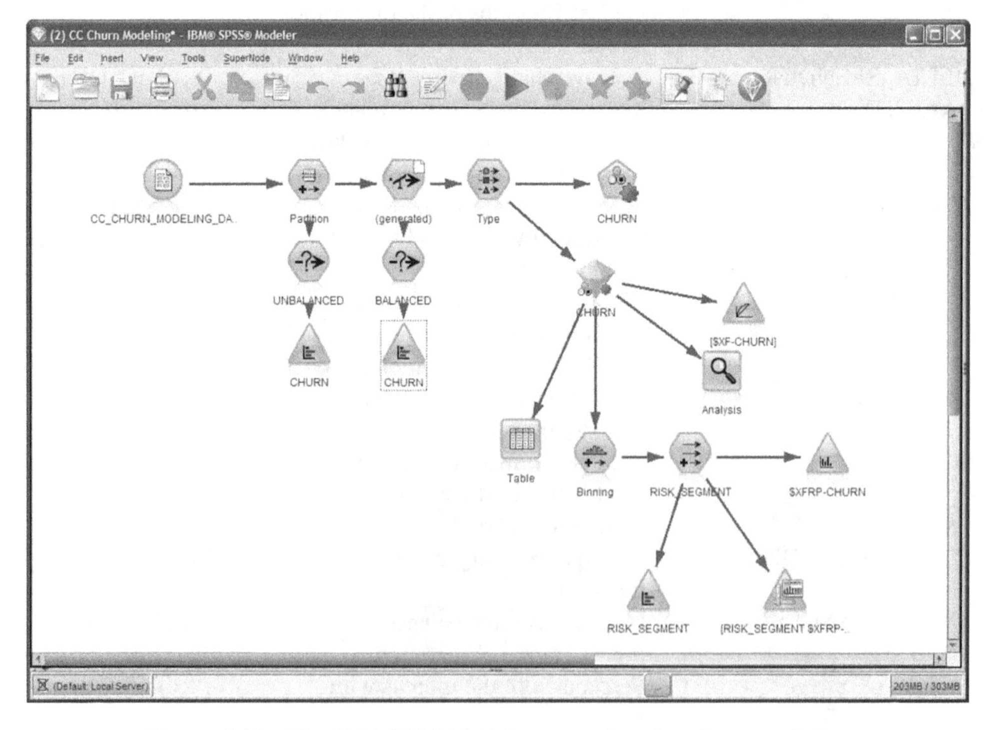

Figure 6.16 The IBM SPSS Modeler procedure for churn modeling

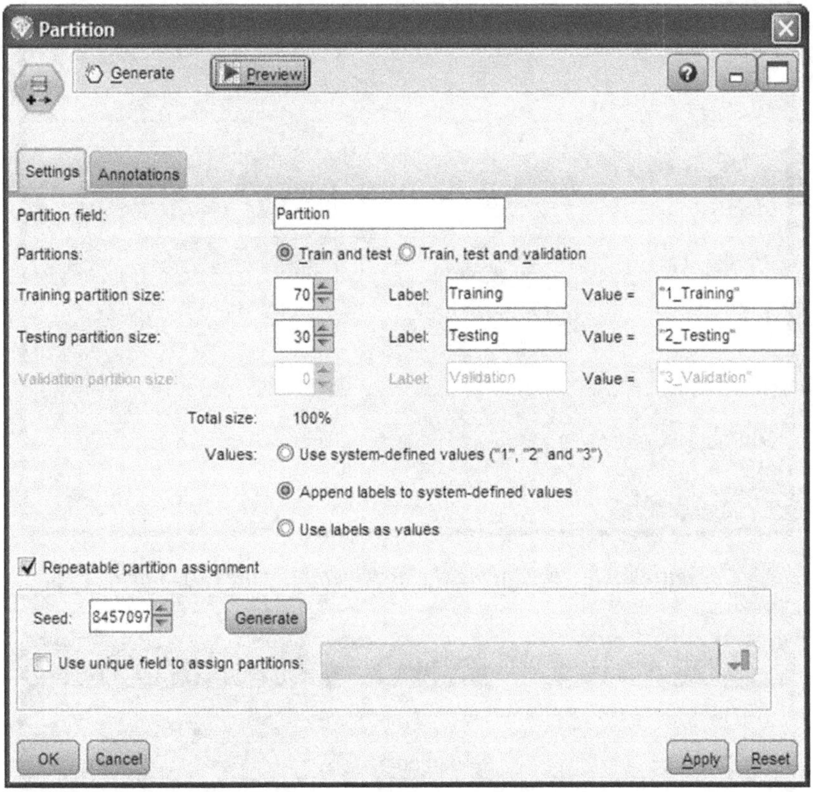

Figure 6.17 Partitioning the modeling dataset for evaluation purposes

propensity modeling in the case of rare target events. Many propensity models do not behave well in the case of imbalanced distribution of the target field. The weighting of outcomes corrects distribution imbalances and facilitates the identification of more refined patterns.

The Balance node (Figure 6.18) applies a stratified random sampling on the training dataset. A sample ratio of 1.0 was specified for the records belonging to the class of churners. Thus, all churned customers were retained in the training dataset. On the other hand, a sample ratio of 0.23 was specified for nonchurners; thus, about one every four of nonchurners was retained.

This disproportionate sampling, called undersampling, reduced the proportion of the more frequent class and boosted the percentage of churners in the training dataset from the initial 7% to approximately 25% after balancing. The balance effect and the final distribution of the CHURN field in the training partition is shown in Figure 6.19. As you can see, a more balanced distribution of 25–75% was achieved by reducing the number of nonchurners.

The testing partition was not balanced, allowing a simpler and more straightforward evaluation of the model results in the next steps.

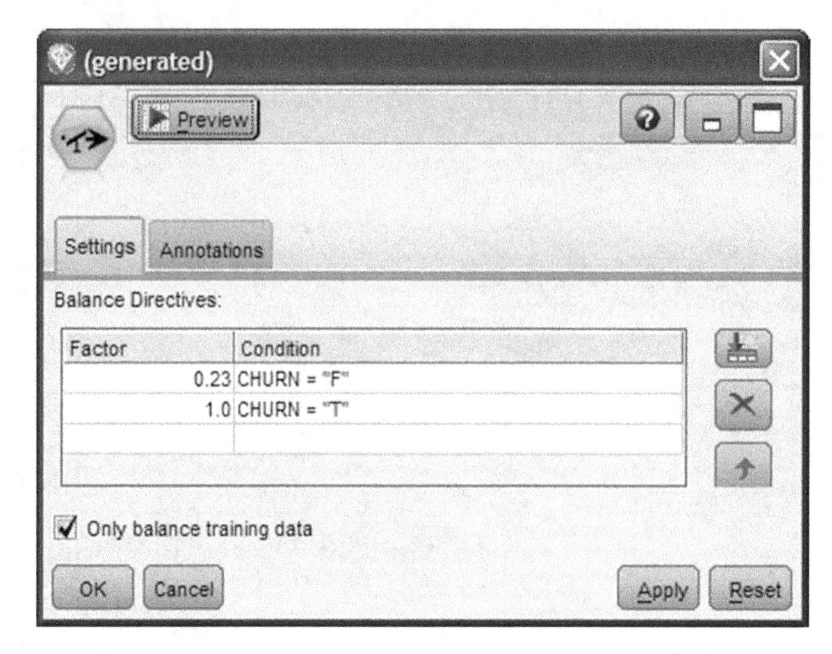

Figure 6.18 Balancing the distribution of the target field

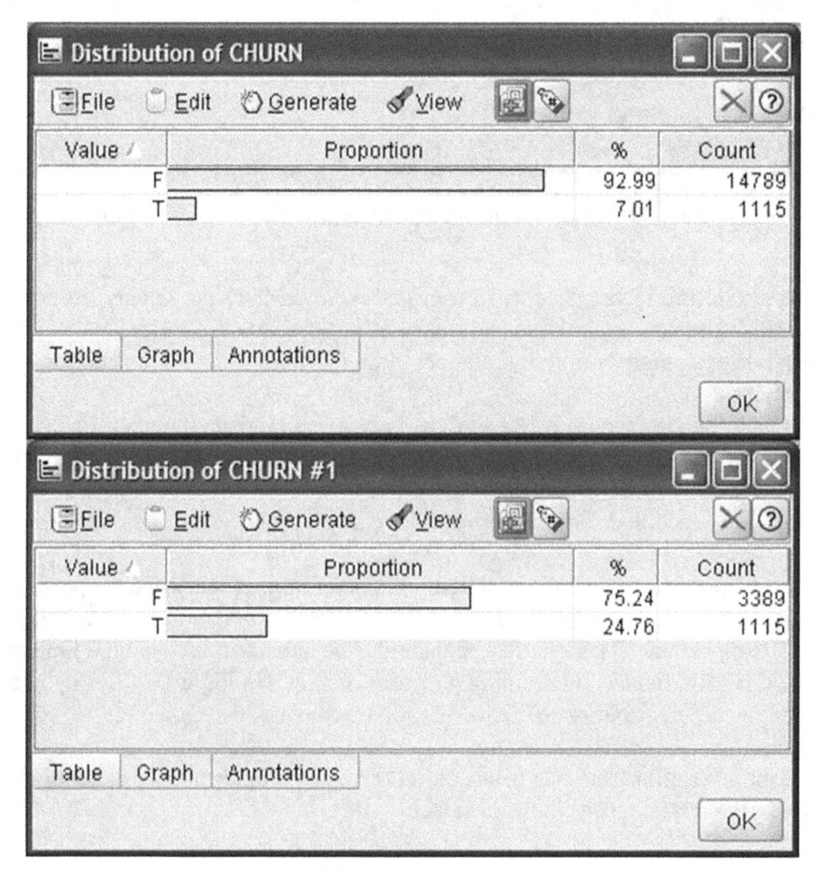

Figure 6.19 The initial and the "balanced" distribution of the CHURN field in the training partition

A Cache was enabled at the Balance node, saving the data that pass through the node in a temporary folder on disk. This option ensured that the Balance sampling was stored and the training partition was not altered with each data pass.

6.6.3 Setting the role of the fields in the model

Each propensity model tries to learn the input data patterns associated with the occurrence of the target event. In Modeler, the role of each field, whether to be used as an input (predict) or as an output (be predicted), is set with a Type node (Figure 6.20).

In our case study, the CHURN field was the target field to be predicted and was set with an output role (Direction Out). The fields denoted as predictors in Table 6.2 were designated as inputs (Direction In). The rest of the fields were set with direction None and were omitted from model training.

6.6.4 Training the churn model

After selecting the predictors, it was time for model training. The Auto-Classifier node enabled the building and the comparison of multiple models in a single modeling run. As shown in Figure 6.21, the first criteria selected for ranking the models was the Lift at the top 5 percentile. Therefore, the Lift of each model, the increase in predictive performance compared

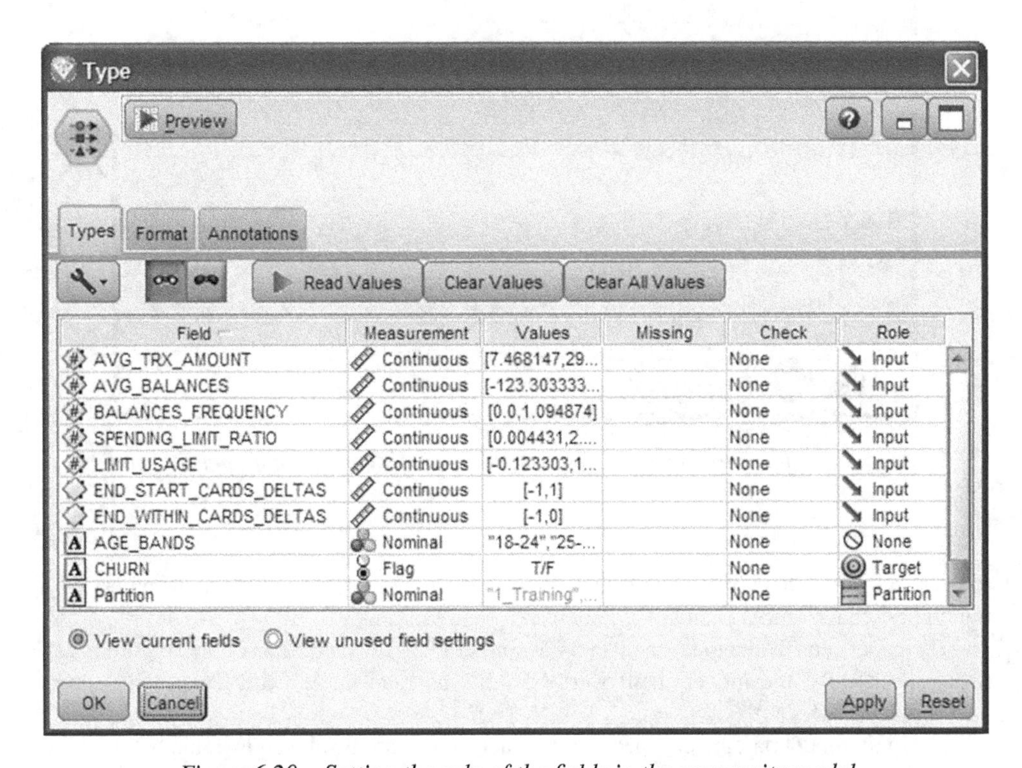

Figure 6.20 Setting the role of the fields in the propensity model

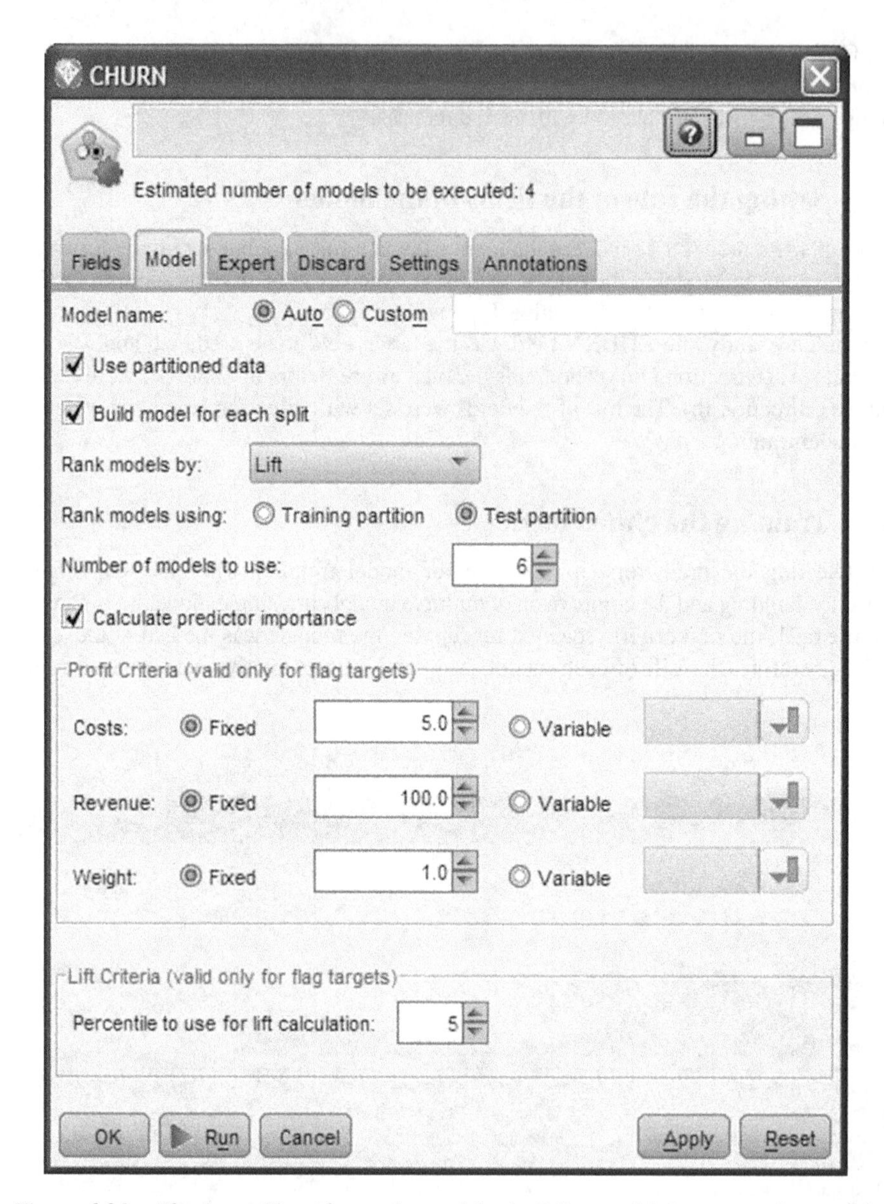

Figure 6.21 The Auto-Classifier node used for building multiple propensity models

to the baseline "model" of random selection, was calculated for the top 5% of customers with the highest churn scores and used for comparing and ranking.

Three Decision Tree models and an SVM model were selected for training (Figure 6.22). More specifically, the models built were CHAID, boosted C5.0, C&R Tree, and support vector machine (SVM).

The main model parameters specified for each model are displayed in Table 6.3.

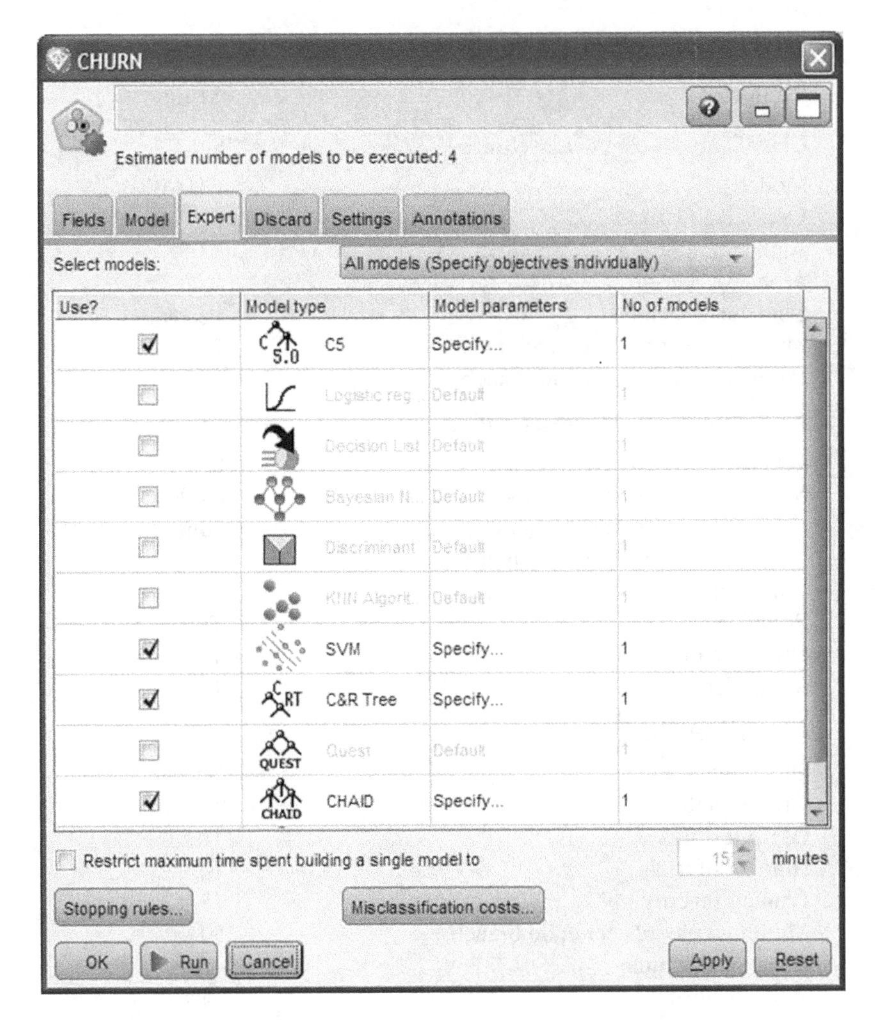

Figure 6.22 The propensity models trained for churn prediction

6.7 Understanding and evaluating the models

By browsing the generated model, we have access (in the Model tab) to an initial output which provides valuable information for assessing the accuracy of the models (Figure 6.23). The performance measures of the candidate models include Lift, Maximum Profit, Overall Accuracy, and Area Under Curve. All these metrics are based on the testing dataset which was not used for the model training and was not balanced.

The CHAID model is ranked first since it yielded the highest Lift (4.3) at the top 5% percentile. Since the overall percentage of churners was about 7%, a Lift of 4 denotes that the concentration of actual churners among the top 5% of customers with the highest churn scores was 4 times higher, about 28%. Thus, a list of the top 5% scored customers is expected to be 4 times denser in churners compared to randomness. The Lifts for the other classifiers were equally good, ranging from 3.8 for C5.0 to 4.2 for C&R Tree. Of course, the

Table 6.3 The churn models parameter settings

Parameter	Setting
CHAID model parameter settings	
Model	CHAID
Levels below root	6
Alpha for splitting	0.05
Alpha for merging	0.05
Chi-square method	Pearson
Minimum records in parent branch	200
Minimum records in child branch	100
C&R Tree model parameter settings	
Levels below root	8
Minimum change in impurity	0.0001
Impurity measure for categorical target	Gini
Minimum records in parent branch	80
Minimum records in child branch	40
Prune tree	True
Use standard error rule	True
Multiplier	1.0
C5.0 model parameter settings	
Output type	Decision Tree
Group symbolics	True
Use Boosting	True
Number of trials	10
Pruning severity	75
Minimum records per child branch	50
Use global pruning	True
Winnow attributes	False
SVM model parameter settings	
Stopping criteria	$1.0e^{-3}$
Regularization parameter (C)	10
Kernel type	RBF
RBF gamma	0.25

concentration of churners is expected to rise even more among the higher percentiles. CHAID and C5.0 performed best in terms of Area Under Curve, followed closely by the rest of the models. The overall conclusion is that all models performed well, presenting comparable evaluation measures. In terms of complexity, the SVM model appears as the most complex model since, due to the lack of an internal pruning mechanism, it had used all predictors. On the other end, CHAID only used 8 of the initial 19 predictors.

The Gains chart, displayed in Figure 6.24, provides a visual comparison of the models' predictive power. It is generated by selecting the option Evaluation chart(s) from the Generate menu of the Model tab.

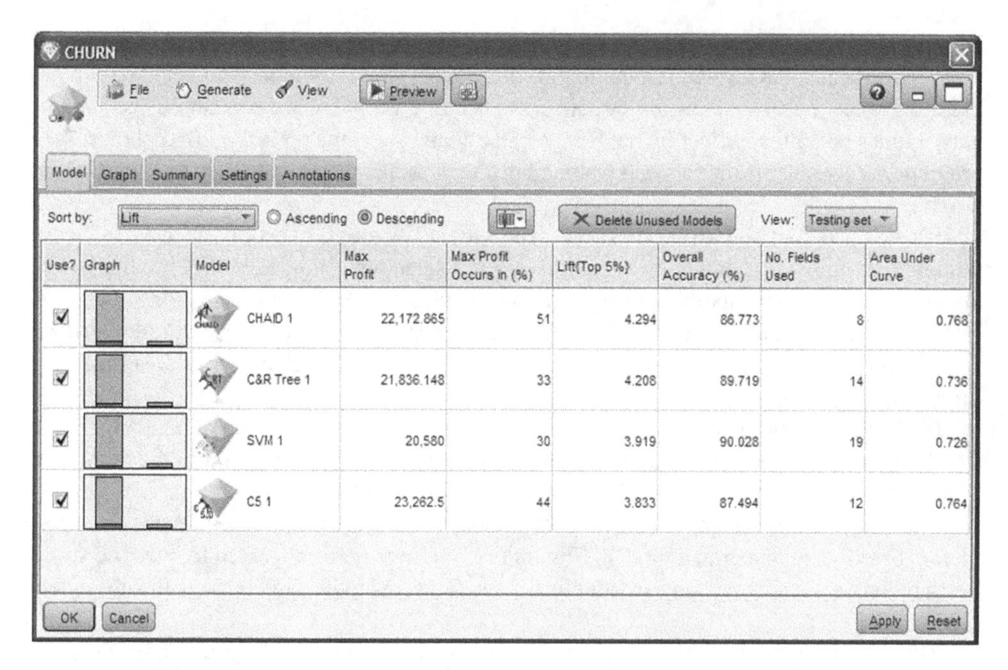

Use?	Graph	Model	Max Profit	Max Profit Occurs in (%)	Lift{Top 5%}	Overall Accuracy (%)	No. Fields Used	Area Under Curve
☑		CHAID 1	22,172.865	51	4.294	86.773	8	0.768
☑		C&R Tree 1	21,836.148	33	4.208	89.719	14	0.736
☑		SVM 1	20,580	30	3.919	90.028	19	0.726
☑		C5 1	23,262.5	44	3.833	87.494	12	0.764

Figure 6.23 Performance metrics for the candidate models

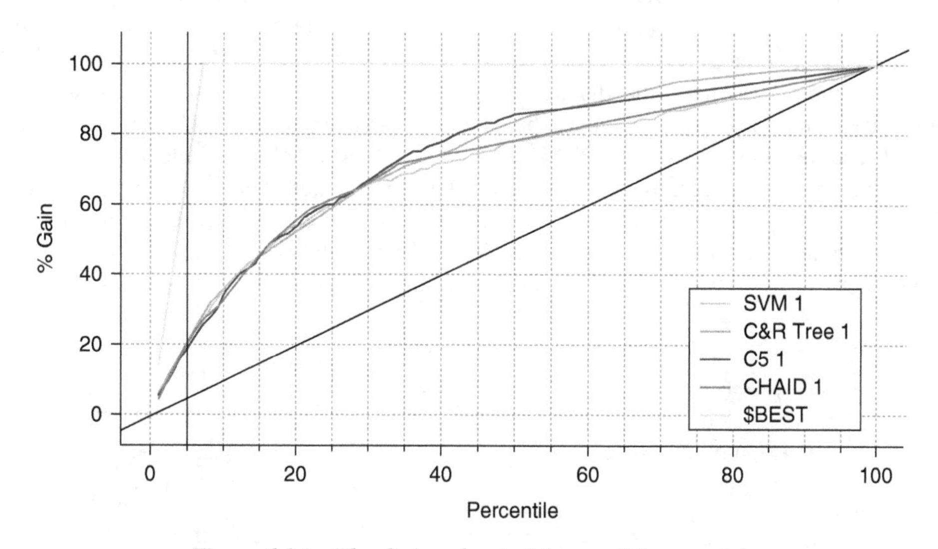

Figure 6.24 The Gains chart of the candidate models

A separate Gains curve is plotted for each model. The x-axis represents the percentiles of customers after being sorted according to their churn scores. The added vertical reference line corresponds to the top 5% percentile. The diagonal line represents the baseline model of randomness. The $BEST line corresponds to a hypothetical ideal model which classifies correctly all cases. A good model is expected to show a gradual but steep incline at the top percentiles before finally converging to the diagonal line of randomness.

As we can see in the graph, the Gain % of all models at the top 5% percentile reaches 20%. This means that a retention campaign list with those customers is expected to capture the 20% of all churners. Thus, by targeting the 5% of customers, we contact 20% of all targets, 4 times better than selecting randomly. Once again, we come upon the Lift of 4 discussed previously. Overall, all models behaved equally well at their top percentiles (left part of the x-axis).

So which model should be used for deployment? A tough question since in this case models are competing head-to-head. But instead of selecting a single classifier, why not use all of them? The analysts of the organization decided to employ an ensemble procedure. Instead of relying on an individual model, they chose to combine the predictions of all models through a voting procedure. The ensemble method selected was average raw propensity.

In the next paragraphs, we'll examine the performance of the ensemble model which was finally used by the bank officers for scoring customers. But before examining the combined model, let's have a look at the results of the individual models.

The Graph tab of the generated CHURN model node displays the importance of the predictors based on all generated models (Figure 6.25). Usage fields appeared to have the strongest predictive power since an early decline in usage seems to be associated with subsequent churn.

The CHAID model with its intuitive results in tree format can help us understand the churn patterns.

In Figure 6.26, the CHAID tree is displayed in tree format. The bar with the darkest shade of gray represents the proportion of churners. The overall percentage of churners in the balanced training dataset was approximately 25%. Note that due to the balancing and the oversampling of churners, the displayed churn rates do not represent the actual churn percentages, but they can nevertheless be examined comparatively in order to interpret the model. The spending frequency (SPENDING_FREQUENCY field) was selected for the first split, and customers were partitioned accordingly. Customers with the lowest spending frequency (below 0.417) landed on Node 1 presenting an increased churn rate of about 61%. Those customers were further divided according to their monthly average number of spending transactions (AVG_TRX field). The churn rate among customers with small number of

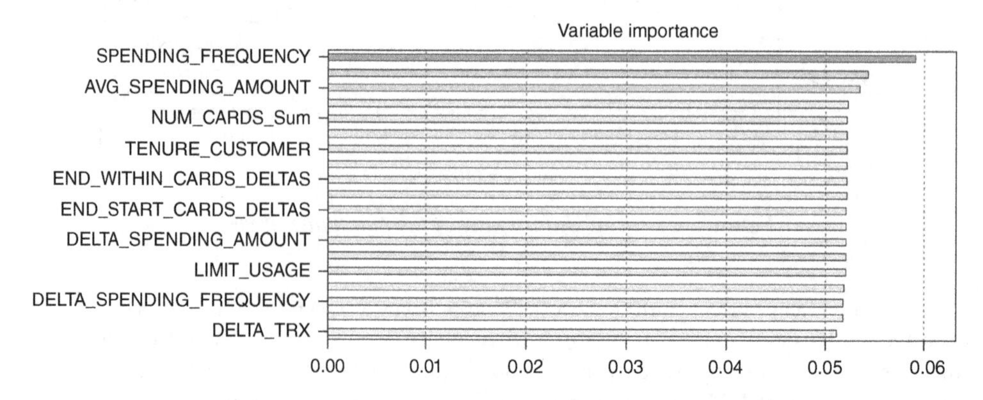

Figure 6.25 The importance of the predictors

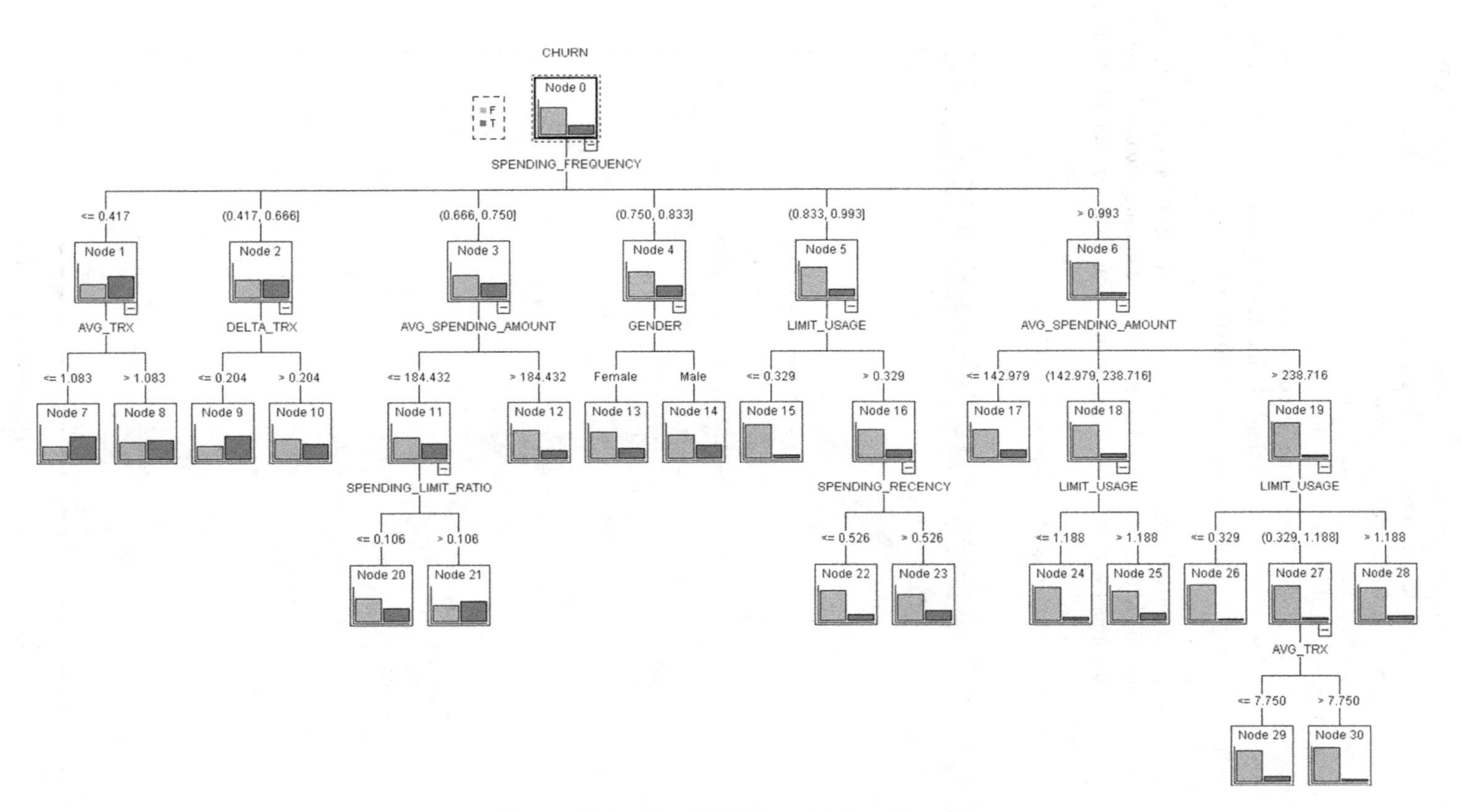

Figure 6.26 The CHAID model (tree format)

transactions (lower than 1 per month) further increased to 65%. These customers landed on Node 7 of the tree and seem to comprise a target group of customers with increased churn propensity. Likewise, by studying Node 9, we see that customers with low spending frequency which was further decreased during the most recent months (DELTA_TRX below 0.2) also had a high proportion of actual churners.

The case of Node 21 is also interesting. It contains customers with low spending in terms of frequency and amount but with relatively increased spending to limit ratio. The observed churn rate was quite high among those customers (about 55%) who may face a problem with their limit.

The right "branches" of the CHAID model were dominated by heavier users and included a limited number of actual churners. The lowest percentage of churners was found in Node 26 which contained customers with relatively high spending on a regular monthly basis and didn't have problems with their limit.

In Figure 6.27, the CHAID model is displayed in rules format.

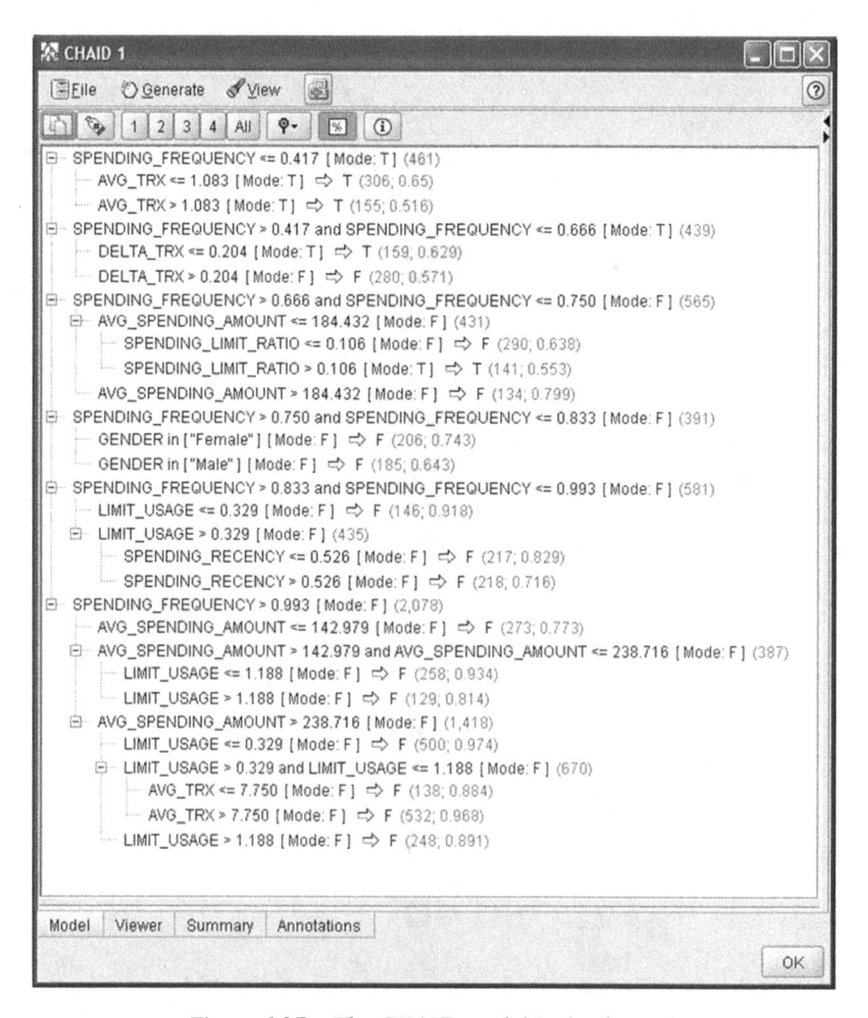

Figure 6.27 The CHAID model (rules format)

Now, let's go back to meta-modeling and examine the combined model produced by the four Decision Trees and the SVM model. The separation of the generated ensemble model (CHURN model) was evaluated with an Evaluation node.

The ensemble model presented an acceptable performance, yielding a Gain of about 21% (Figure 6.28) and a Lift of about 4.3 (Figure 6.29) at the top 5 percentile. Obviously, the Lift was larger at higher cutoffs. The results of the Analysis node, shown in Figure 6.30, present the combined model's Accuracy and Error (misclassification) rate. It also presents the Confusion (Coincidence) matrix which can be used for computing the Sensitivity, Specificity, Precision, and *F*-measures. Due to the imbalanced distribution of the target field, the analysts mainly studied the latter measures instead of the overall accuracy and misclassification rate. The ensemble model yielded an *F*-measure of approximately 30% compared to the 28% of the individual CHAID model. Its AUC (Area under the ROC curve) and Gini mesures were also increased (0.776 and 0.552 respectively).

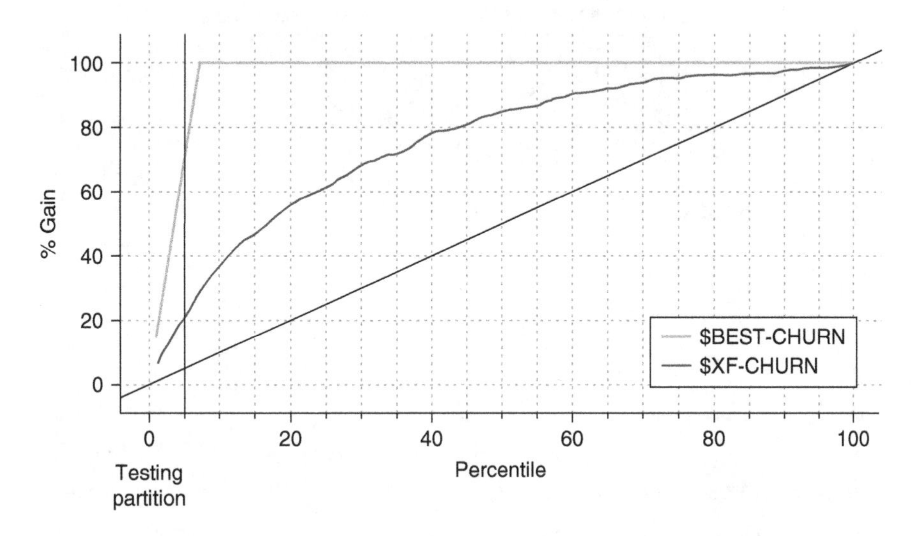

Figure 6.28 The Gains chart of the ensemble model

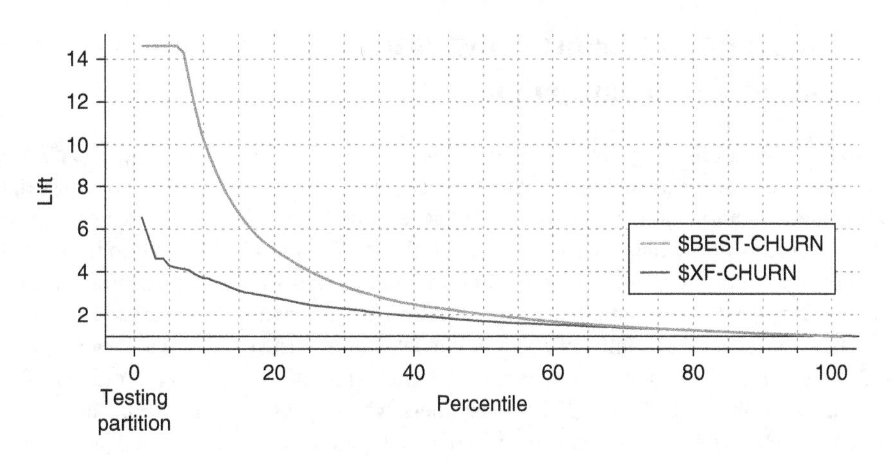

Figure 6.29 The Lift chart of the ensemble model

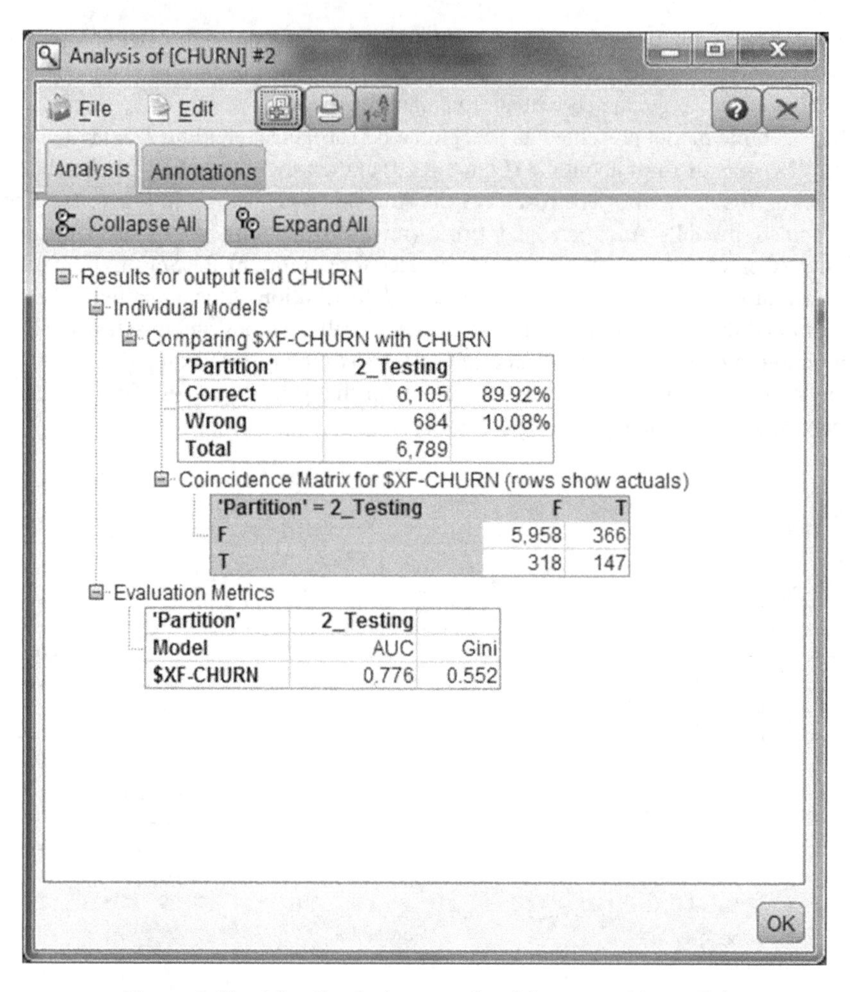

Figure 6.30 The Confusion matrix of the ensemble model

6.8 Model deployment: using churn propensities to target the retention campaign

After training and evaluating the classifiers, it was time for model deployment. The ensemble model was used for scoring and identifying the customers with increased churn likelihood, and the scheduled retention campaign was targeted according to the calculated churn propensities.

The marketers of the bank started the preparation of the retention campaign in July 2013, right after the end of the modeling procedure. In order to refresh the churn scores, the historical data had to be updated, and the data preparation process had to be repeated for currently active customers. Twelve months of observation data were retrieved summarizing usage from July 2012 to June 2013 as shown in Figure 6.31. The retention campaign included all customers active at the end of June 2013. Customers with at least one open card at that point were scored, and their churn propensities were calculated. Churn, was defined as the closing of all cards, and it was predicted 4 months ahead, from September to December 2013.

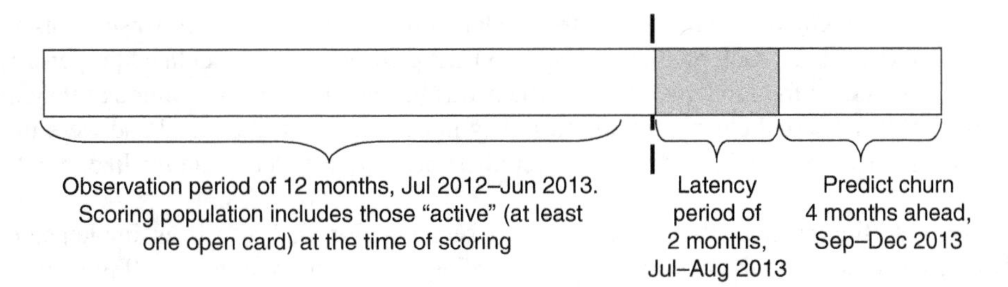

Figure 6.31 The voluntary churn model scoring procedure

Table (3 fields, 11.385 records)

File Edit Generate

	CUSTOMER_ID	$XF-CHURN	$XFRP-CHURN
1	C100002	F	0.085
2	C100004	F	0.096
3	C100006	F	0.101
4	C100007	F	0.246
5	C100010	T	0.772
6	C100012	F	0.264
7	C100013	F	0.076
8	C100014	T	0.515
9	C100018	F	0.152
10	C100022	F	0.147
11	C100023	F	0.138
12	C100027	F	0.072
13	C100028	F	0.141
14	C100029	F	0.139
15	C100032	F	0.099
16	C100034	F	0.092
17	C100035	F	0.233
18	C100037	F	0.097
19	C100038	F	0.316
20	C100039	F	0.103

Table Annotations

OK

Figure 6.32 Scoring active customers and calculating churn propensities

In Modeler, a generated model also serves as a scoring engine which assigns a score at each record passed through it. A sample of scored customers is shown in Figure 6.32. For each customer, two new model-generated fields were derived. The $XF-CHURN field (the $ prefix designates a model-generated field) is the prediction field and denotes the assignment to a CHURN class. Since the model used for scoring was an ensemble model with the average raw propensity method used for voting, the Modeler estimated the churn propensities instead of the confidences for each prediction (field $XFRP-CHURN). The calculated propensities were raw, based on the modeling dataset which was balanced. Therefore, although they did not represent the actual churn probabilities, they indicated churn likelihood and were used as churn scores for the rank ordering of customers according to their relative likelihood to leave.

Using a Binning node, a propensity-based segmentation had also been implemented, dividing customers into three distinct pools according to their churn propensity (Figure 6.33):

1. Top Churn Risk: Top 5% percentile including customers with highest churn propensities

2. High Churn Risk: High 15% percentile including customers with relatively increased churn propensities

3. Medium–Low Churn Risk: Customers with lower churn propensities, belonging in the lowest 70% percentiles

The churn propensity information was then cross-examined with each customer's present value. This combined segmentation scheme which was derived (value-based segments and propensity-based segments) was used to prioritize the retention campaign. The marketing department decided to roll out a small and targeted campaign. Therefore, their final campaign population included customers from the Top Churn Risk and Value segments.

The bank's next steps included the evaluation of the campaign's results in respect to the effectiveness of the retention offer as well as the model's accuracy. Additionally, a separate churn model focused only on the high value segments was scheduled for the near future.

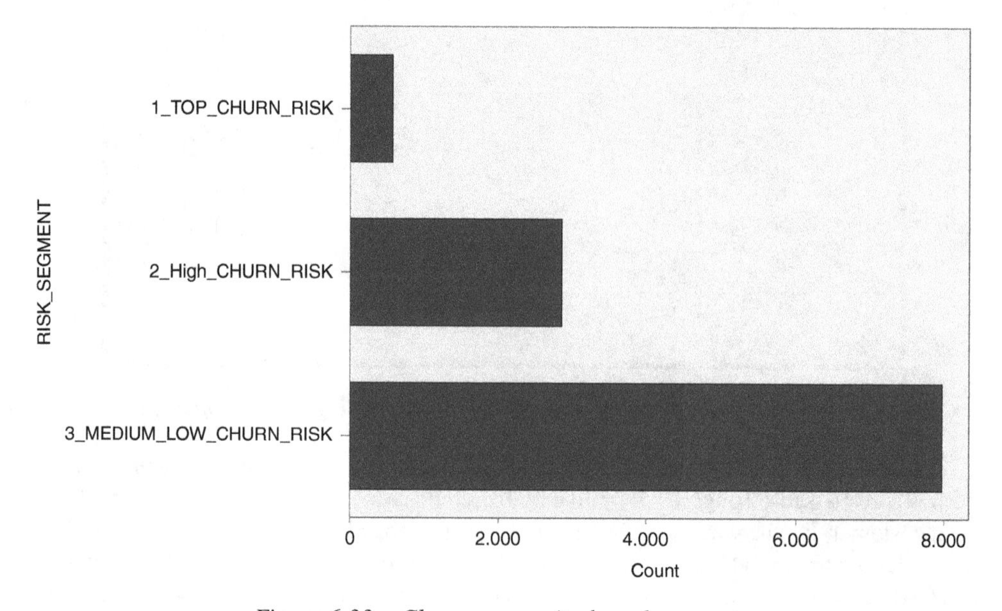

Figure 6.33 Churn propensity-based segments

6.9 The voluntary churn model revisited using RapidMiner

The churn modeling procedure is presented again, this time using RapidMiner's modeling algorithms. The modeling process is shown in Figure 6.34.

6.9.1 Loading the data and setting the roles of the attributes

After loading the modeling dataset, analysts retained in the process only the target field (the label field, in RapidMiner's terminology) and the predictors listed in Table 6.2. A Select Attribute operator was used to keep those fields and filter out the rest which did not contribute to the model.

In the next step of the process, a Set Role operator was used to assign a label (target) role to the CHURN field. The rest of the fields kept their default regular role and participated as predictors in the subsequent propensity model. The Set Role settings are presented in Figure 6.35.

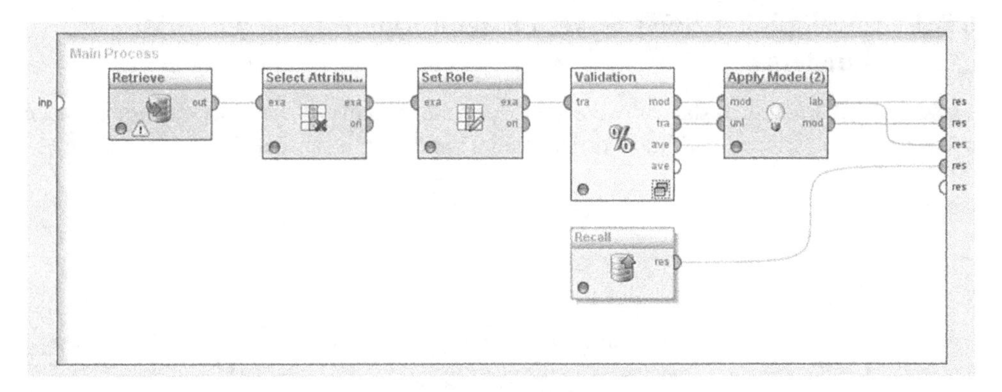

Figure 6.34 The RapidMiner modeling process

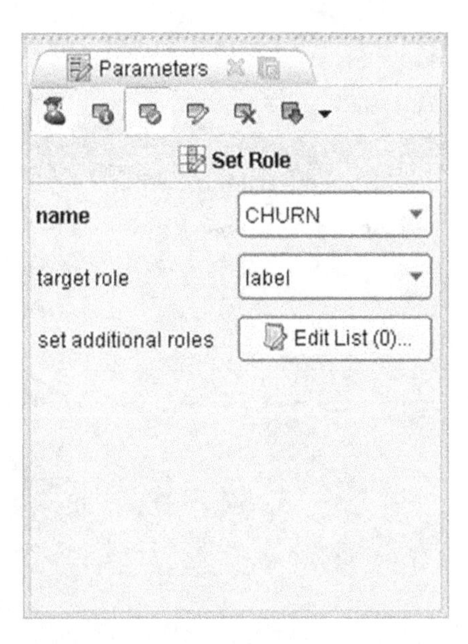

Figure 6.35 The Set Role operator used for defining the role of the attributes in the model

6.9.2 Applying a Split (Holdout) validation and adjusting the imbalance of the target field's distribution

Before training the classification model, a Split (Holdout) validation method has been applied through a Split Validation operator. This operator partitioned the modeling dataset into training and testing parts of 70 and 30%, respectively, as shown in Figure 6.36.

The partitioning was based on random sampling. A split ratio of 0.7 randomly assigned the 70% of the training dataset to the training sample and the remaining records to the testing sample. The Split Validation operator, is comprised of two subprocesses (Figure 6.37).

The left subprocess corresponds to the training sample and covers the model training phase. The right subprocess corresponds to the testing dataset and wraps the evaluation actions.

6.9.3 Developing a Naïve Bayes model for identifying potential churners

The balanced training instances were then fed into a Naïve Bayes model with Laplace correction and the model was trained. The model validation is presented in Section 6.9.4.

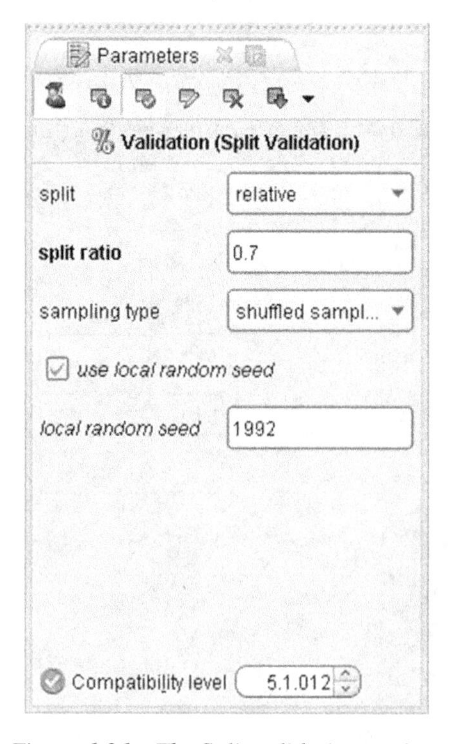

Figure 6.36 The Split validation settings

6.9.4 Evaluating the performance of the model and deploying it to calculate churn propensities

As mentioned previously, the model evaluation procedure is executed in the right subprocess of the Split Validation operator. The subprocess receives the testing sample which is then passed through a Performance (Binomial Classification) operator to calculate a set of performance measures. Evaluation measures require scored cases. Therefore, before the Performance operator, an Apply Model operator has been inserted to classify the unseen records of the testing sample and score (label) the cases according to the generated model.

The Confusion matrix of the model is presented in Figure 6.38. As we can see, 323 churners and 4422 nonchurners were classified correctly (TP, true positives, and TN, true negatives, respectively). The overall accuracy was 69.7%. That means that 69.7% of the testing cases were correctly classified, while the model failed to classify correctly the 30.3% of the cases (misclassification rate). Since the data were imbalanced, it is recommended to focus on more adequate measures such as Sensitivity, Specificity, *F*-measure, and Area under the ROC curve.

The ROC curve plots the model's sensitivity in the *y*-axis against the (1-specificity) values in the *x*-axis, in simple words the trade-off between the true-positive rate (proportion of churners classified correctly) and the false-positive rate (proportion of nonchurners incorrectly labeled as churners) at different propensity cutoffs. The ROC curve of the Naïve Bayes model is depicted in Figure 6.39. It shows a steep increase at the left of the *x*-axis (higher cutoffs) where the curve substantially diverges from the diagonal line of random guessing, indicating

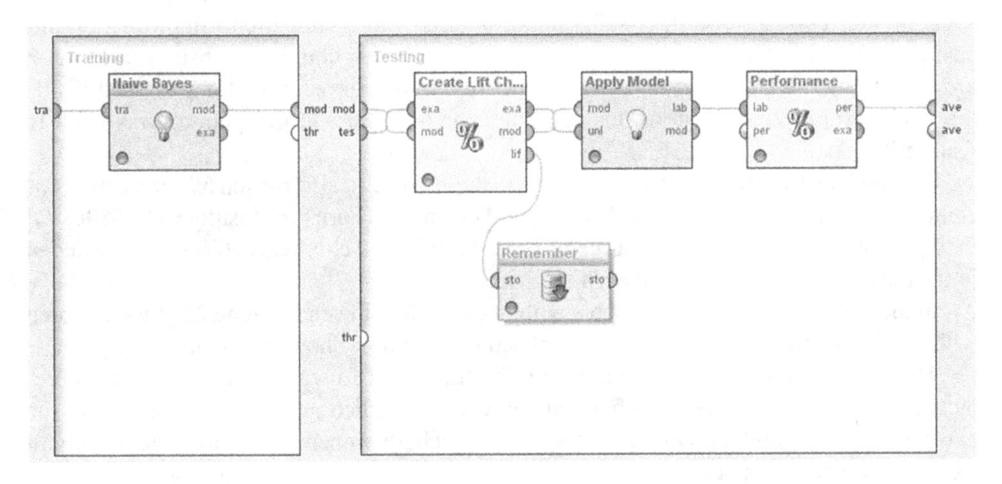

Figure 6.37 The Split Validation operator for partitioning the modeling dataset

Table View Plot View

accuracy: 69.70%

	true F	true T	class precision
pred. F	4422	143	96.87%
pred. T	1920	323	14.40%
class recall	69.73%	69.31%	

Figure 6.38 The RapidMiner model Confusion matrix

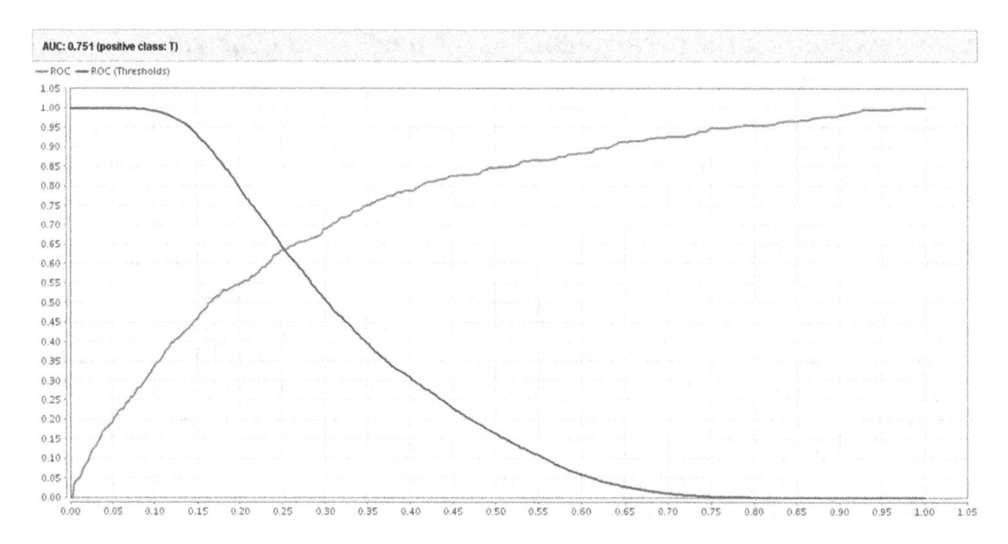

Figure 6.39 The Naïve Bayes model's ROC curve

acceptable model performance. The model presents a decent Area under the ROC curve measure of 0.75 and an F-measure of 23.85%.

The model's Gains chart, created through a Create Lift Chart operator, is displayed in Figure 6.40. It is based on the 6808 validation cases (30% validation sample) and ranks customers into 20 equal-sized bins (tiles) according to their churn propensity (note that in order to apply the Lift chart on the testing cases, a Remember operator was used inside the testing subprocess to store the Lift chart, and then a Recall operator was used to restore it).

The model's Lift was 3.5 at the top 5% percentile. Specifically, the top 5% percentile was comprised of the 341 customers with the highest estimated churn propensities (341/6808). Of those customers, 81 were actual churners. Thus, 17.4% of all churners (81/466) was included in the top 5% percentile, yielding a Lift of 3.5.

In the deployment phase, customers with open cards at the end of June 2013 were scored with an Apply Model operator. A sample of scored records is shown in Figure 6.41.

Three new estimated fields were derived by the model for each case. The model's predicted class (prediction(CHURN) field) along with the prediction confidence for each of the two classes. The confidence(T) field is the estimated churn propensity and indicates the churn likelihood for each customer.

6.10 Developing the churn model with Data Mining for Excel

The next sections present the development of the churn model using the algorithms of Data Mining for Excel. Two models were examined, one using the original data and one after balancing the distribution of the target. Since the two approaches yielded similar results, the first approach was finally selected.

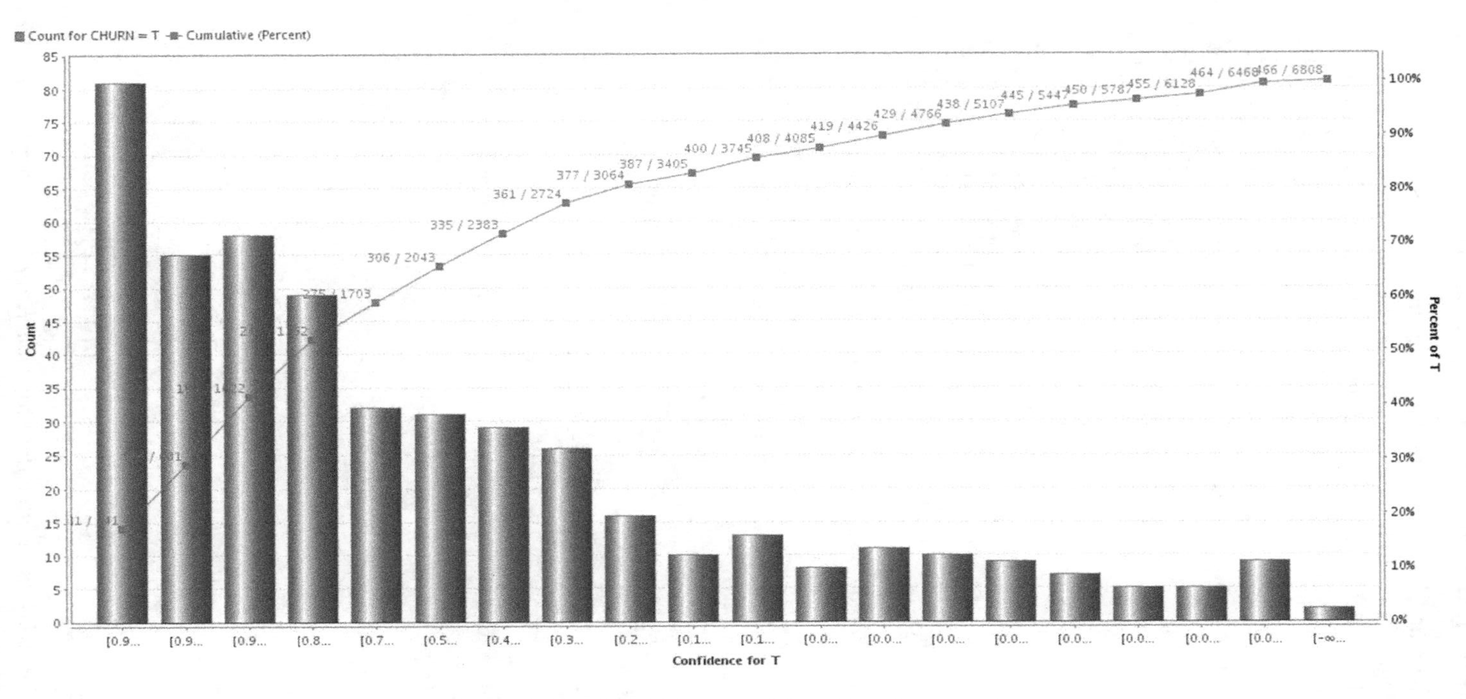

Figure 6.40 The Lift chart of the Naïve Bayes model

6.10.1 Building the model using the Classify Wizard

In the first step of the Classify Wizard, the source training data had been selected and loaded as shown in Figure 6.42. In this example, a specific data range of the active datasheet was specified as the modeling dataset. In general, training data can also be loaded using an external data source and/or an SQL query.

ExampleSet (22693 examples, 4 special attributes, 19 regular attributes) View Filter (22693 / 22693): all

Row No.	CHURN	confidence(F)	confidence(T)	prediction(CHURN)	GENDER	MARITAL_S...	AGE	NUM_CAR...
1	F	0	1	T	Female	Married	33.836	1
2	F	0.927	0.073	F	Female	Married	55.540	1
3	F	0.000	1.000	T	Female	Married	23.937	1
4	F	0.741	0.259	F	Female	Married	24.170	1
5	F	0.083	0.917	T	Male	Single	29.444	1
6	F	0.003	0.997	T	Female	Married	42.419	1
7	F	0.960	0.040	F	Female	Married	27.419	3
8	F	0.688	0.312	F	Female	Married	25.236	1
9	F	0.147	0.853	T	Female	Married	38.395	1
10	F	0.000	1.000	T	Male	Married	25.915	2
11	F	0.996	0.004	F	Male	Married	58.195	3

Figure 6.41 The RapidMiner model predictions and estimated propensities for each customer

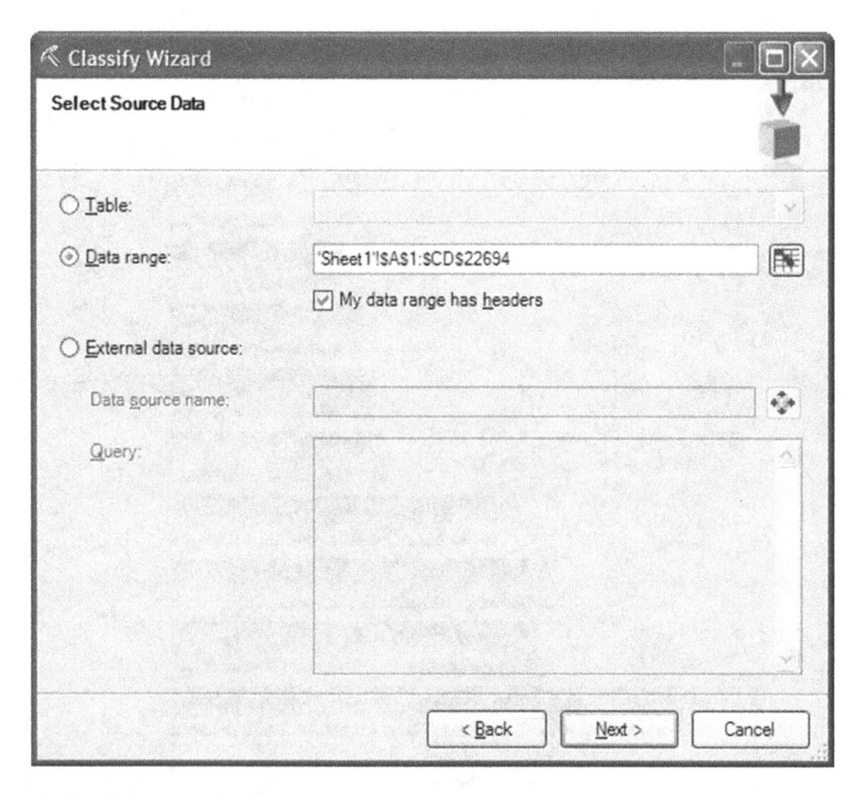

Figure 6.42 Selecting the source data for model training in the Classify Wizard of Data Mining for Excel

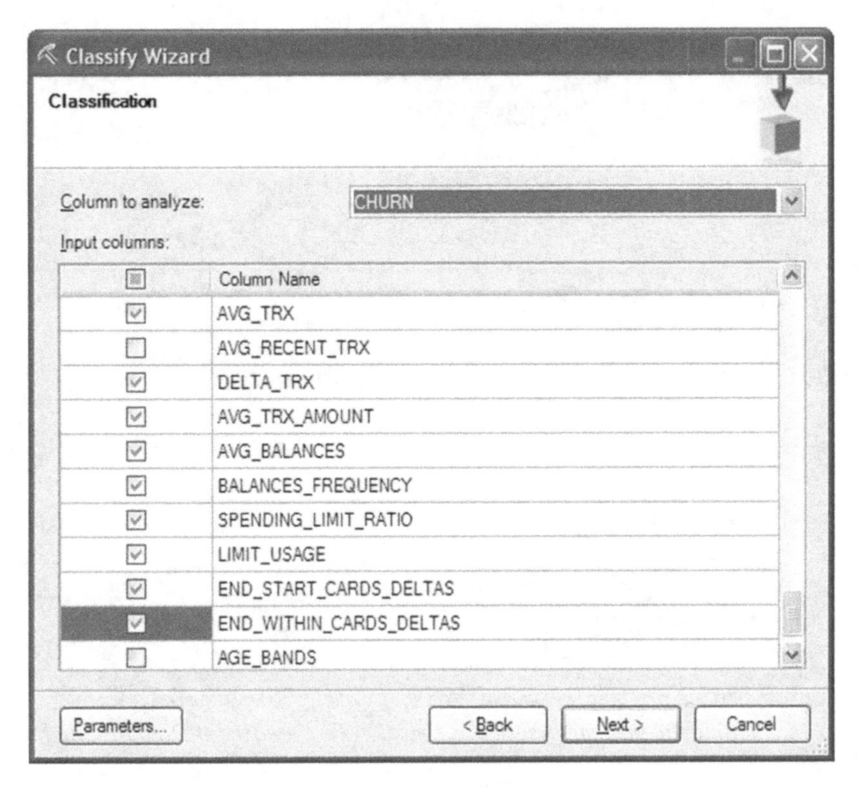

Figure 6.43 Choosing the predictors and the target in the Classify Wizard of Data Mining for Excel

The second step of the wizard involved the selection of the predictors ("Inputs") and the target ("Column to analyze") (Figure 6.43). The complete catalogue of predictors is listed in Table 6.2.

6.10.2 Selecting the classification algorithm and its parameters

After experimentation with different classification algorithms, the modeling team decided to proceed with a Decision Tree algorithm with the default BDE (Bayesian Dirichlet Equivalent with Uniform prior) attribute selection method ("Score method"). The selected algorithm and its parameters were defined in the "Parameters..." menu as displayed in Figure 6.44.

The default "Split method" of "'Both," which optimally combines multiway and binary splits of the predictors, was applied. To obtain a more refined solution and a tree with more levels, the default "Complexity penalty" value was decreased to 0.1. Similarly, the "Minimum Support" value (minimum acceptable size of the leaf nodes) was set to a minimum of 10 instances.

6.10.3 Applying a Split (Holdout) validation

In order to avoid an optimistic validation of the model performance, it was decided to apply a Split (Holdout) validation. A random percentage of 30% of the training instances had been hold out to be used as the testing dataset as shown in Figure 6.45.

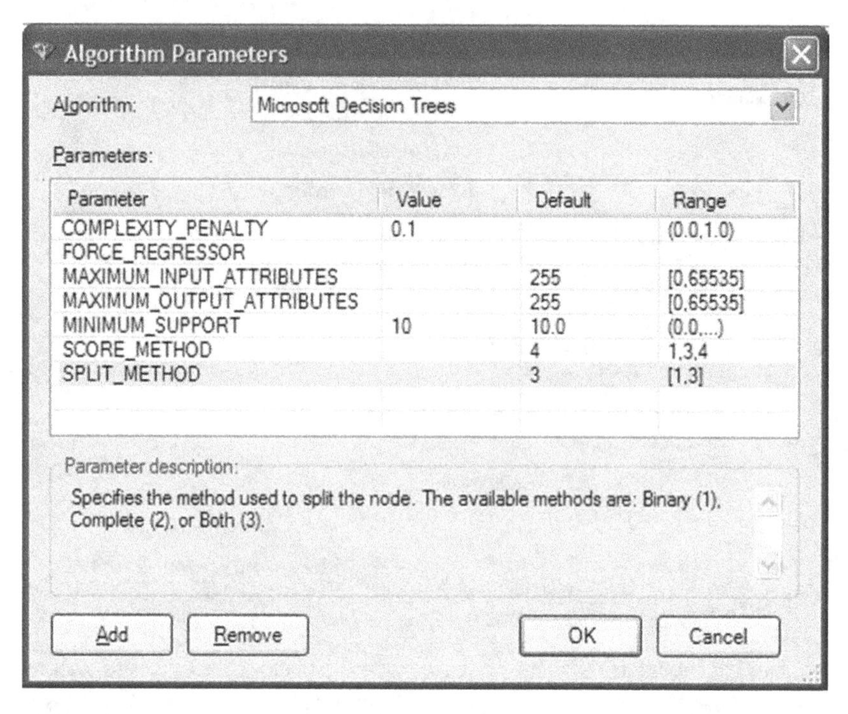

Figure 6.44 Setting the parameters for the Decision Tree models in Data Mining for Excel

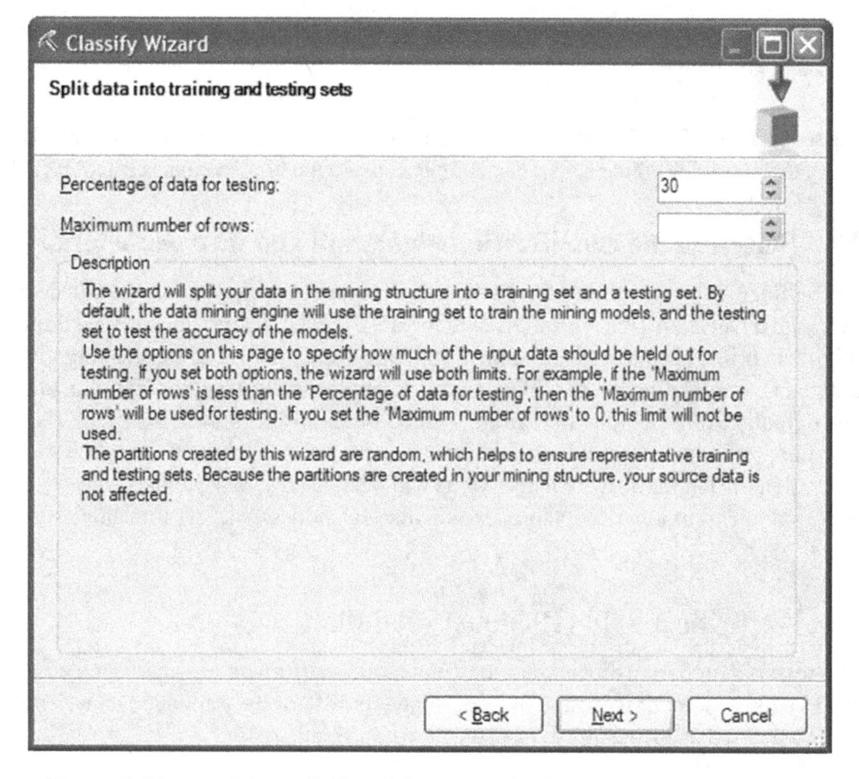

Figure 6.45 Applying a Split validation method in Data Mining for Excel

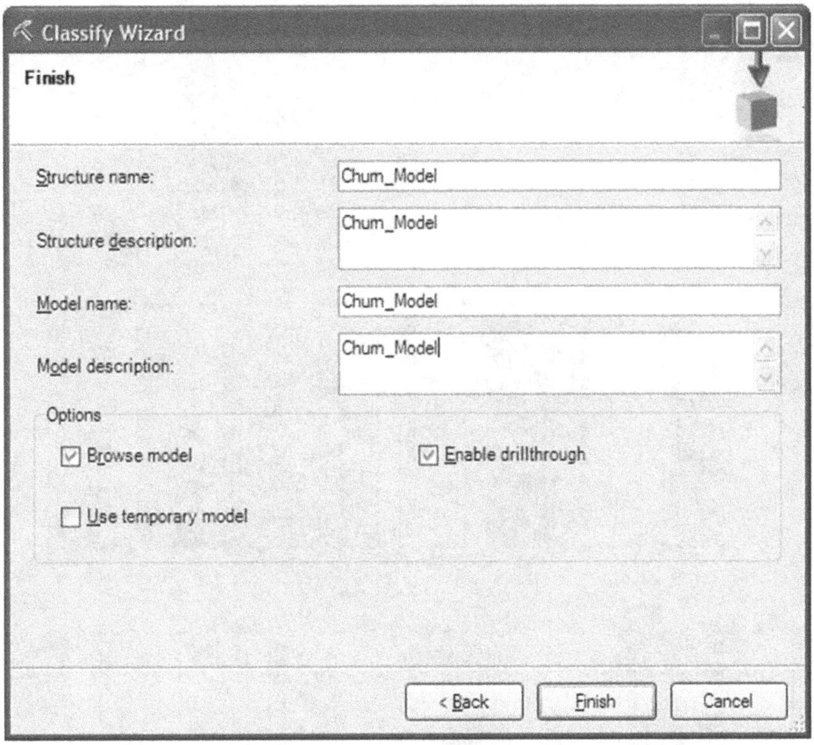

Figure 6.46 Storing the mining structure and the model

In the last step of the wizard, the created mining structure, including the model and the testing (holdout) dataset, was stored as shown in Figure 6.46. The stored testing dataset was subsequently used for model validation.

6.10.4 Browsing the Decision Tree model

The generated Decision Tree model is presented in Figure 6.47. By browsing the tree and by following the recursive partitioning and its branches, from the root down to the leaf nodes, we can gain insight on the data patterns associated with increased churn rate. In each node, a darker background color designates more dense concentrations of churners.

The spending frequency was selected for the first partition. Remember that the same predictor was selected for the first split in the Modeler CHAID tree. By studying the tree, we can infer that churn seems to be associated with low spending frequency and a decrease in transactions.

6.10.5 Validation of the model performance

The model validation was performed using the Classification Matrix Wizard and the Accuracy Chart Wizard of the Data Mining for Excel.

The Classification Matrix wizard evaluates the accuracy and the error rate of the classification model through a Confusion (misclassification) matrix. In the first step of the Classification wizard, the stored BDE Decision Tree model was selected for the validation as shown in Figure 6.48.

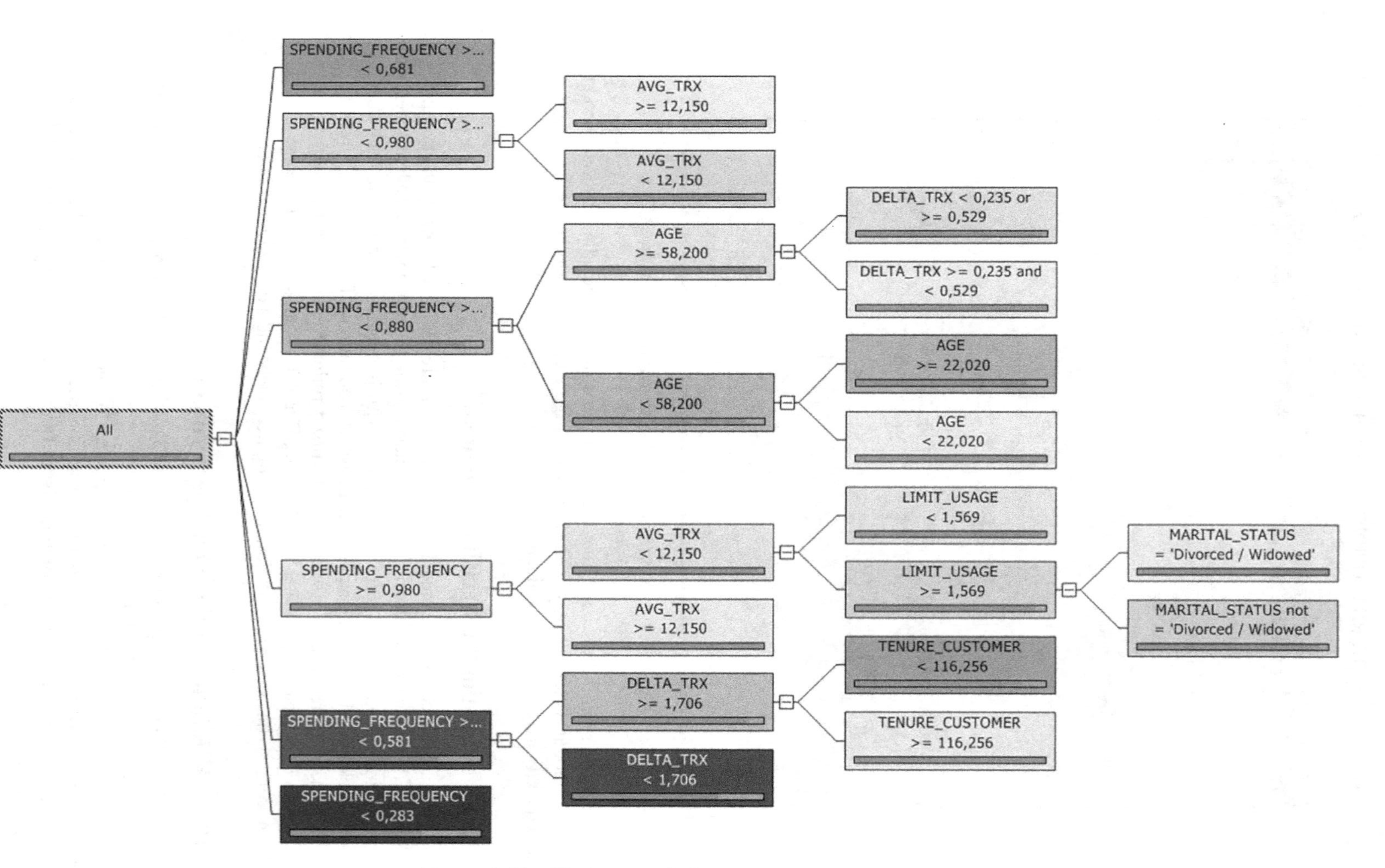

Figure 6.47 The Microsoft decision tree churn model

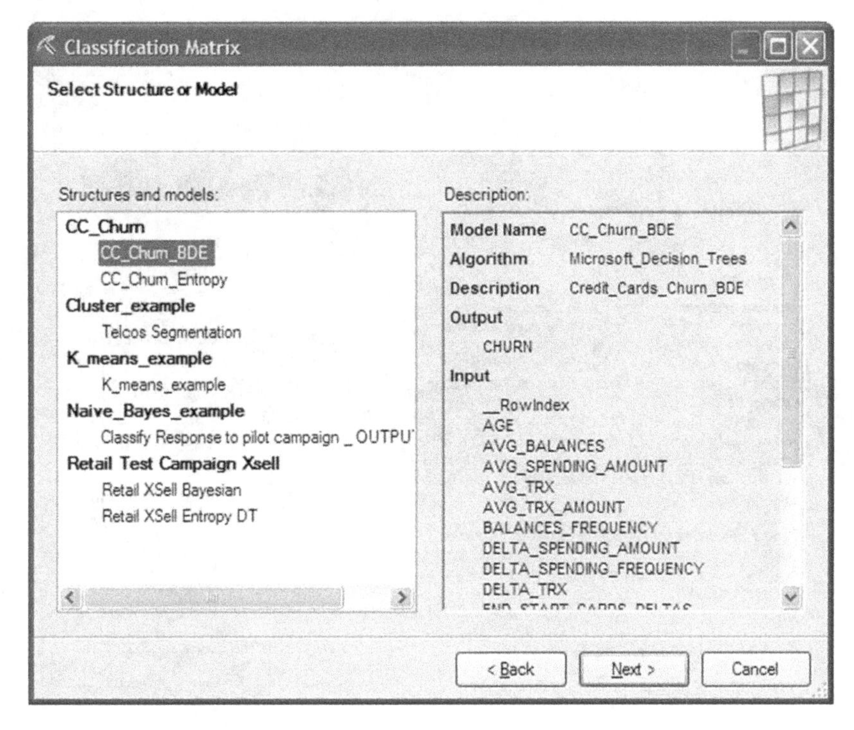

Figure 6.48 Selecting the model to evaluate in Data Mining for Excel

The target field as well as the desired format of the results (in counts and/or percentages) was specified in the second step of the wizard as shown in Figure 6.49.

Obviously, since a Holdout validation technique had been applied, the model was selected to be tested on the testing (holdout) dataset of the mining structure (Figure 6.50).

The accuracy and the error rate of the Decision Tree model are presented in Table 6.4. The model presented an overall accuracy of 93.11% and an error rate of 6.89%.

The detailed Confusion matrix of the model is presented in Table 6.5. The actual classes are listed in the columns of the matrix and are cross-examined with the predicted classes which are listed in the rows of the matrix.

The model classified all actual churners as nonchurners because the estimated churn propensities were lower than the threshold value of 50% for classifying an instance as churner. This was an anticipated result due to the sparse target class and the imbalanced distribution of the target, and it does not mean that the model was not useful. The discrimination effectiveness of the model was the main criterion for the evaluation of its usability, and this effectiveness was examined with a Gains chart, produced through the Accuracy Chart wizard (Figure 6.51).

The Gains chart of the model is displayed in Figure 6.52.

Table 6.6 lists the cumulative Gain %, that is, the cumulative percentage of actual churners, for the top 20 churn propensity percentiles.

A random 5% sample would have included 5% of the total actual churners. After ranking the customers according to their estimated churn propensities, the relevant proportion in the top 5% percentile has been raised to 20.68%, yielding a Lift of 4.14 (20.68/5).

Figure 6.53 presents the cumulative distribution of churners and nonchurners across the estimated propensity percentiles. The maximum separation (equal to the KS statistic) was 41.17%, and it was observed in the 32% percentile.

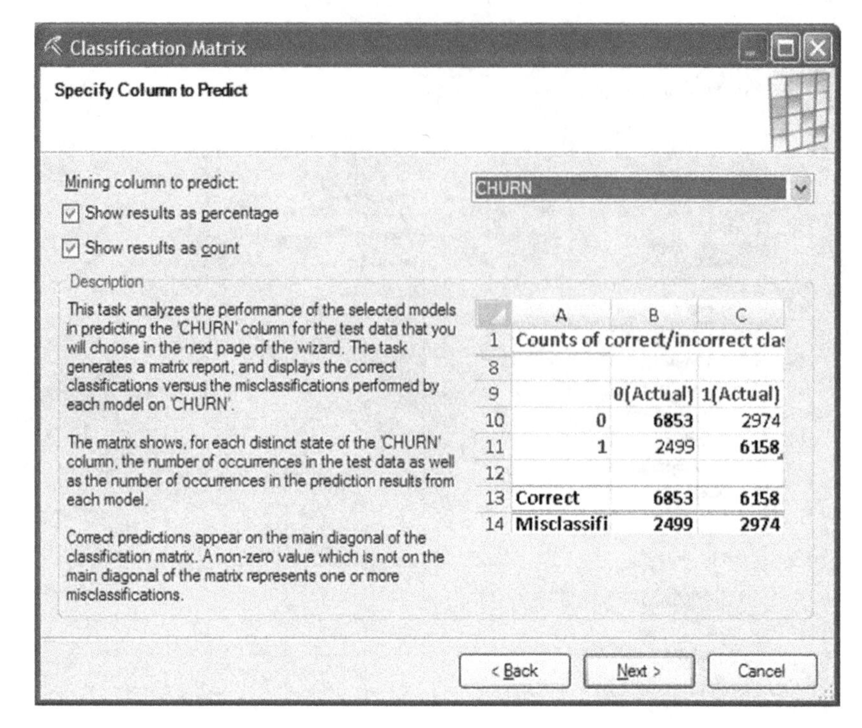

Figure 6.49 Selecting the target field for validation in Data Mining for Excel

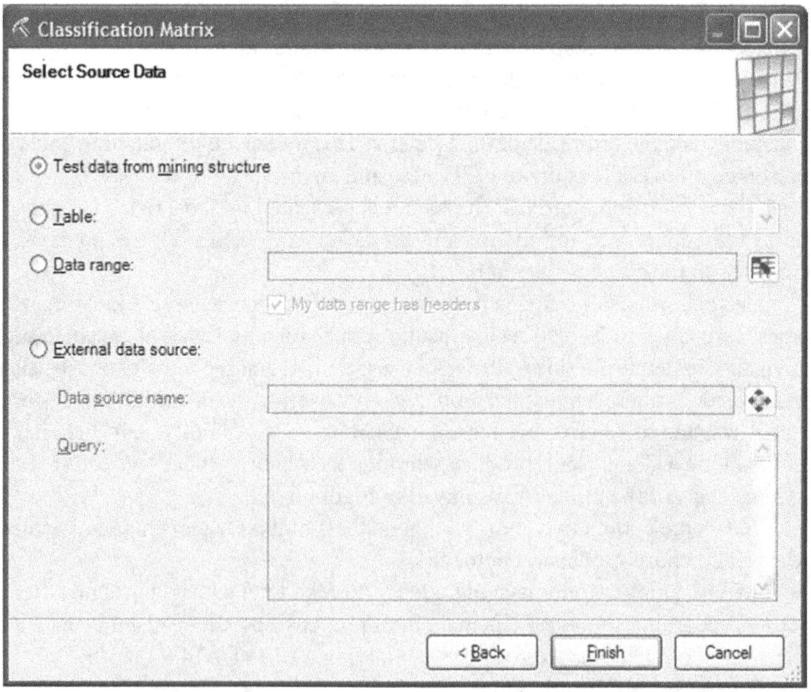

Classification Matrix

Select Source Data

Test data from mining structure

Table:

Data range:

My data range has headers

External data source:

Data source name:

Query:

< Back   Finish   Cancel

Figure 6.50 Selecting the validation dataset in the Classification Matrix wizard of Data Mining for Excel

Table 6.4 The accuracy and the error rate of the churn Decision Tree model

Model name	CC_Churn_BDE	CC_Churn_BDE
Total correct	93.11%	6338
Total misclassified	6.89%	469

Table 6.5 The Confusion matrix of the Decision Tree model

Results as counts for model "CC_Churn_BDE"

	F(Actual)	*T*(Actual)
F	6338	469
T	0	0
Correct	6338	0
Misclassified	0	469

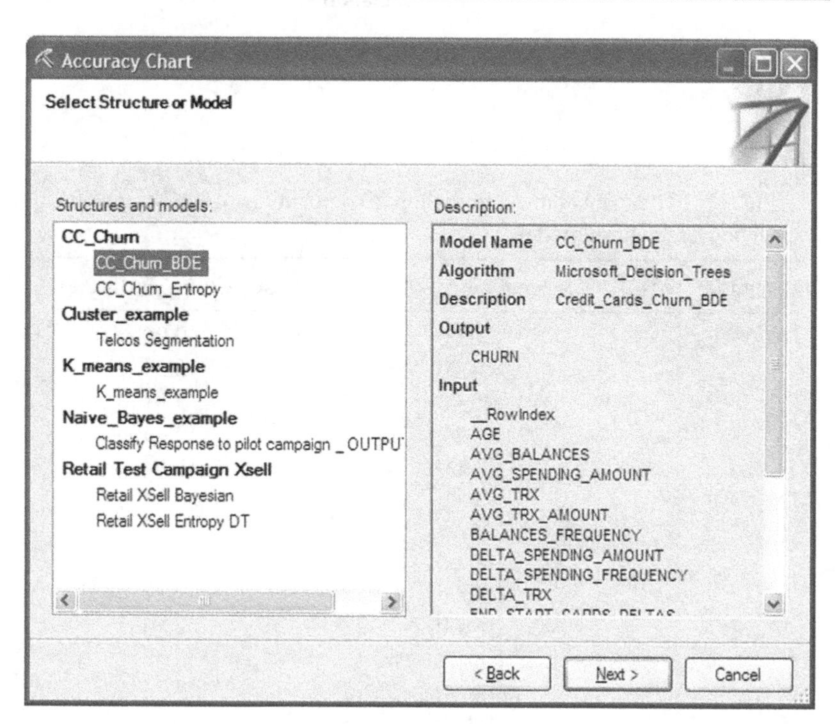

Figure 6.51 Requesting a Gains chart through the Accuracy Chart wizard of Data Mining for Excel

6.10.6 Model deployment

In the deployment phase, the "Query" wizard was used to deploy the stored and validated model on the scoring dataset which contained customers active at the time of scoring. The model estimates which were selected to be produced included the predicted class (based on the 50% propensity threshold), the prediction confidence (the estimated probability of the prediction), and the churn propensity (the estimated probability of churn) as shown in Figure 6.54.

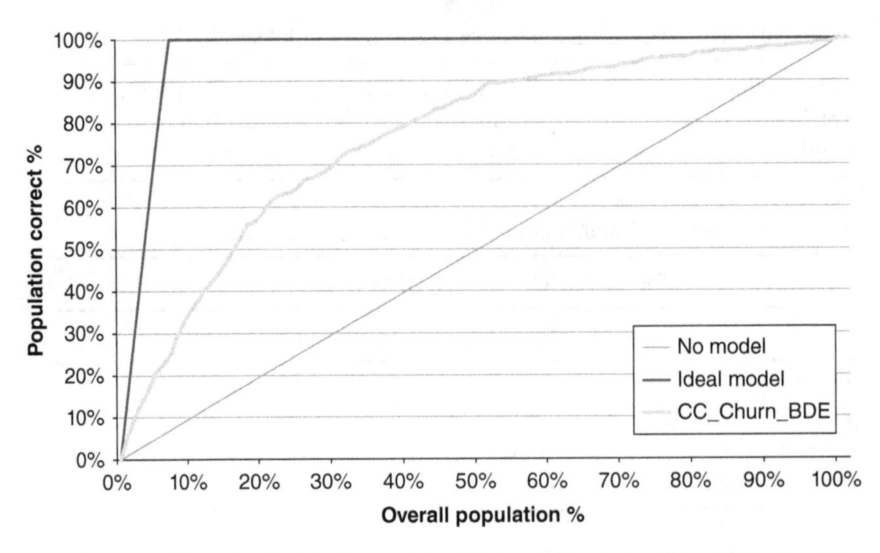

Figure 6.52 The Gains chart of the Microsoft Decision Tree churn model

Table 6.6 The Gains table and the top 20 percentiles of the Microsoft Decision Tree churn model

Percentile (%)	Ideal model (%)	CC_Churn_BDE (%)
0	0.00	0.00
1	14.50	5.54
2	29.00	9.81
3	43.50	13.43
4	58.00	16.63
5	72.49	20.68
6	86.99	22.60
7	100.00	24.95
8	100.00	29.21
9	100.00	32.84
10	100.00	35.39
11	100.00	37.74
12	100.00	40.30
13	100.00	42.22
14	100.00	44.35
15	100.00	46.91
16	100.00	49.68
17	100.00	52.67
18	100.00	55.86
19	100.00	56.50
20	100.00	58.21

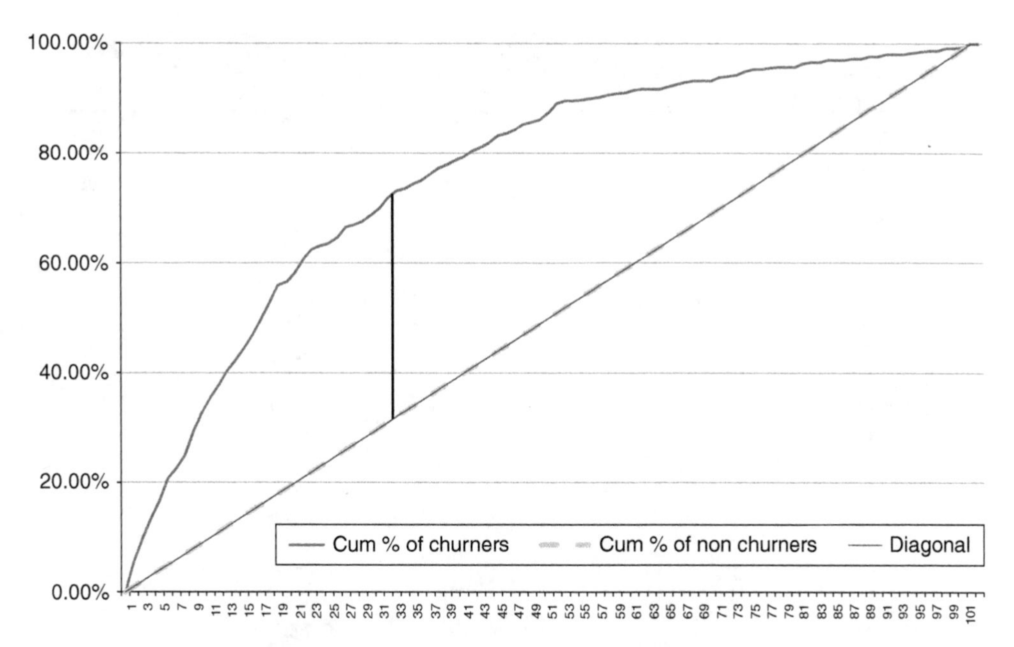

Figure 6.53 The cumulative distribution of churners and nonchurners across the propensity percentiles

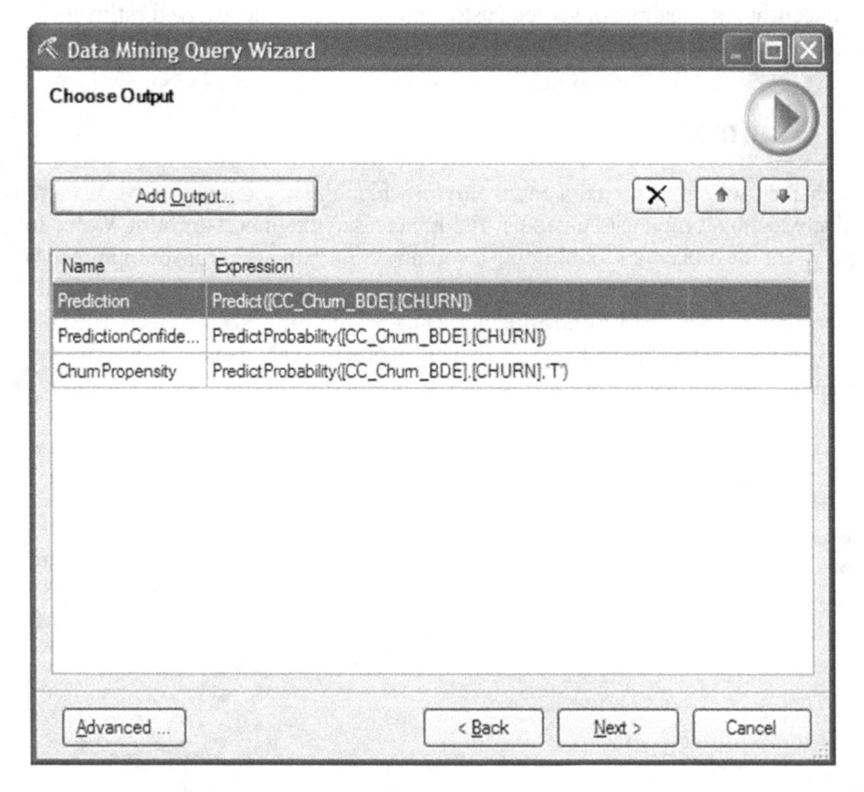

Figure 6.54 Using the Query wizard to deploy the model and score customers in Data Mining for Excel

CUSTOMER_ID	Prediction	PredictionConfidence	ChurnPropensity
C100001	F	0.758	0.242
C100002	F	0.975	0.025
C100003	F	0.894	0.106
C100004	F	0.975	0.025
C100005	F	0.861	0.139
C100006	F	0.949	0.051
C100007	F	0.894	0.106
C100008	F	0.975	0.025
C100009	F	0.894	0.106
C100010	F	0.758	0.242
C100011	F	0.941	0.059
C100012	F	0.894	0.106
C100013	F	0.989	0.011
C100014	F	0.861	0.139
C100015	F	0.758	0.242
C100017	F	0.975	0.025
C100018	F	0.975	0.025
C100019	F	0.975	0.025
C100020	F	0.894	0.106

Figure 6.55 The model estimates and the scored file

A screenshot of a sample of scored customers and their model derived estimates is shown in Figure 6.55.

6.11 Summary

In this chapter, we've presented a case study in which the marketers of a bank attempted to proactively identify credit card customers with increased likelihood to churn. We've followed all the steps of the procedure for building a voluntary churn model, from defining the business objective and the mining approach to data preparation and building, evaluating, and deploying the model. We've used IBM SPSS Modeler, Data Mining for Excel, as well as RapidMiner, and we've trained and evaluated various models in order to calculate the churn propensity for each customer.

7

Value segmentation and cross-selling in retail

7.1 The business background and objective

In retail, competition is hard. The lack of a formal commitment and the easiness with which shoppers can switch to competitors make the process of building a loyal customer base tougher for retailers. Quality of offered commodities and competitive pricing are often not enough for standing out from the competition. In such a dynamic environment, a competitive enterprise should try to understand its customers by gaining insight on their attitudes and behaviors.

To achieve this, retailers have to collect and analyze usage data. Fortunately, this type of information is usually stored at a very detailed level at the point of sales. Transactional data which typically record the detailed information of every transaction are logged with every purchase. A prerequisite for summarizing transactional information at a customer level and for tracking the purchase history of each customer is that every transaction is identified to a customer. This issue is usually tackled with the introduction of a loyalty scheme which assigns a unique identification number, card ID, to each loyalty card. Customers use their loyalty cards for their purchases, and this way, the anonymous transactional records are associated with a specific card ID and a customer.

In this chapter, we'll focus on the efforts of a retailer (chain of department stores) to analyze purchases' data and gain insight on its customers. More specifically, the marketers and the analysts of the retailer decided to explore raw transactional data in order to:

- Segment the customer base according to their value (purchase amount) to discern high value from medium- and low-value customers and incorporate prioritization strategies for handling each customer accordingly.

Effective CRM using Predictive Analytics, First Edition. Antonios Chorianopoulos.

Companion website: www.wiley.com/go/chorianopoulos/effective_crm

- Go a step ahead and apply RFM cell segmentation. Combine the amount of purchases with frequency and recency of transactions (RFM analysis) to gain a 360° view of the customer relationship with the enterprise.
- Use the above information along with other demographical and behavioral data to promote their new department of House and furniture products. They've scheduled a cross-selling campaign for their new department, and they decided to build a relevant model to better target their campaign. In the past, their cross-selling lists were only based on RFM cells. They recorded the response rates of RFM cells in previous campaigns and targeted the best-performing cells. This time, they've planned to train a propensity model using all the available information in order to refine their campaign lists and increase their response rates.

7.2 An outline of the data preparation procedure

Five months of transactional data was retrieved for the needs of the mining applications planned (value-based segmentation, RFM cell segmentation, cross-selling model) covering the 5 first months of 2012. The original data, as we'll see in the next paragraphs, contained multiple lines per order transaction, recording the specifics of each transaction: transaction date, invoice number, store ID, payment type, amount, and item code. However, customer-level data were required for all mining applications. Therefore, through extensive data preparation, the detailed raw information had to be transformed and aggregated to summarize the purchase habits of each cardholder, providing a customer "signature," a unified view of the customer. The usage aspects summarized at a customer level included:

- The frequency and recency of purchases
- The total spending amount
- Relative spending amount per product group
- Size of basket (average spending amount per transaction)
- Preferred payment method
- Preferred period (day/time) of purchases
- Preferred store
- Tenure of each customer

Table 7.1 presents the initial data for two fictional cardholders (Card IDs C1, C2) and three purchase transactions (Invoices INV1, INV2, INV3). Each transactional record (line) contains information about, among other things, the invoice's number, the date and time of transaction, the code of the product purchased, and the paid amount.

The data preparation procedure is presented in detail in the next paragraphs. Its main steps are outlined below.

Table 7.1 Initial transactional data retrieved

Card id	Invoice num	Transaction date	Item code	Amount ($)
C1	INV1	2012-2-20 09:10	PR1001	20
C1	INV1	2012-2-20 09:10	PR2002	58
C1	INV2	2012-3-17 18:12	PR1002	110
C1	INV2	2012-3-17 18:12	PR1003	20
C2	INV3	2012-4-17 19:17	PR2001	58
...				

Step 1. Pivoting the transactional data

Before aggregating the detailed information to summarize customer behavior, records were categorized according to the product group of the purchased item, the time zone (morning–afternoon), and the day (weekdays–weekends) of transaction. The data had then been pivoted and new fields were generated denoting spent amount for each category. This approach enables to capture relative spending (per product group and date/time zone) with a single subsequent aggregation command. Table 7.2 presents the pivoting for the transactional records of Table 7.1. We assume that the third character of each product item code represents its product group. The time zone grouping discerns morning transactions which took place from 09:00 to 16:00, from afternoon transactions, done after 16:00. The second transaction of cardholder C1 (INV2) had occurred on Saturday, 17th of March 2012; the rest of transactions took place on weekdays.

Step 2. Aggregating at a transaction level

Individual records were then aggregated (grouped) per transaction (invoice number, since in this example we consider each order's invoice as a separate transaction) as shown in Table 7.3. This intermediate aggregation was applied in order to measure the number of transactions per customer at the subsequent customer-level aggregation.

Step 3. Aggregating at a customer level

The next step involved the grouping of records at a customer level as shown in Table 7.4. After this final aggregation, input records conveyed the overall purchase behavior of each cardholder: the total number of transactions (invoices, orders), the most recent date of transaction, necessary for computing the R component of the RFM cell segmentation, the total paid amount and the relative spending per product group, and day/time zone.

Was the data preparation phase over after this step? Not yet, the final customer "signature" file had to be enriched with derived KPIs.

Table 7.2 Pivoting transactional data to generate new fields per product group, day and time zone of transaction

				Amount ($)					
Card id	Invoice num	Transaction date	Item code	Product group pr1	Product group pr2	Morning (09:00–16:00)	Afternoon (16:00–)	Weekdays	Weekends
C1	INV1	2012-2-20 09:10	PR1001	20		20		20	
C1	INV1	2012-2-20 09:10	PR2002		58	58		58	
C1	INV2	2012-3-17 18:12	PR1002	110			110		110
C1	INV2	2012-3-17 18:12	PR1003	20			20		20
C2	INV3	2012-4-17 19:17	PR2001		58		58	58	
…									

Table 7.3 Aggregating at a transaction (invoice) level

			Amount ($)					
Card id	Invoice num	Transaction date	Product group pr1	Product group pr2	Morning (09:00–16:00)	Afternoon (16:00–)	Weekdays	Weekends
C1	INV1	2012-2-20 09:10	20	58	78		78	
C1	INV2	2012-3-17 18:12	130			130		130
C2	INV3	2012-4-17 19:17		58		58	58	
…								

Table 7.4 Aggregating at a customer (Card ID) level

		Amount ($)						
Card id	Transaction date_max	Product group pr1	Product group pr2	Morning (09:00–16:00)	Afternoon (1600–)	Weekdays	Weekends	Number of transactions
C1	2012-3-17 18:12	150	58	78	130	78	130	2
C2	2012-4-17 19:17		58		58	58		1
…								

Step 4. Deriving new fields to enrich customer information

Customer-level data were then enriched with informative attributes such as:

- Monthly average purchase amount, in total and per product group
- Monthly average number of transactions
- Time since last transaction
- Basket size: average transaction amount per transaction
- Relative spending (% of spending) per product group
- Total number of distinct product groups with purchases
- Relative spending (% of spending) per day/time zone

The constructed attributes were used as inputs in the mining applications which followed.

7.3 The data dictionary

In the next paragraphs, we'll examine in detail the data preparation procedure with the use of IBM SPSS Modeler. The data dictionary of the transactional input data is listed in Table 7.5.

Note: only purchase data of cardholders (transactions associated with a specific card ID) were retained and used in the applications. Unmatched purchase data not associated with customers were discarded. After all, only registered customers could be monitored, addressed, and handled according to the results of the marketing applications planned.

The demographics of the cardholders, collected upon application for the loyalty card, were also retrieved and merged with the transactional data to enrich the profile of each customer. These data are listed in Table 7.6.

7.4 The data preparation procedure

The data preparation process outlined in Section 7.2 is presented in detail below using the IBM SPSS Modeler data management functions. Those interested only for the modeling applications could skip to Section 7.5 (although it is strongly recommended to follow the data transformations applied).

7.4.1 Pivoting and aggregating transactional data at a customer level

The first steps of the data preparation process included a series of IBM SPSS Modeler Derive nodes for the construction of new attributes denoting categorization information for each transaction. More specifically, each transactional record was decoded and categorized according to date/time zone (weekday–weekend, morning–afternoon), product group, store, and payment type. Although in the real world product grouping would have been done by mapping the Universal Product Code (UPC) of each purchased item with a reference (lookup table) containing the product grouping hierarchy, for simplification, we assume in our case study that the first two characters of the ITEM_CODE field indicate the relevant product group.

Table 7.5 The transactional input data

Data file: RETAIL_DP_DATA.TXT	
Field name	Description
CARD_ID	The ID of the customer loyalty card
INVOICE_NO	The invoice number of the order transaction
TRANS_DATE	The date/time of the transaction
STORE	The store code in which the transaction took place
ITEM_CODE	The item code of the purchased product
PAYMENT_TYPE	The payment type (card/cash)
AMOUNT	The amount for the item purchased

Table 7.6 The demographical input data

Data file: RETAIL_DEMOGRAPHICS_DATA.txt	
Field name	Description
CARD_ID	The ID of the customer's loyalty card. To be used as the key field for joining demographics with purchase data
GENDER	Gender of customer
REGISTRATION_DATE	The registration date, the date when customer applied for the loyalty card
CHILDREN	Number of children of cardholder, when available
BIRTH_DATE	Cardholder's birth date

Comparing customer behaviors in different date/time periods

Retailers are usually interested in comparing customer behaviors in different date/time periods, for instance, in sales periods, in holidays, in periods of special offers, etc. To achieve this, an approach similar to the one described here could be followed: mapping each order to a period based on date/time information of the transaction, pivoting and aggregation of the relevant information to monitor each customer's behavior in the period of interest.

At first, the weekday information was extracted from the transaction's timestamp field (TRANS_DATE) using the Modeler datetime_weekday() function. Then, with the datetime_hour() function, the transaction hour has been extracted and grouped into time zones. This was achieved by using a Modeler Flag Derive node as shown in Figure 7.1. All transactions that took place before 4 p.m. were flagged as morning transactions and the rest as afternoon.

After recoding the information, it was time for pivoting the data to summarize the purchase amount (field AMOUNT) per product group, date/time zone, payment type (card/cash), and store. A series of Restructure nodes were used which created numerical fields based on the values of the grouping attributes.

The Restructure node for pivoting based on the payment type is displayed in Figure 7.2.

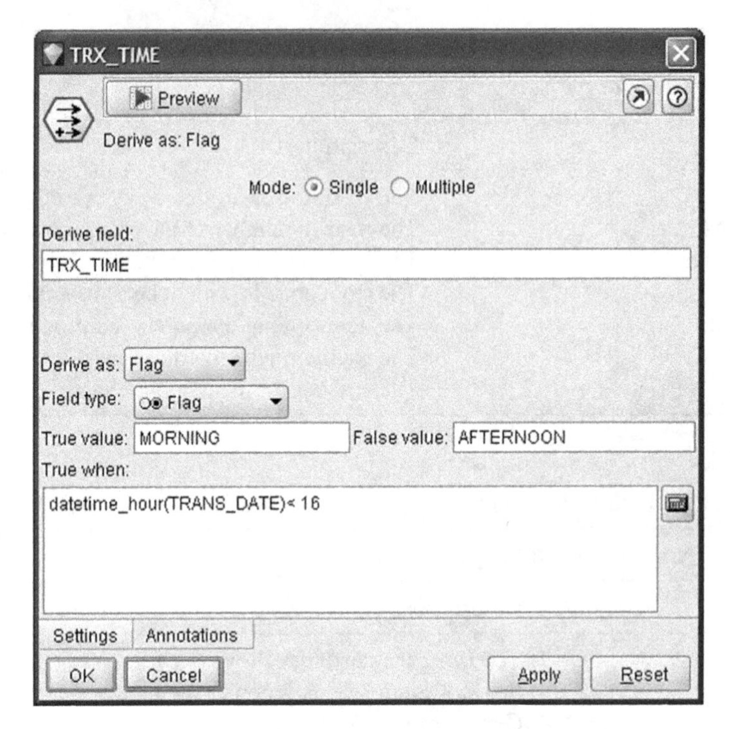

Figure 7.1 Categorizing transactions into time zones

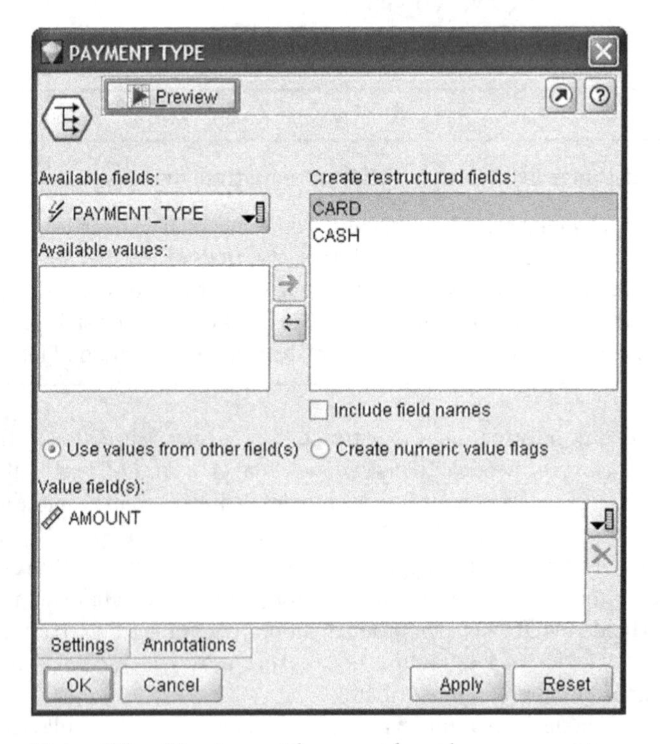

Figure 7.2 Pivoting paid amount based on payment type

Similarly, a series of Restructure nodes were also applied to generate a set of numeric flag fields with a value of 1 for our categorizations. This way, in the subsequent aggregation, the sum of the 1s simply denotes the number of transactions per store, date/time period, etc.

The aggregation was carried out in two steps in order to count the transactions per customer. The first aggregation summarized the numeric fields per transaction/invoice (field INVOICE_NO). In the second step, the invoice information was further aggregated at a cardholder/card ID (field CARD_ID) level as shown in Figure 7.3.

The number of aggregated records per card ID designates the number of order transactions (invoices) in the period examined.

The total spending amount and the sum of spending amount per store, product group, date/time zone, and payment type were computed for each cardholder. Similarly, the aggregated fields summarized customer behavior in terms of the number of transactions. The most recent transaction date was also captured as the maximum of all transaction dates (maximum of TRX_DATE field). This field will be used later for the calculation of the recency for each customer.

Defining transactions as visits

In this case study, we considered each invoice (order) as a different transaction and aggregated date per customer and invoice ID. An alternative approach would be to consider each "visit" as a transaction. To do this, the only modification required is to use the customer ID, the transaction day, and the store ID as the key fields for the aggregation of transactional records instead of the invoice number. In this approach, transaction corresponds to a "visit" at a store, and multiple orders in the same day (same purchases in the same visit) are considered as a single transaction.

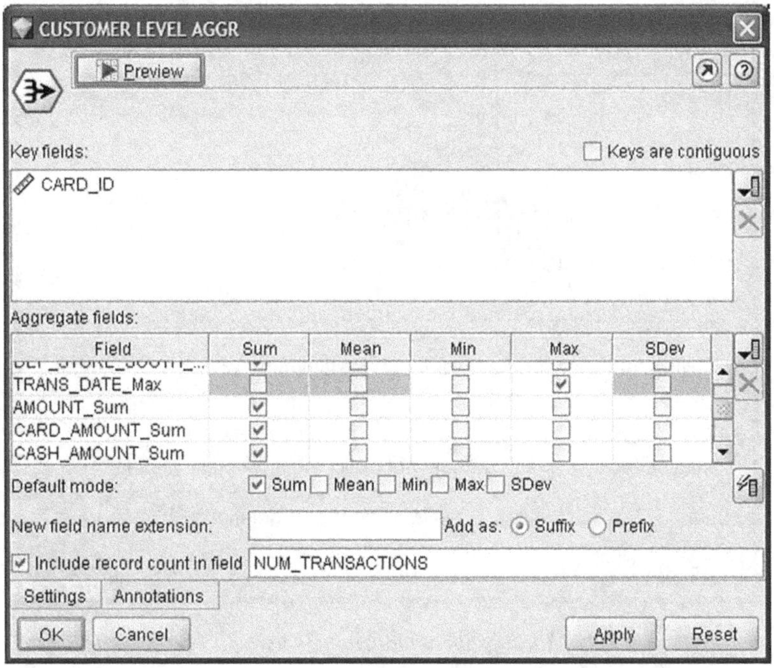

CUSTOMER LEVEL AGGR

Preview

Key fields: ☐ Keys are contiguous

CARD_ID

Aggregate fields:

Field	Sum	Mean	Min	Max	SDev
TRANS_DATE_Max	☐	☐	☐	☑	☐
AMOUNT_Sum	☑	☐	☐	☐	☐
CARD_AMOUNT_Sum	☑	☐	☐	☐	☐
CASH_AMOUNT_Sum	☑	☐	☐	☐	☐

Default mode: ☑ Sum ☐ Mean ☐ Min ☐ Max ☐ SDev

New field name extension: Add as: ◉ Suffix ○ Prefix

☑ Include record count in field NUM_TRANSACTIONS

Settings Annotations

OK Cancel Apply Reset

Figure 7.3 Aggregating transactional data at a customer level

The demographics, listed in Table 7.6, were then retrieved and joined with the purchase data using the card ID (field CARD_ID) as the key for merge. A Modeler Merge node had been used and an inner join had been applied as shown in Figure 7.4. The inner join merge retained only customers with demographics since for the rest of the customers, no contact data were available and hence any interaction and marketing action was not feasible.

7.4.2 Enriching customer data and building the customer signature

The summarized fields were then used for the construction of informative KPIs which meant to be used as inputs in the subsequent data mining applications. The derived attributes included:

Customer tenure

Based on the recorded registration date, the time in months of the relationship of the card-holder (TENURE field) with the retailer has been calculated (months on books).

Relative spending

To derive the relative spending for the categorizations of the transactional records (by product group, date/time zone, store, and payment type), each individual field was divided with the total spending amount. A Modeler multiple Conditional Derive node

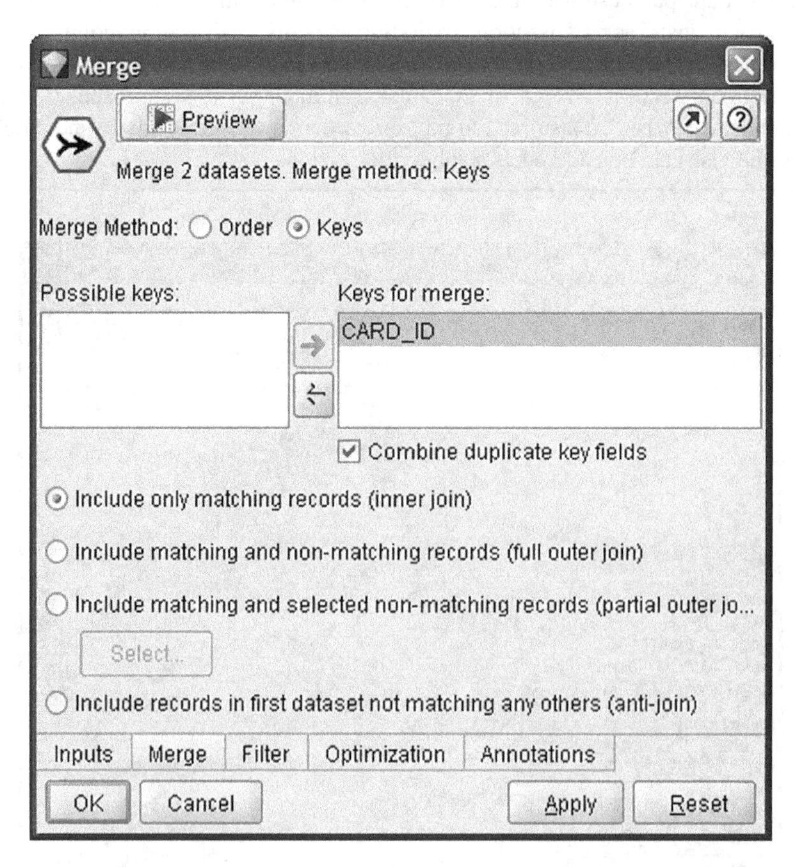

Figure 7.4 Adding the demographics using a Merge node

was applied for the calculation as shown in Figure 7.5. The derived fields were named with a PRC_ prefix.

Monthly average purchase amount per product group

To compute the monthly average purchase amount per product group, each spending amount was divided with 5 (months) for old customers or with the tenure months for new cardholders. Remember that the transactional data retrieved covered a 5-month period. The derived fields have been labeled with an AVG_ prefix.

Ratio of transactions per date/time zone, store, and payment type

In a way similar to the one applied for calculating the relative spending, the ratio of the number of transactions per date/time zone, store, and payment type has been derived. A Modeler multiple Conditional Derive node has been applied and each individual field was divided with the total number of transactions.

Average basket size

The average basket size indicates the average spending amount per transaction (in our example order/invoice). The logic of this field is to discern those customers that spend a lot at each visit to the stores (rare visits perhaps yet it remains for investigation after calculating the frequency of the visits) from those which tend to check out small baskets. Through a

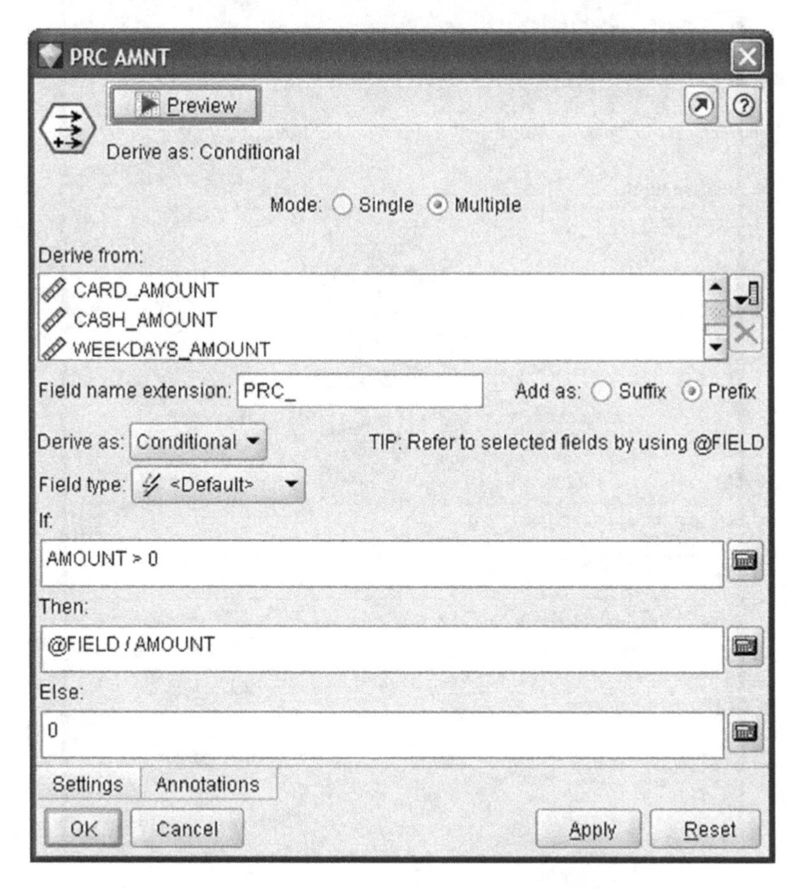

Figure 7.5 Calculating the relative spending KPIs

Conditional Derive node (Figure 7.6), the average basket size attribute (AVG_BASKET_SIZE field) was derived as the ratio of total spending amount to the total number of transactions in the period examined.

Basket diversity

The next derived field's scope (NUM_PRODUCT_GROUPS field) was to identify customers who tend to purchase items from multiple product groups. The number of distinct product groups with purchases has been counted for each customer. The values of this attribute range from 1 for customers who only bought products from a single group to 7 for super buyers who have bought something from every group.

Flags of product groups with purchases

A series of flag (binomial) fields have also been created indicating spending for each product group. The derived flags have a FLAG_ prefix. They are ready to be used as inputs in Association modeling for market basket analysis and for the identification of products that tend to be purchased together.

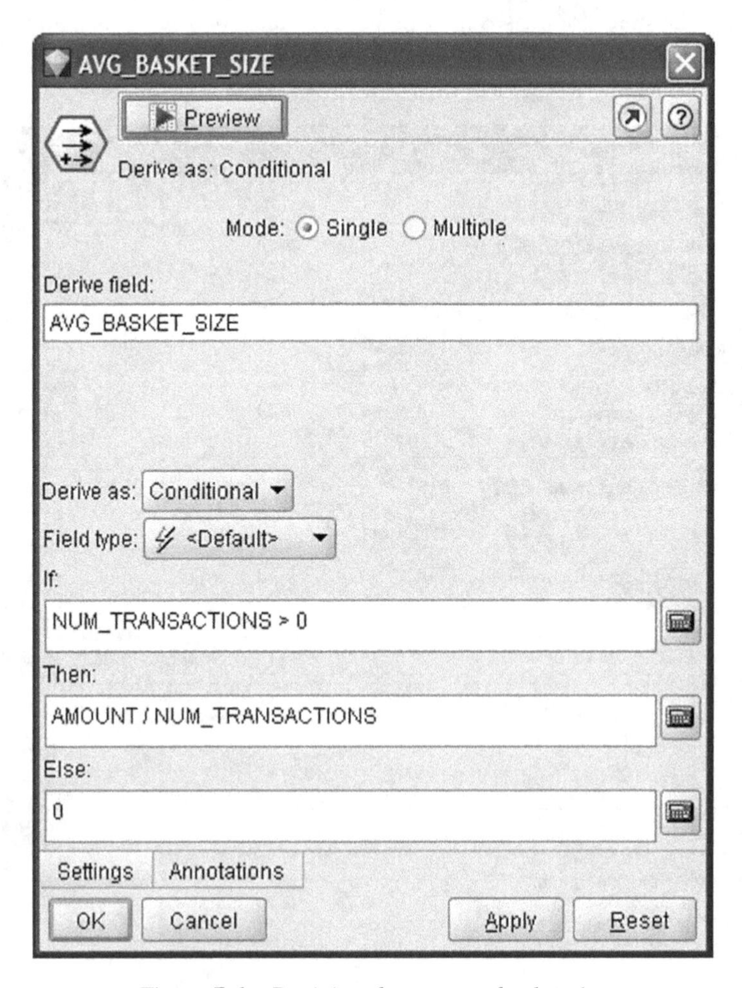

Figure 7.6 Deriving the average basket size

Recency

The RECENCY field was the first RFM component attribute derived. It denotes the time (in days) from the last transaction (TRANS_DATE_Max field) to the analysis as-of (reference) date (June 1, 2012). A Modeler Derive node and the date_days_difference() function was applied for the computation and is displayed in Figure 7.7.

Frequency

The *F* part of the RFM stands for the frequency of transactions. It was calculated as the monthly average number of order transactions with a Modeler Conditional Derive node as shown in Figure 7.8.

Monetary value

The M component of RFM stands for the monetary value, and it was derived as the monthly average purchase amount by dividing the total spending amount with 5 (months) or with the tenure months for new customers as shown in Figure 7.9.

7.5 The data dictionary of the modeling file

Finally, the enriched customer signature file, summarizing customer purchases in terms of frequency, recency, volume, and type, was ready to be used for the needs of the mining applications which followed. The data dictionary of the modeling file is listed in Table 7.7. The

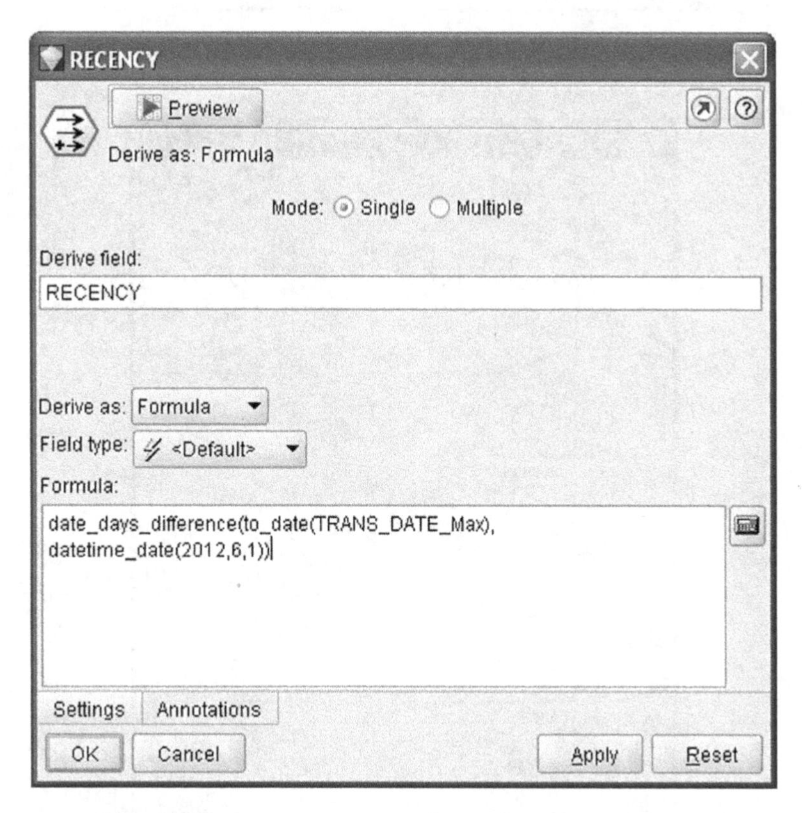

Figure 7.7 Constructing recency, the first RFM component

FREQUENCY
Preview
Derive as: Conditional
Mode: Single Multiple
Derive field:
FREQUENCY
Derive as: Conditional
Field type: <Default>
If:
TENURE > 5
Then:
NUM_TRANSACTIONS / 5
Else:
NUM_TRANSACTIONS / TENURE
Settings
Annotations
OK
Cancel
Apply
Reset

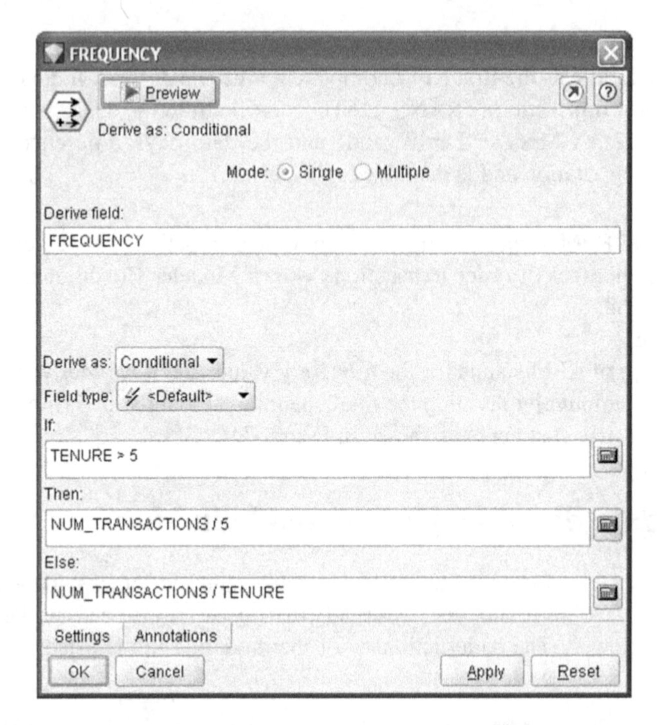

Figure 7.8 Constructing frequency, the second RFM component

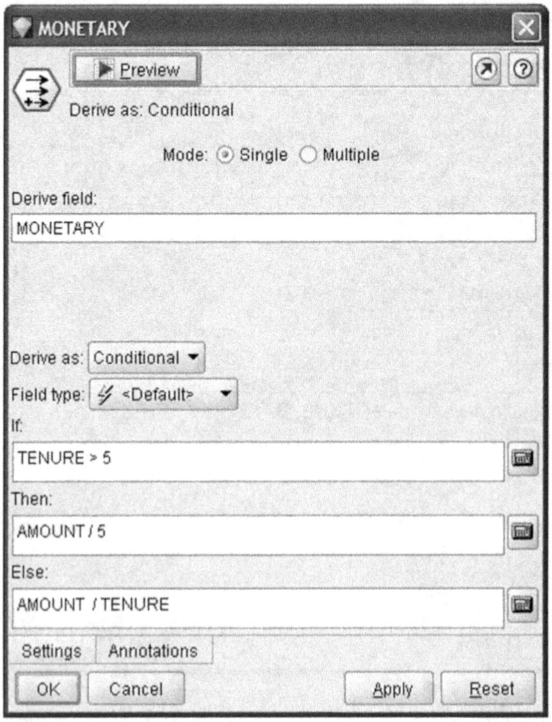

MONETARY
Preview
Derive as: Conditional
Mode: Single Multiple
Derive field:
MONETARY
Derive as: Conditional
Field type: <Default>
If:
TENURE > 5
Then:
AMOUNT / 5
Else:
AMOUNT / TENURE
Settings
Annotations
OK
Cancel
Apply
Reset

Figure 7.9 Constructing the monetary component of RFM

Table 7.7 The data dictionary of the modeling file

Data file: RETAIL_MODELING_DATA.txt		
Cardholders' demographics		
Field name	Description	Role in the cross-sell model
CARD_ID	The card identification number	
GENDER	Gender of the cardholder	PREDICTOR
CHILDREN	Number of children of the cardholder	
BIRTH_DATE	Cardholder's birth date	
AGE	Age of the customer	
AGE_BANDS	Age bins (18–24, 25–34, 35–44, 45–54, 55+)	
REGISTRATION_DATE	Cardholder's registration date	
TENURE	Time (in months) since a cardholder	PREDICTOR
Purchase data		
Field name	Description	Role in the model
TRANS_DATE_Max	Most recent date of order transaction	
AMOUNT	Total amount of purchases for each customer for the period of 5 months examined	
CARD_AMOUNT	Amount of purchases done with a credit card	
CASH_AMOUNT	Amount of purchases with cash	
WEEKDAYS_AMOUNT	Amount of purchases on weekdays	
WEEKEND_AMOUNT	Amount of purchases at weekends	
AFTERNOON_AMOUNT	Amount of afternoon (after 4 p.m.) purchases	
MORNING_AMOUNT	Amount of morning (before 4 p.m.) purchases	
BEAUTY_AMOUNT	Amount of purchases at the Beauty department	
CHILDRENWEAR_AMOUNT	Amount of purchases at the Childrenwear department	
GROCERY_AMOUNT	Amount of purchases at the Food and Grocery department	
HOME_AMOUNT	Amount of purchases at the House and furniture department	
WOMEN_APPAREL_AMOUNT	Amount of purchases at the Women apparel department	
ACCESSORIES_AMOUNT	Amount of purchases at the Accessories department	
MEN_APPAREL_AMOUNT	Amount of purchases at the Men apparel department	

(*Continued*)

Table 7.7 (**Continued**)

Purchase data

Field name	Description	Role in the model
DOWNTOWN_AMOUNT	Amount of purchases at the Downtown Store	
SUBURBS_AMOUNT	Amount of purchases at the Suburbs store	
DEP_STORE_NORTH_ AMOUNT	Amount of purchases at the North store	
DEP_STORE_SOUTH_ AMOUNT	Amount of purchases at the South store	
NUM_TRANSACTIONS	Total number of order transactions. Each invoice is considered as a transaction	
CARD_TRANSACTIONS	Total number of transactions using a credit card	
CASH_TRANSACTIONS	Total number of transactions with cash	
WEEKDAYS_ TRANSACTIONS	Number of transactions on weekdays	
WEEKEND_ TRANSACTIONS	Number of transactions at weekends	
AFTERNOON_ TRANSACTIONS	Number of afternoon (after 4 p.m.) transactions	
MORNING_ TRANSACTIONS	Number of morning (before 4 p.m.) transactions	
DOWNTOWN_ TRANSACTIONS	Number of transactions at the Downtown Store	
SUBURBS_ TRANSACTIONS	Number of transactions at the Suburbs store	
DEP_STORE_NORTH_ TRANSACTIONS	Number of transactions at the North store	
DEP_STORE_SOUTH_ TRANSACTIONS	Number of transactions at the South store	
PRC_CARD_AMOUNT	Relative spending amount (percentage of total spending amount) using a credit card	PREDICTOR
PRC_CASH_AMOUNT	Relative spending amount using cash	PREDICTOR
PRC_WEEKDAYS_ AMOUNT	Relative spending amount on weekdays	PREDICTOR
PRC_WEEKEND_ AMOUNT	Relative spending amount at weekends	PREDICTOR
PRC_AFTERNOON_ AMOUNT	Relative spending amount at afternoons	PREDICTOR

Table 7.7 (**Continued**)

Purchase data		
Field name	Description	Role in the model
PRC_MORNING_AMOUNT	Relative spending amount at mornings	PREDICTOR
PRC_BEAUTY_AMOUNT	Relative spending amount at the Beauty department	PREDICTOR
PRC_CHILDRENWEAR_AMOUNT	Relative spending amount at the Childrenwear department	PREDICTOR
PRC_GROCERY_AMOUNT	Relative spending amount at the Food and Grocery department	PREDICTOR
PRC_HOME_AMOUNT	Relative spending amount at the House and furniture department	
PRC_WOMEN_APPAREL_AMOUNT	Relative spending amount at the Women apparel department	PREDICTOR
PRC_ACCESSORIES_AMOUNT	Relative spending amount at the Accessories department	PREDICTOR
PRC_MEN_APPAREL_AMOUNT	Relative spending amount at the Men apparel department	PREDICTOR
PRC_DOWNTOWN_AMOUNT	Relative spending amount at the Downtown Store	PREDICTOR
PRC_SUBURBS_AMOUNT	Relative spending amount at the Suburbs store	PREDICTOR
PRC_DEP_STORE_NORTH_AMOUNT	Relative spending amount at the North store	PREDICTOR
PRC_DEP_STORE_SOUTH_AMOUNT	Relative spending amount at the South store	PREDICTOR
PRC_CARD_TRANSACTIONS	Ratio of order transactions using a credit card	
PRC_CASH_TRANSACTIONS	Ratio of transactions with cash	
PRC_WEEKDAYS_TRANSACTIONS	Ratio of transactions on weekdays	
PRC_WEEKEND_TRANSACTIONS	Ratio of transactions at weekends	
PRC_AFTERNOON_TRANSACTIONS	Ratio of afternoon (after 4 p.m.) transactions	
PRC_MORNING_TRANSACTIONS	Ratio of morning (before 4 p.m.) transactions	
PRC_DOWNTOWN_TRANSACTIONS	Ratio of transactions at the Downtown store	
PRC_SUBURBS_TRANSACTIONS	Ratio of transactions at the Suburbs store	

(Continued)

Table 7.7 (**Continued**)

Purchase data		
Field name	Description	Role in the model
PRC_DEP_STORE_NORTH_TRANSACTIONS	Ratio of transactions at the North store	
PRC_DEP_STORE_SOUTH_TRANSACTIONS	Ratio of transactions at the South store	
AVG_BEAUTY_AMOUNT	Monthly average spending amount at the Beauty department	PREDICTOR
AVG_CHILDRENWEAR_AMOUNT	Monthly average spending amount at the Childrenwear department	PREDICTOR
AVG_GROCERY_AMOUNT	Monthly average spending amount at the Food and Grocery department	PREDICTOR
AVG_HOME_AMOUNT	Monthly average spending amount at the House and furniture department	
AVG_WOMEN_APPAREL_AMOUNT	Monthly average spending amount at the Women apparel department	PREDICTOR
AVG_ACCESSORIES_AMOUNT	Monthly average spending amount at the Accessories department	PREDICTOR
AVG_MEN_APPAREL_AMOUNT	Monthly average spending amount at the Men apparel department	PREDICTOR
AVG_BASKET_SIZE	Average basket size indicating average amount per transaction	PREDICTOR
NUM_PRODUCT_GROUPS	Number of distinct product groups with purchases	PREDICTOR
FLAG_BEAUTY	A flag (binomial) field indicating purchases from the product group	PREDICTOR
FLAG_CHILDRENWEAR	A flag (binomial) field indicating purchases from the product group	PREDICTOR
FLAG_GROCERY	A flag (binomial) field indicating purchases from the product group	PREDICTOR
FLAG_HOME	A flag (binomial) field indicating purchases from the product group	
FLAG_WOMEN_APPAREL	A flag (binomial) field indicating purchases from the product group	PREDICTOR
FLAG_ACCESSORIES	A flag (binomial) field indicating purchases from the product group	PREDICTOR
FLAG_MEN_APPAREL	A flag (binomial) field indicating purchases from the product group	PREDICTOR
RECENCY	Time (in days) since last transaction	PREDICTOR—also used as RFM component

Table 7.7 (**Continued**)

Purchase data		
Field name	Description	Role in the model
FREQUENCY	Monthly average number of order transactions	PREDICTOR—also used as RFM component
MONETARY	Monthly average purchase amount	PREDICTOR—also used as RFM component

last column of the table indicates the attributes which were used as predictors in the cross-selling model developed for promoting the House department products.

7.6 Value segmentation

All customers are not equal. Some are more valuable than others. Identifying those valuable customers and understanding their importance should be considered as a top priority for any organization.

Value-based segmentation is one of the key tools for developing customer prioritization strategies. It can enable service-level differentiation and prioritization. Marketers of the retailer decided to develop a value-based segmentation scheme in order to assign each customer to a segment according to his value. Their plan was to use this segmentation in order to separate valuable customers from the rest and gain insight on their differentiating characteristics.

7.6.1 Grouping customers according to their value

The value-based segmentation was applied to the customers with a loyalty card, and it was based on the monthly average spent amount of each customer (the derived MONETARY field). Thus, all cardholders were ranked according to their purchases. The ordered records were grouped (binned) to chunks of equal frequency referred to as quantiles. These tiles were then used to construct the value segments.

What if a value index is not available?

In the case of absence of a calculated valid value index, usage fields, known by business users to be highly related to value, could be used as temporary substitutes of the value measure, in order to proceed with value segmentation. For instance, fields like monthly average balances in banking or monthly average number of call minutes in telecommunications may not be an ideal component for value segmentation, but they could be used as a work-around.

In value-based segmentation, the number of binning tiles to be constructed depends on the specific needs of the organization. The derived segments are usually of the following form: highest $n\%$, medium–high $n\%$, medium–low $n\%$, and low $n\%$. In general, it is recommended to select a sufficiently rich and detailed segmentation level, especially in the top groups, in order to discern the most valuable customers. On the other hand, a detailed segregation level may not be required in the bottom of the value pyramid.

The segmentation bands selected by the retailer were the following:

1. *CORE: bottom 50% of customers with lowest purchase amount*
2. *BRONZE: 30% of customers with medium–low purchase amount*
3. *SILVER: 15% of customers with high purchase amount*
4. *GOLD: top 5% of customers with highest purchase amount*

Since customers with no purchases in the 5-month period examined were excluded from the procedure, a fifth segment not listed but implied is the one comprised of inactive customers.

Before beginning any data mining task, it is necessary to perform a health check in the data to be mined. Initial data exploration typically involves looking for missing data and checking for inconsistencies, identifying outliers, and examining the field distributions with basic descriptive statistics and charts like bar charts and histograms. IBM SPSS Modeler offers a tool called Data Audit that performs all these preliminary explorations and allows users to have a first look at the data, examine them, and spot potential abnormalities.

A snapshot of the results of the Data Audit node applied to the modeling file is presented in Figure 7.10.

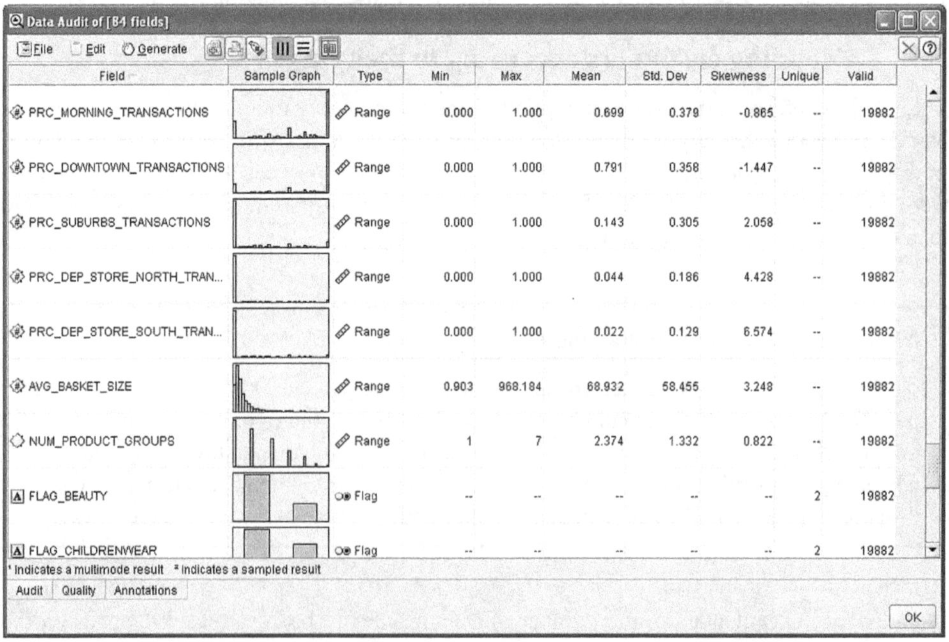

Data Audit of [84 fields]

File Edit Generate

Field	Sample Graph	Type	Min	Max	Mean	Std. Dev	Skewness	Unique	Valid
PRC_MORNING_TRANSACTIONS		Range	0.000	1.000	0.699	0.379	-0.865	--	19882
PRC_DOWNTOWN_TRANSACTIONS		Range	0.000	1.000	0.791	0.358	-1.447	--	19882
PRC_SUBURBS_TRANSACTIONS		Range	0.000	1.000	0.143	0.305	2.058	--	19882
PRC_DEP_STORE_NORTH_TRAN...		Range	0.000	1.000	0.044	0.186	4.428	--	19882
PRC_DEP_STORE_SOUTH_TRAN...		Range	0.000	1.000	0.022	0.129	6.574	--	19882
AVG_BASKET_SIZE		Range	0.903	968.184	68.932	58.455	3.248	--	19882
NUM_PRODUCT_GROUPS		Range	1	7	2.374	1.332	0.822	--	19882
FLAG_BEAUTY		Flag	--	--	--	--	--	2	19882
FLAG_CHILDRENWEAR		Flag	--	--	--	--	--	2	19882

' Indicates a multimode result * Indicates a sampled result

Audit Quality Annotations

OK

Figure 7.10 Using a Data Audit node to perform an initial exploration of the data

The binning procedure for value-based segmentation is graphically depicted in Figure 7.11.

The Modeler stream used for the value (and RFM) segmentation is presented in Figure 7.12.

A Binning node was used for grouping cardholders in groups of equal frequency. In order to be able to discern the top 5% customers, binning into vingitiles, 20 tiles of 5% each, was selected as shown in Figure 7.13.

The "tiles (equal count)" was selected as the binning method, and customers were assigned to vingitiles according to their monthly average purchase amount by specifying the MONETARY field as the bin field. The derived bands have then been regrouped with a Set Derive node, in order to refine the grouping and construct the VBS field which assigns each customer to one of the desired segments: Core, Bronze, Silver, and Gold. The conditions for regrouping are displayed in Figure 7.14.

7.6.2 Value segments: exploration and marketing usage

The next step before starting to make use of the derived segments was the investigation of their main characteristics.

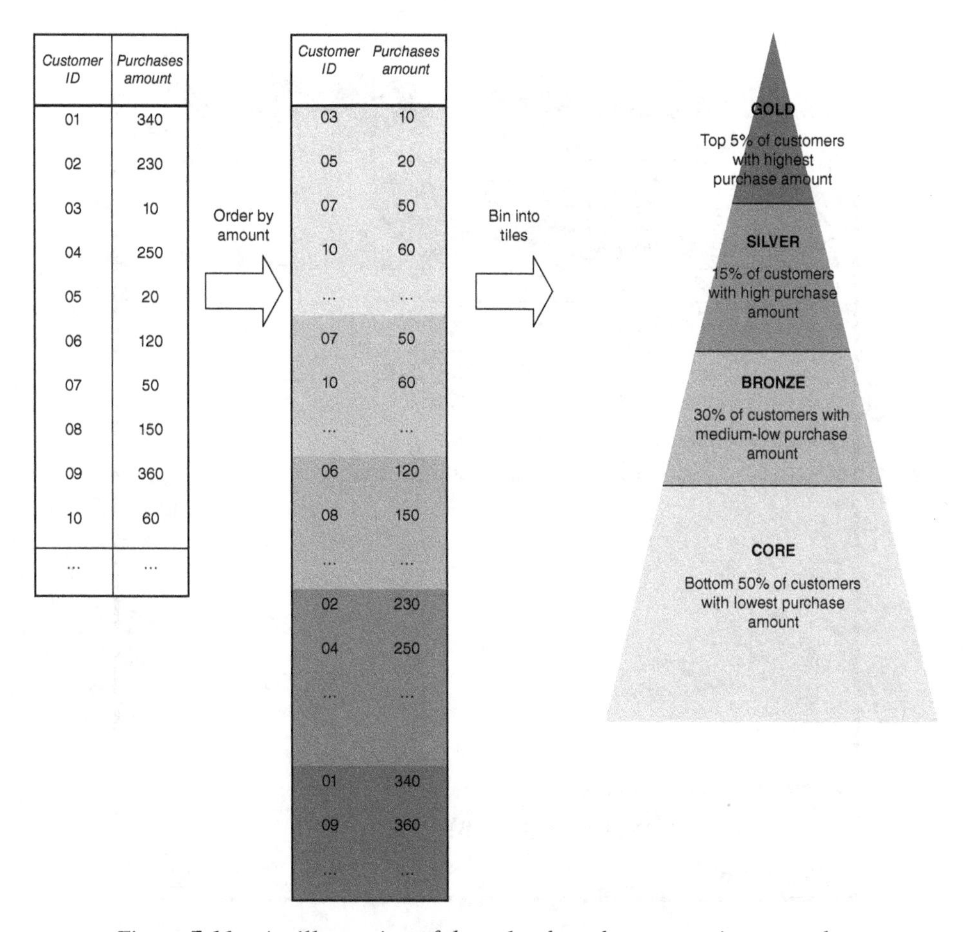

Figure 7.11 An illustration of the value-based segmentation procedure

(2) Retail RFM_VBS* - IBM® SPSS® Modeler
File
Edit
Insert
View
Tools
SuperNode
Window
Help
84 Fields
RETAIL_MODELING_DATA
Binning VBS
VBS
VBS
Binning R
REVERSE R
Binning F
Binning M
RFM
RFM
(Default: Local Server)
220MB / 315MB

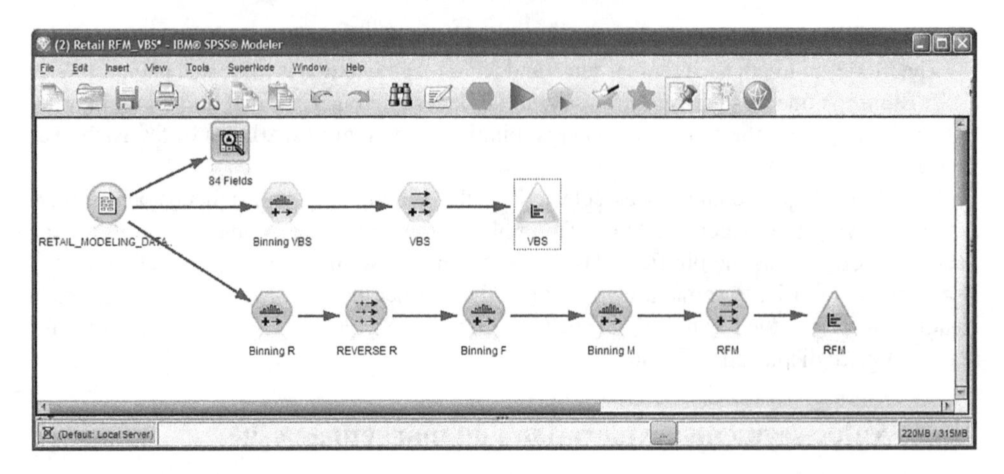

Figure 7.12 The IBM SPSS Modeler stream for value and RFM segmentation

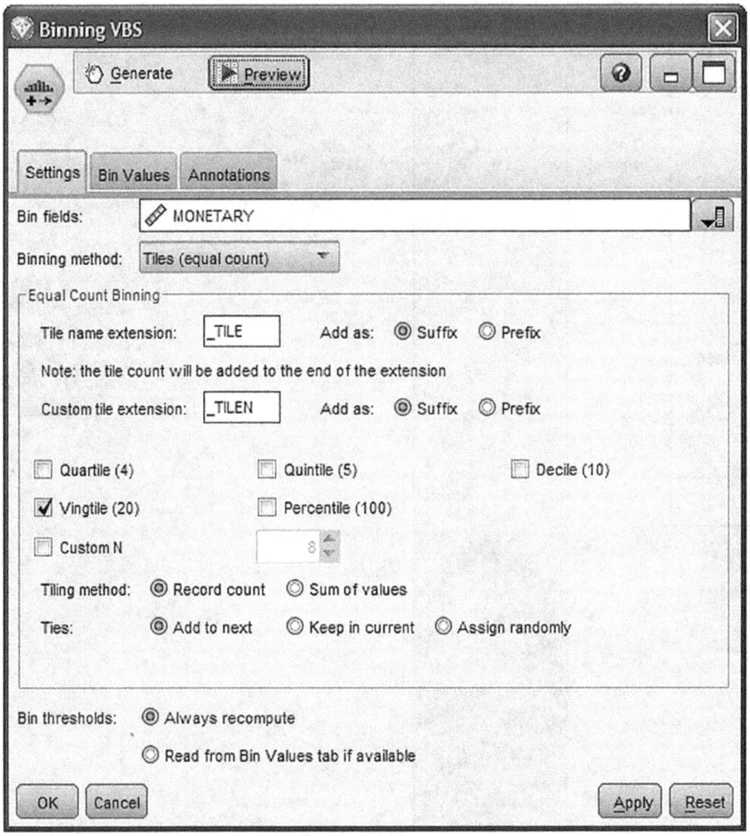

Binning VBS
Generate
Preview
Settings
Bin Values
Annotations
Bin fields:
MONETARY
Binning method:
Tiles (equal count)
Equal Count Binning
Tile name extension:
_TILE
Add as:
Suffix
Prefix
Note: the tile count will be added to the end of the extension
Custom tile extension:
_TILEN
Add as:
Suffix
Prefix
Quartile (4)
Quintile (5)
Decile (10)
Vingtile (20)
Percentile (100)
Custom N
8
Tiling method:
Record count
Sum of values
Ties:
Add to next
Keep in current
Assign randomly
Bin thresholds:
Always recompute
Read from Bin Values tab if available
OK
Cancel
Apply
Reset

Figure 7.13 Using a Binning node for grouping customers in groups of 5%

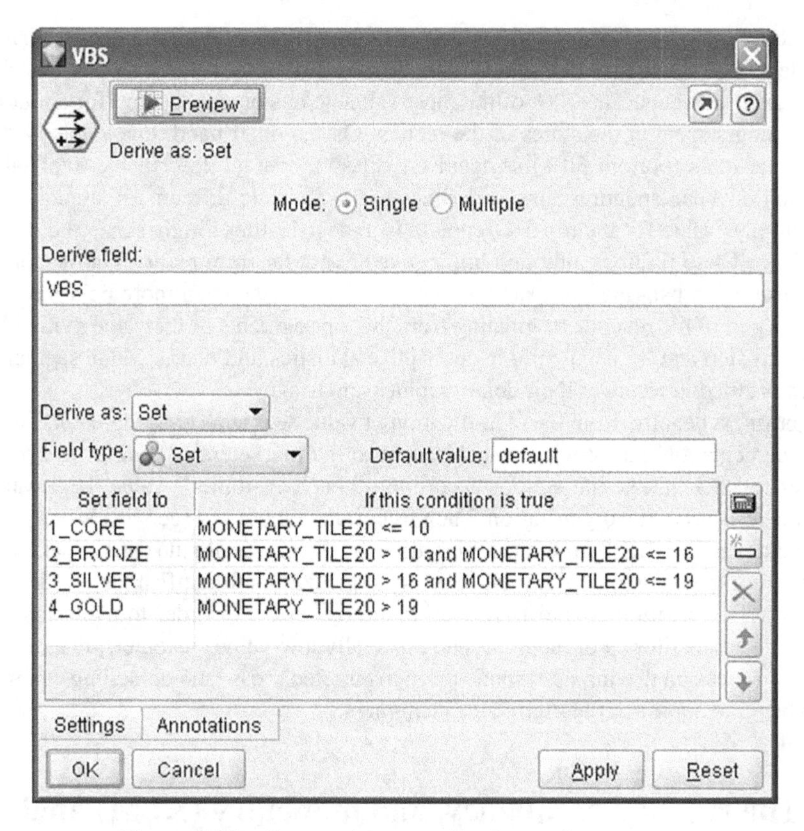

Figure 7.14 Regrouping quantiles into value segments

Table 7.8 Value-based segments and total purchase amount

		Percentage of customers	Sum percentage of total purchase amount	Mean purchase amount
Value-based segments	1_CORE (Bottom 50%)	50	15	56
	2_BRONZE (Medium–low 30%)	30	30	186
	3_SILVER (High 15%)	15	32	394
	4_GOLD (Top 5%)	5	23	841
Total		100	100	185

The initial goal of this segmentation was to capture the assumed large-scale differences between customers, in terms of spending. Thus, marketers begun to investigate this hypothesis by examining the contribution of each segment to the total sales (purchase amount). Table 7.8 presents the percentage of customers and the percentage of the total sales amount for the 5-month period examined by value segment.

The above table shows that a substantial percentage of the total sales comes from a disproportionately small number of high-value users. Almost half of the retailer's sales arise from the top two value segments. The 20% of the most valuable customers account for about 45% of the total amount spent at the stores of the retailer. On the other hand, low-value users which comprise the mass (bottom 50%) segment provide a moderate 23% of the total sales. The comparison of mean spending also underlines the large-scale differences among segments. The mean sales value for the top 5% segment is almost 17 times higher than the one for the bottom 50%. These findings although impressive are not far from reality. On the contrary, in other industries, in banking, for example, the situation may be even more polarized, with an even larger part of the revenue originating from the top segments of the value pyramid.

In a next step and by using simple descriptive statistics and charts, value segments have also been profiled in terms of their demographics and usage.

The business benefits from the identification of value segments are prominent. The implemented segmentation can provide valuable help to the marketers in setting the appropriate objectives for their marketing actions according to each customer's value. High-value customers are the heart of the organization. Their importance should be recognized and rewarded. They should perceive their importance every time they interact with the enterprise. Prevention of defection of these customers is vital for every organization. Identification of valuable customers at risk of attrition should trigger an enormous effort in order to avoid losing these customers to competitors. For medium- and especially low-value customers, marketing strategies should focus on driving up revenue through targeted cross- and up-selling campaigns in order to begin to approach the high-value customers.

7.7 The recency, frequency, and monetary (RFM) analysis

The marketers of the retailer extended the value segmentation by implementing an RFM analysis. In RFM cell segmentation, monetary value information is examined along with recency and frequency of purchases, and customers are assigned to cells according to "since when," "how much," and "how often" they purchase. Before presenting the RFM procedure in detail, let's have a look at RFM basics.

7.7.1 RFM basics

RFM analysis is a common approach for understanding and monitoring consuming behaviors. It is quite popular, especially in the retail industry. As its name implies, it involves the calculation of three core KPIs, recency, frequency, and monetary value which summarize the corresponding dimensions of the customer relationship with the enterprise. The recency measurement indicates the time since the last purchase transaction. Frequency denotes the number and the rate of the purchase transactions. Monetary value measures the purchase amount. These indicators are typically calculated at a customer (cardholder) level through simple data processing of the recorded transactional data.

RFM analysis can be used to identify the good customers with the best scores in the relevant KPIs, who generally tend to be good prospects for additional purchases. It can also identify other purchase patterns and respective customer types of interest, such as infrequent big spenders or customers with small but frequent purchases who might also have sales perspectives, depending on the market and the specific product promoted.

In the retail industry, the RFM dimensions are usually defined as below:

Recency: time (in units of time, typically in days or in months) since the most recent purchase transaction or shopping visit.

Frequency: total number of purchase transactions or shopping visits in the period examined. An alternative, and more "normalized" approach that also takes into account the tenure of the customer, calculates frequency as the average number of transactions per unit of time, for instance, the monthly average number of transactions.

Monetary value: the total or the average per time unit (e.g., monthly average value) amount of purchases within the examined period. According to an alternative yet not so popular definition, the monetary value indicator is defined as the average transaction value (average amount per purchase transaction). Since the total value tends to be correlated with the frequency of the transactions, the reasoning of this alternative definition is to capture a different and supplementing aspect of the purchase behavior.

The construction of the RFM indicators is a simple data management task which does not involve any data mining modeling. It does involve through a series of aggregations and simple computations that transform the raw purchase records into meaningful scores. In order to perform RFM analysis, each transaction should be linked with a specific customer (card ID) so that each customer's purchase history can be tracked and traced over time. Fortunately, in most situations, the usage of a loyalty program makes the collection of "personalized" transactional data possible.

RFM components should be calculated on a regular basis and stored along with the other behavioral indicators in the organization's mining datamart. They can be used as individual fields in subsequent tasks, for instance, as inputs, along with other predictors, in cross-selling models. They can also be included as clustering fields for the development of a multiattribute behavioral segmentation scheme. Typically, they are simply combined to form a single RFM measure and a respective cell segmentation scheme.

The RFM analysis involves the grouping (binning) of customers into chunks of equal frequency, named quantiles, in a way similar to the one presented for value-based segmentation. This binning procedure is applied independently on the three RFM component measures. Customers are ranked according to the respective measure, and they are grouped in classes of equal size. For instance, the break in four groups results quartiles of 25% and the break in five groups, quintiles of 20%. In the case of binning into quintiles, for example, the *R*, *F*, and *M* measures are converted into rank scores ranging from 1 to 5. Group 1 includes the 20% of customers with the lowest values and group 5 the top 20% of customers with the top values in the corresponding measure. Especially for the recency measure, the scale of the derived ordinal score should be reversed so that larger scores indicate the most recent buyers.

The derived *R*, *F*, and *M* bins become the components for the RFM cell assignment. These bins are combined with a simple concatenation to provide the RFM cell assignment. Customers with top frequency, recency, and monetary values are assigned to the RFM cell 555. Similarly, customers with average recency (quintile 3), top frequency (quintile 5), and lowest monetary values (quintile 1) form the RFM cell 351 and so on.

The aforementioned procedure for the construction of the RFM cells is illustrated in Figure 7.15.

When grouping customers in quintiles (groups of 20%), the procedure results a maximum of $5 \times 5 \times 5 = 125$ RFM cells as displayed in Figure 7.16.

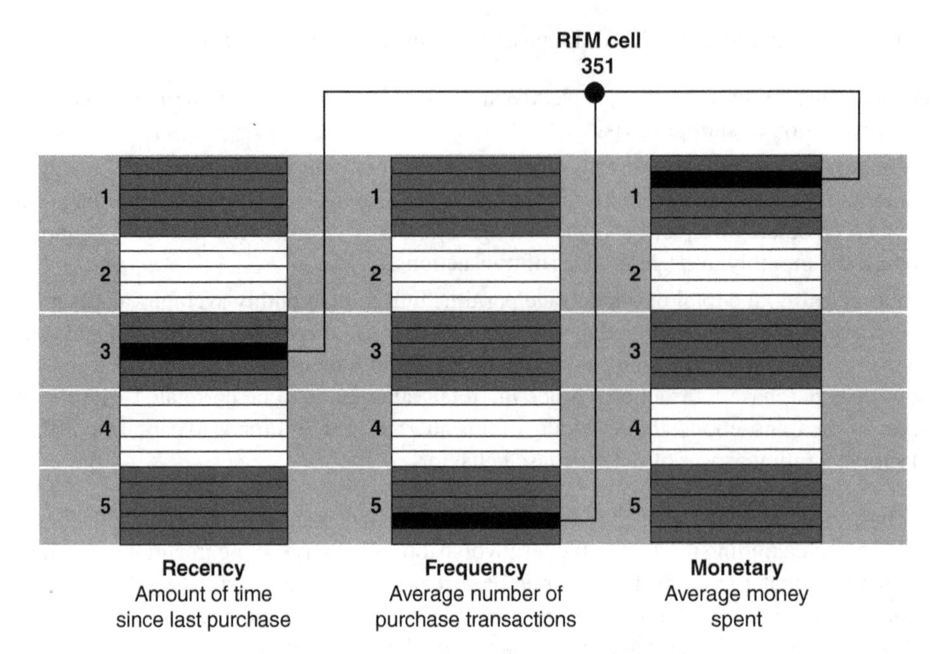

Figure 7.15 Assignment to the RFM cells. Source: Tsiptsis and Chorianopoulos (2009). Reproduced with permission from Wiley

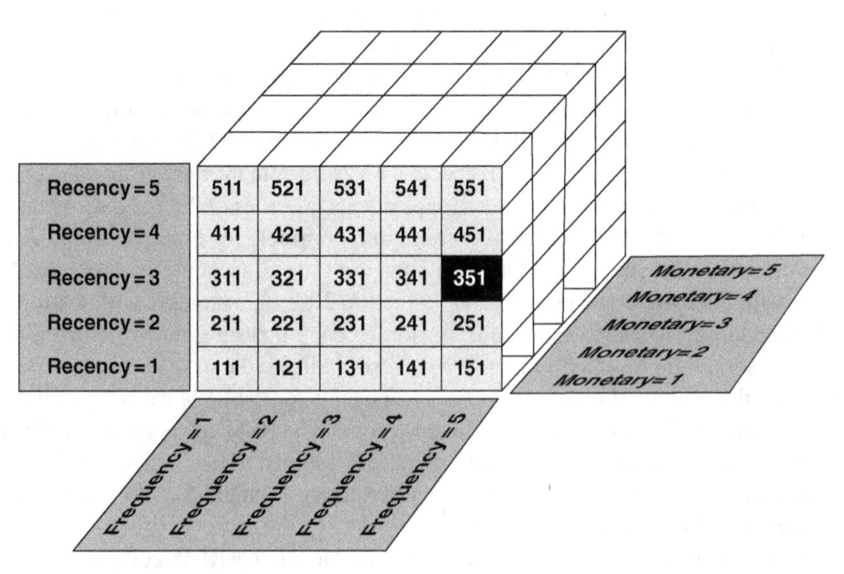

Figure 7.16 The total RFM cells in the case of binning into quintiles (groups of 20%). Source: Tsiptsis and Chorianopoulos (2009). Reproduced with permission from Wiley

This combination of the *R*, *F*, and *M* components into RFM cells is widely used; however, it has a certain disadvantage. The large number of the derived cells makes the procedure quite cumbersome and hard to manage. An alternative method for segmenting customers according to their RFM patterns is to use the respective components as inputs in a clustering model and let the algorithm reveal the underlying natural groupings of customers.

7.8 The RFM cell segmentation procedure

RFM analysis was based on the same modeling dataset used for value segmentation with 5 months of aggregated purchase transactions. The procedure involved the calculation of the RFM components and the assignment of customers into RFM segments. As noted before, only active customers with a loyalty card were included in the project. Customers with no purchases during the period examined were identified as candidates for inclusion in upcoming retention and reactivation campaigns. We remind you that the three RFM components had already been calculated during the data preparation phase:

RECENCY: derived as the number of days since the most recent purchase transaction
FREQUENCY: denoting the monthly average number of distinct purchase transactions
MONETARY: indicating monthly average spent amount

Customers were then binned into five groups of 20% (quintiles) through three separate binnings, one for each RFM component. This discretization procedure resulted in a set of three ordinal attributes (RECENCY_TILE5, FREQUENCY_TILE5, MONETARY_TILE5) which were then concatenated to form the RFM cell segmentation. In each of the derived bins, top customers (in terms of RFM) were assigned to bin 5 while worst customers to bin 1. In terms of recency, good customers are those that had recently visited store and hence present a low recency figure. That's why the analysts of the retailer reversed the recency scale so that a value of 1 corresponds to "distant" customers and a value of 5 to "recent" customers. The formula used for this adjustment ((5-RECENCY_TILE5)+1) was applied through a Filler node (a Filler node overwrites an existing field's values instead of creating a new attribute).

The resulted quintiles were finally concatenated by using a simple Formula Derive node to form the RFM cell assignment (field RFM). The distribution of customers into RFM cells is shown in Figure 7.17.

Figure 7.18 presents a jittered scatter plot of the monetary and frequency quintiles in *X* and *Y* axes respectively. The gray tone (transparency) for each dot represents the recency tile. There is a dispersion of customers across all RFM cells, indicating the vast range of different consumer behaviors.

RFM: benefits, usage, and limitations

The individual RFM components and the derived segments convey useful information in respect to the purchase habits of the consumers. Undoubtedly, any retail enterprise should monitor the purchase frequency, intensity, and recency as they represent significant dimensions of the customer's relationship with the enterprise.

Moreover, by following RFM migrations over time, an organization can keep track of changes in the purchase habits of each customer and use this information to proactively trigger appropriate marketing actions. For instance, specific events, like the decline of the total value of purchases, a sudden drop on the frequency of visits, or the nonshowing for an unusually large period of time, may indicate the beginning of the decaying of the relationship with the organization. These signals, if recognized on time, should initiate event-triggered reactivation campaigns.

RFM analysis was originally developed for the retailers, but with proper modifications, it can also be applied in other industries. It originated from the catalogue industry in the 1980s and was proved quite useful in targeting the right customers in direct marketing campaigns. The response rates of the RFM cells on past campaigns were recorded, and the best-performing cells were targeted in the next campaigns. An obvious drawback of this approach is that it usually ends up with almost the same target list of good customers, which could finally get annoyed by repeating contacts. Although useful, the RFM approach, when not combined with other important customer attributes, such as the product preferences, fails to provide a complete understanding of the customer behavior. An enterprise should have a 360° view of the customer, and it should use all the available information to guide business decisions.

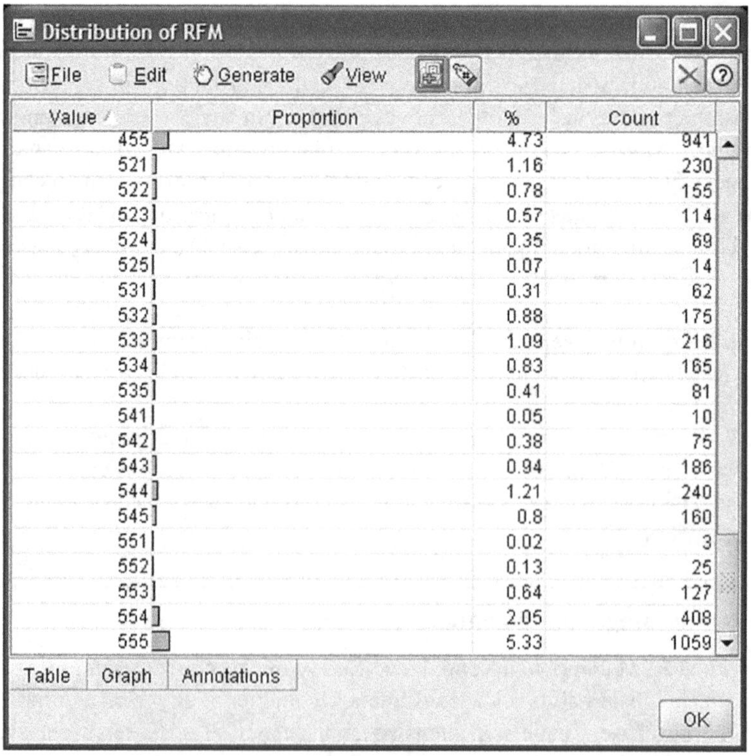

Distribution of RFM

File Edit Generate View

Value	Proportion	%	Count
455		4.73	941
521		1.16	230
522		0.78	155
523		0.57	114
524		0.35	69
525		0.07	14
531		0.31	62
532		0.88	175
533		1.09	216
534		0.83	165
535		0.41	81
541		0.05	10
542		0.38	75
543		0.94	186
544		1.21	240
545		0.8	160
551		0.02	3
552		0.13	25
553		0.64	127
554		2.05	408
555		5.33	1059

Table Graph Annotations

OK

Figure 7.17 The distribution of the constructed RFM cells

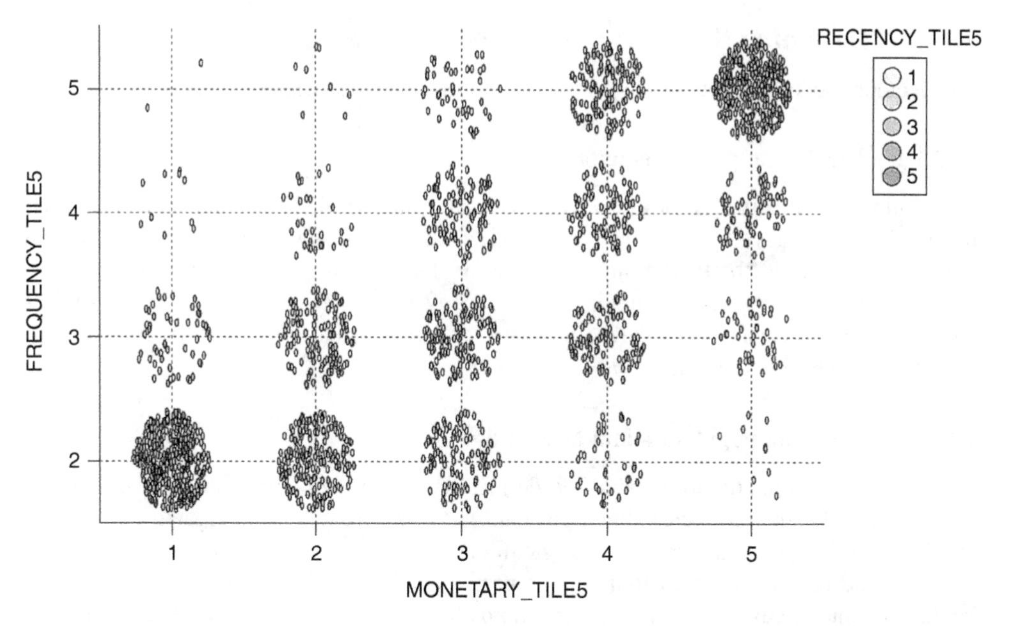

Figure 7.18 An RFM scatter plot

7.9 Setting up a cross-selling model

The development of a cross-selling model for the House and furniture department was the final mining application. The department was new, with moderate penetration, and the marketers of the retailer wanted to make it known to customers likely to show interest. Therefore, they planned a promotional campaign with a discount offer. Yet they didn't want to communicate this offer to each and every customer. They wanted to roll out a rather small-scale but well-targeted campaign, addressed to customers with a profile that suits the specific products. To identify their target group, they carried out a pilot campaign and used the collected responses to build a cross-selling model.

7.10 The mining approach

The marketers of the retailer had scheduled a campaign to promote their new House and furniture department. To improve the campaign's effectiveness and minimize nonresponses, waste of resources, and unsolicited communication, they planned to use analytics to refine their target list. An issue that had to be tackled was that due to the fact that the particular department was new, the number of shoppers was relatively small. Moreover, the marketers wanted to simulate and test the campaign before its actual rollout. Hence, the mining approach decided was to run a pilot campaign with exactly the same characteristics (same promoted product, same channel, same incentive, same initial target group) and collect and analyze responses to identify customers with a purchase potential.

7.10.1 Designing the cross-selling model process

The decided modeling process is outlined in the following paragraphs.

7.10.1.1 The data and the predictors

The mining file constructed for the needs of the value and RFM segmentation was also used for the development of the cross-selling model. Through extensive data transformations on transactional records, the final mining file listed in Table 7.7 provided the customer "signature" information for each cardholder, summarizing consumer behaviors in terms of frequency, intensity, recency, and type of purchases. The role of each attribute in the cross-selling model is also presented in Table 7.7.

7.10.1.2 Modeling population and level of data

The type of the campaign to be launched defines the modeling population and the required level of data. In this case, customer-level data were required. The aggregated mining file provided the necessary information. But the mining file included all active cardholders. The model would be trained on customers who participated in the test campaign. Therefore, a random sample of customers (from those who hadn't yet visited the House department) had been selected and approached with the special offer, a discount voucher with a traceable promotion code. Redeeming of vouchers had been tracked, relevant information had been collected, and the propensity model had finally been trained on response data. In deployment, the generated model was applied to the entire customer base to score customers according to their likelihood to accept the offer.

7.10.1.3 Target population and definition of target attribute

The target field was derived using the collected campaign response data. The target population included all participants of the pilot campaign who accepted the offer.

7.10.1.4 Time periods and historical information required

A 5-month "snapshot" of behavioral data, covering an observation period from January 1, 2012, to May 31, 2012, was retrieved and analyzed to identify the characteristics of those accepted the offer. The test-campaign approach may have its drawbacks in terms of required resources and delay in collecting the required response data, but it is more straightforward in regard to data preparation for modeling. No latency period was reserved and there was no need to ensure that predictors don't overlap with the target event. And above all, the model is trained on cases which had gone through the exact customer interaction planned to be launched at a larger scale.

An outline of the mining procedure followed is presented in Figure 7.19.

7.11 The modeling procedure

The IBM SPSS Modeler stream (procedure) for the cross-sell model is presented in Figure 7.20. The stream loads and uses the fields of Table 7.7 discussed above.

Step			
1. Modeling file	Active card-holders with no purchases at the House & furniture department Historical data used for modeling: customer “signature” mining file with 5 months of purchases history (2012-1-1 to 2012-5-31)		
2. Random sampling for selecting participants in the pilot campaign	Nonparticipants in the pilot campaign	**Modeling population:** participants in the pilot campaign	
3. Running the pilot campaign. Record responses, analyze responses, and train the model	Nonparticipants in the pilot campaign	Customers who refused the offer	**Target population:** customers who accepted the offer
4. Score those not included in the test campaign	**Scoring population:** Apply the model on customers with no purchases at the House and furniture department who didn't participate in the test campaign to score them and target the large-scale campaign		

Figure 7.19 An outline of the mining procedure followed for the development of the cross-selling model

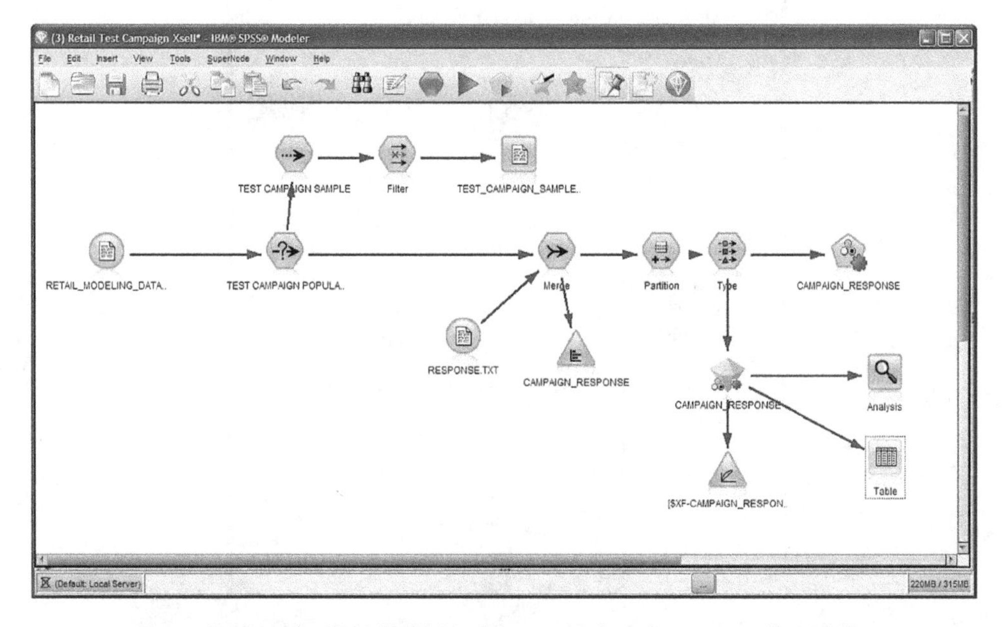

Figure 7.20 The IBM SPSS Modeler procedure for cross-sell modeling

7.11.1 Preparing the test campaign and loading the campaign responses for modeling

Initially, a random sample of customers of the modeling file has been drawn for inclusion in the test campaign. A Sample node has been applied to customers without purchases at the House department, and the campaign list had been stored at an external file. Upon conclusion of the pilot campaign, response data had been collected and stored in a dedicated file. The model had to be trained on known response. Therefore, the relevant data had been retrieved and merged with the rest of the modeling data. An inner join had been applied so that only customers who participated in the campaign were retained for model training. Figure 7.21 presents the Merge node used for joining the two data sources with the CARD_ID field used as the key for merge.

7.11.2 Applying a Split (Holdout) validation: splitting the modeling dataset for evaluation purposes

The loading of response data defined the final modeling population as well as the target attribute, the flag field CAMPAIGN_RESPONSE indicating customers who accepted the offer and used the discount voucher (value "T"). However, not all campaign participants took part in the model training. A random sample of them was left out (Holdout sample) to be used for

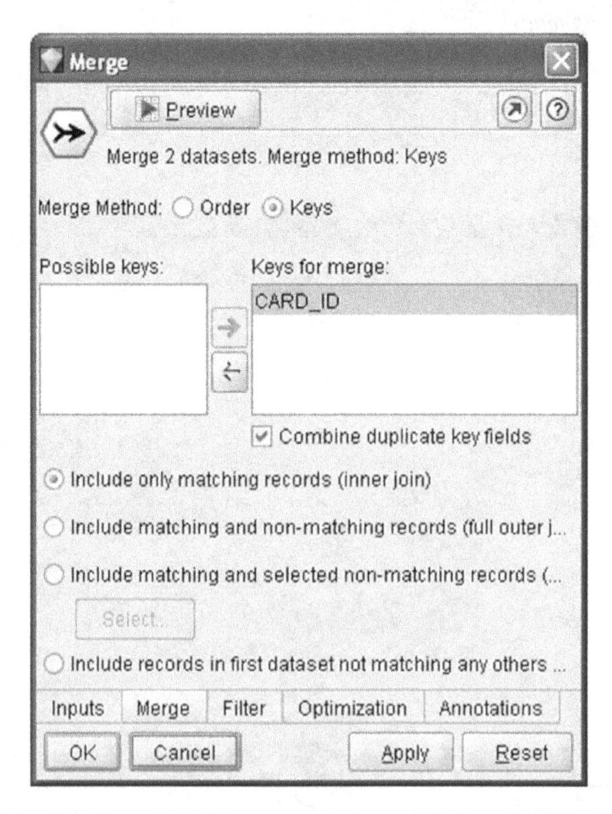

Figure 7.21 Merging campaign response data with the rest of the information

model evaluation. As already stressed before, using the same data for training and testing the model can lead to optimistic measures of accuracy. Therefore, a standard procedure for ensuring realistic model evaluation is to apply a Split (Holdout) validation so that the model is evaluated on unseen instances. In our case study, a Partition node was applied to split the modeling dataset into Training and Testing datasets through random sampling as shown in Figure 7.22. The Training partition size was set to 75%. Therefore, the 75% of the 13 074 campaign participants were used for the training of the model. The remaining 25% were held out to be used for assessing the model accuracy.

Additionally, a random seed was specified in order to "stabilize" the underlying random sampling procedure and ensure the same partitioning on every model run.

7.11.3 Setting the roles of the attributes

The last column of Table 7.7 denotes the attributes used as predictors in the classification model. Obviously, the information regarding the offer acceptance, recorded in the CAMPAIGN_RESPONSE field, defined the model's target. The role of each attribute in the model was set through a Modeler Type node (Figure 7.23). Predictors were assigned an input

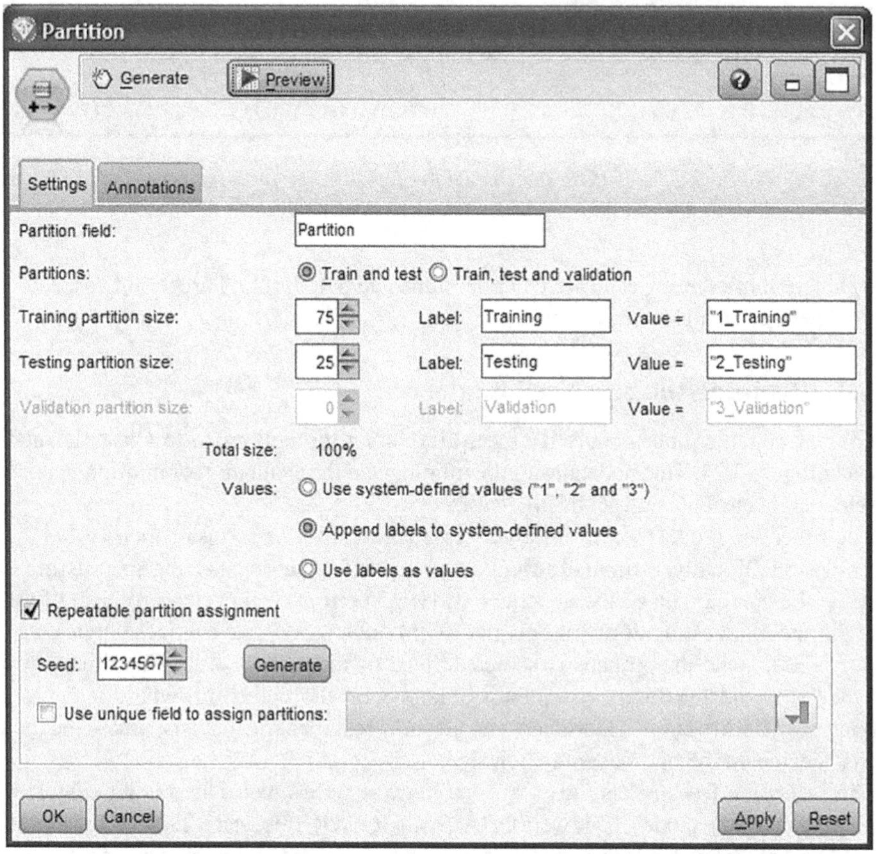

Figure 7.22 Partitioning the modeling dataset for evaluation purposes

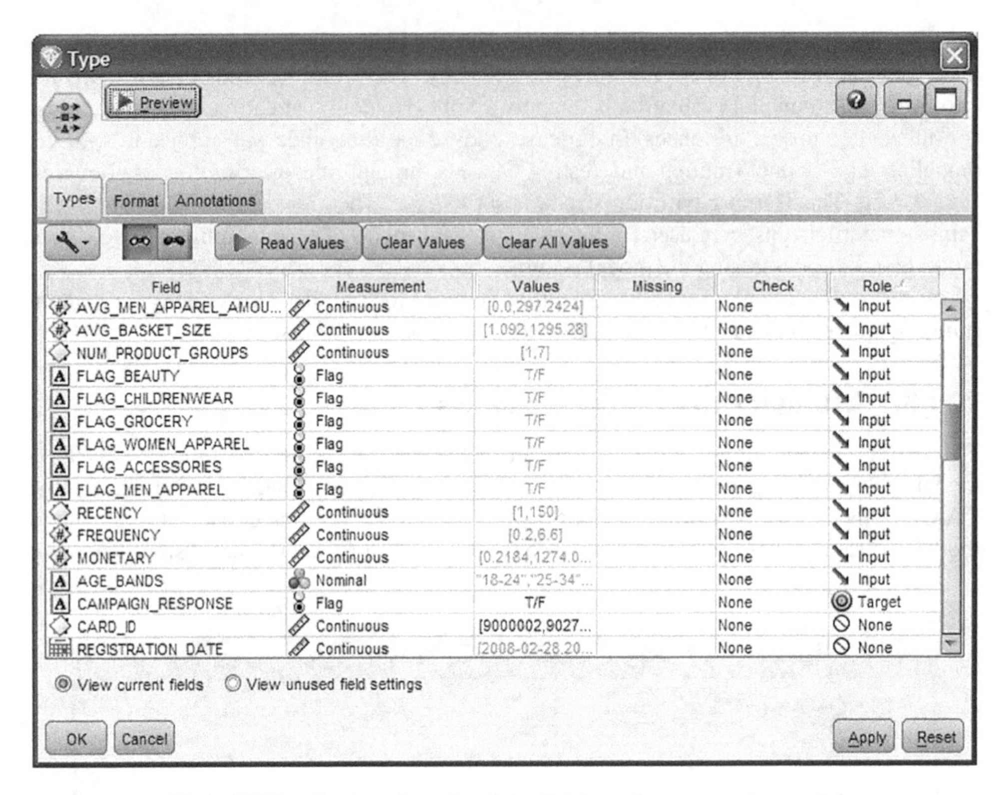

Figure 7.23 Setting the role of the fields in the propensity model

role (Direction In), the target attribute was assigned an output role (Direction Out), and fields omitted were given a None direction.

7.11.4 Training the cross-sell model

A series of classification models has been developed through an Auto Classifier node as shown in Figure 7.24. This node allows the training and the evaluation of multiple models for the same target attribute and set of predictors.

The generated models were initially assessed and ranked in terms of their Lift at the 2% percentile. Additionally, estimated revenue and cost information has been specified in order to enable the comparison of the generated models in terms of expected profit and ROI. By using the specified estimations for cost per offer/contact (50$), revenue if someone accepts the offer (200$), and the propensity scores derived by the models, Auto Classifier can compute the estimated maximum profitability along with the percentile where it occurs. By studying profitability curves, marketers can perform cost–benefit analysis, assess the models, and also decide on the optimal campaign size.

Three Decision Tree models and a Bayesian Network were selected for training (Figure 7.25). More specifically, the models built were CHAID, C5.0, C&R Tree, and a TAN (Tree Augmented Naïve Bayes) model with a feature selection preprocessing step to omit irrelevant predictors.

The parameters specified for each model are displayed in Table 7.9.

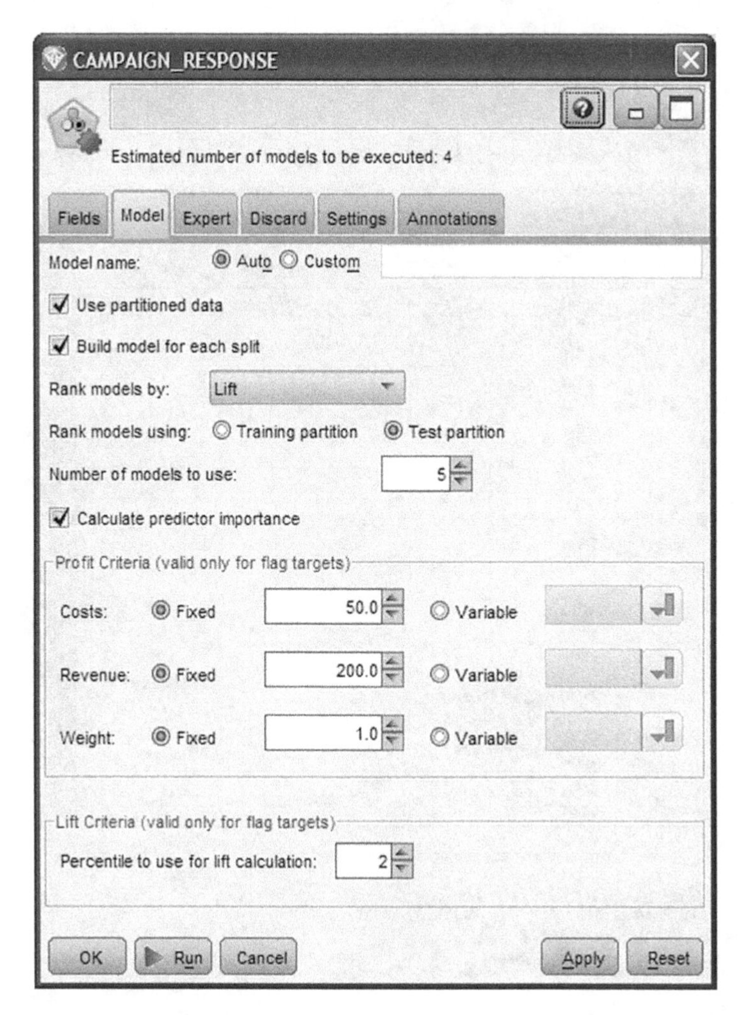

Figure 7.24 Building multiple classification models with Auto Classifier

7.12 Browsing the model results and assessing the predictive accuracy of the classifiers

By double-clicking the generated model nugget in the modeling stream, we can browse the model results. The Model tab provides an initial evaluation of the individual models in terms of Lift, Maximum Profit, Overall Accuracy, and Area Under Curve as displayed in Figure 7.26. All these metrics are based on the testing dataset.

Models are ranked according to their Lift values at the top 2% percentile. Although all models performed similarly well in respect to Lift, with the exception of C&RT which presented relatively lower values, the CHAID model was ranked first with a Lift of 3.35. The overall proportion of responders was about 22%. A Lift value of 3.35 means that the percentage of responders was 3.35 times higher, reaching a proportion of 75% (3.35*22%), among the top 2% of customers with the highest propensities. Thus, the CHAID model managed to triple the

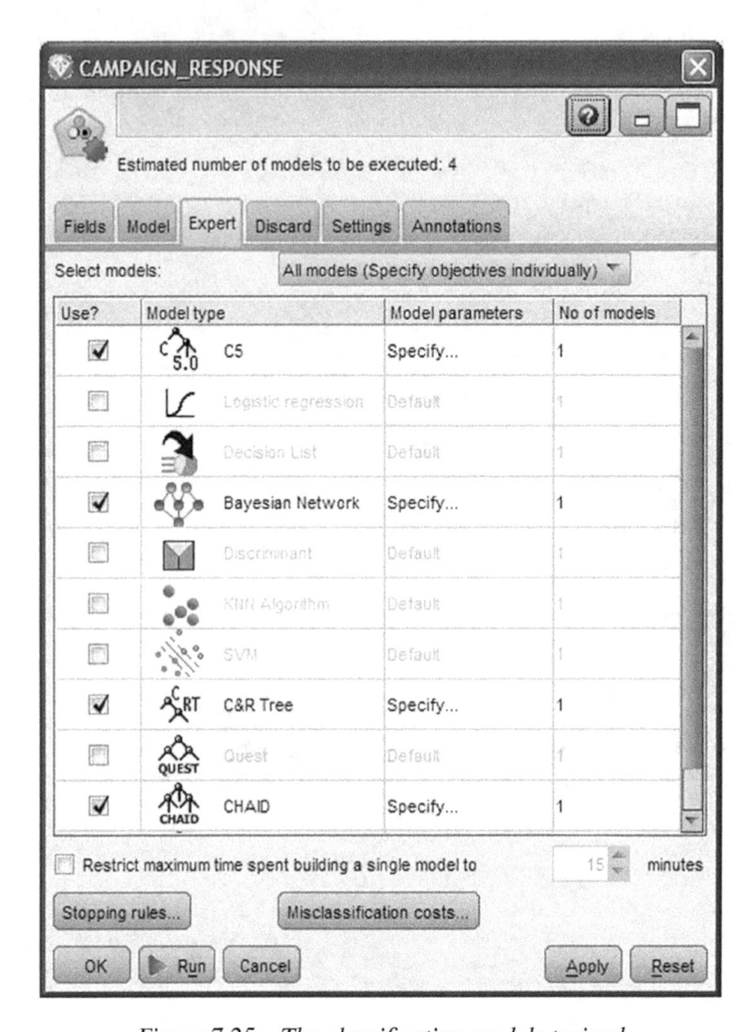

Figure 7.25 The classification models trained

concentration of potential buyers among its highest quantiles. In respect to prediction accuracy, the figures of all models were analogous with the percentage of correct classification around 79% and the misclassification rate around 21%.

The estimated Maximum (cumulative) Profit for the CHAID model was the highest, 31 550$, occurring at the top 30% quantile. C&RT presented the lowest estimated maximum profit.

The Gains chart displayed in Figure 7.27 provides a visual comparison of the models' ability to identify the target customers.

A separate Gains curve is plotted for each model summarizing the percentage of true positives, that is, the percentage of actual responders, for the model percentiles. The diagonal line represents the baseline model of randomness. The $BEST line corresponds to "good's model," a hypothetical ideal model which classifies correctly all cases. A good model is expected to show a gradual but steep increase at the left of the *X*-axis before flattening out. Obviously, all curves reach 100% at the rightmost part of the *X*-axis. But what's interesting

Table 7.9 The parameters of the cross-selling models.

Parameter	Setting
CHAID model parameter settings	
Model	CHAID
Levels below root	6
Alpha for splitting	0.05
Alpha for merging	0.05
Chi-square method	Pearson
Minimum records in parent branch	200
Minimum records in child branch	100
C&R Tree model parameter settings	
Levels below root	7
Minimum change in impurity	0.0001
Impurity measure for categorical target	Gini
Minimum records in parent branch	100
Minimum records in child branch	50
Prune tree	True
Use standard error rule	True
Multiplier	1.0
C5.0 model parameter settings	
Output type	Decision Tree
Group symbolics	True
Use boosting	False
Pruning severity	75
Minimum records per child branch	50
Use global pruning	True
Winnow attributes	False
Bayesian Network model parameter settings	
Structure type	TAN (Tree Augmented Naïve Bayes)
Include feature selection preprocessing step	True
Parameter Learning Method	Maximum Likelihood
Independence test (for the feature selection)	Likelihood ratio
Significance level (for the feature selection)	0.01
Maximum number of inputs (selected by feature selection)	10

is what goes on at the higher cutoffs, among customers with high estimated propensities. The CHAID model seems to behave better with a smooth curve steadily higher than the rest. Differentiations are trivial though at higher percentiles and only begin to become evident after the 20% percentile. What else does the Gains chart tells us? The percentage of potential customers expected to be reached if we use a campaign list ranked by propensities. For

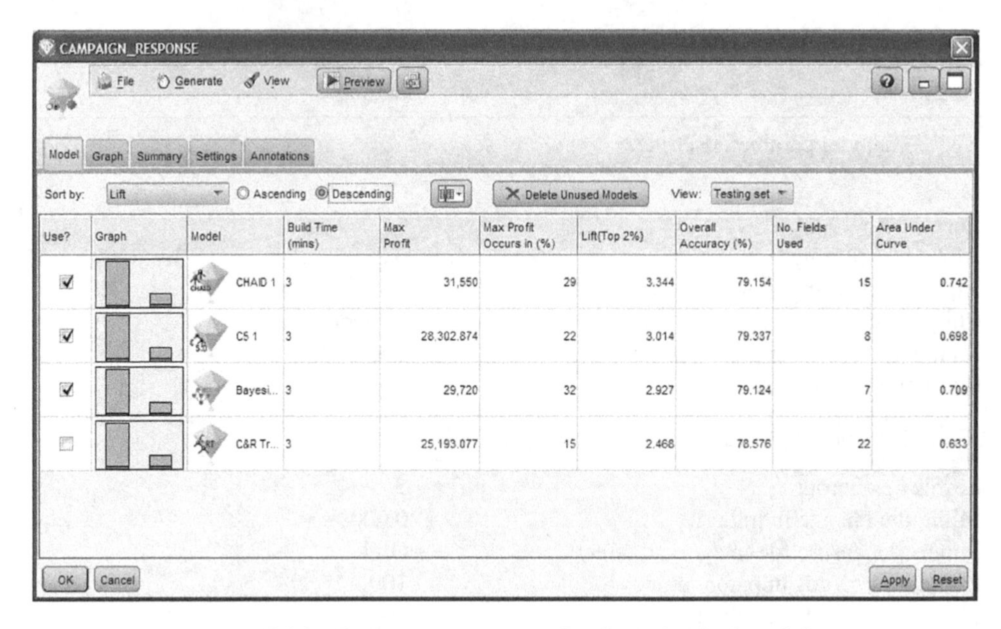

Use?	Graph	Model	Build Time (mins)	Max Profit	Max Profit Occurs in (%)	Lift(Top 2%)	Overall Accuracy (%)	No. Fields Used	Area Under Curve
☑		CHAID 1	3	31,550	29	3.344	79.154	15	0.742
☑		C5 1	3	28,302.874	22	3.014	79.337	8	0.698
☑		Bayesi...	3	29,720	32	2.927	79.124	7	0.709
☐		C&R Tr...	3	25,193.077	15	2.468	78.576	22	0.633

Figure 7.26 Performance metrics for the individual models

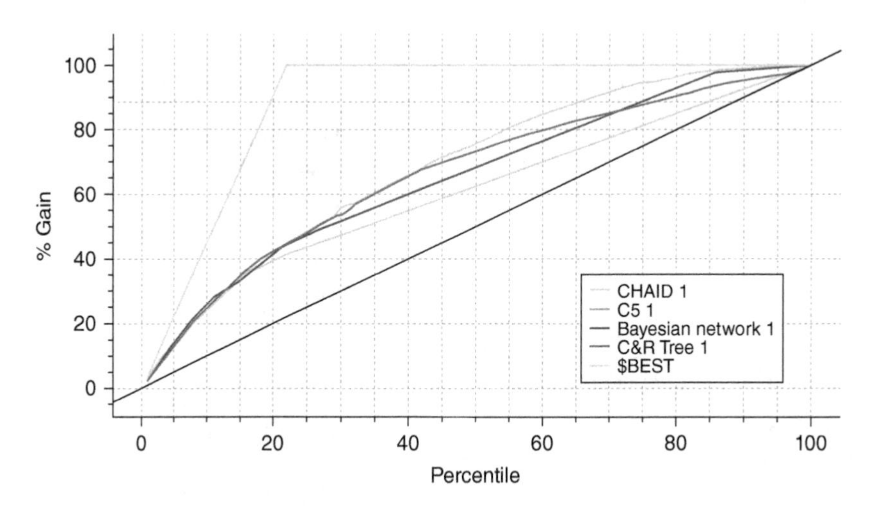

Figure 7.27 The Gains chart for the cross-selling models

example, by contacting the 20% of customers with the higher model propensities, we expect to capture the 40% of the total buyers. The percentage of responders expected to be captured is even larger at higher cutoff values.

To gain insight on the patterns associated with response, let's examine the results of some of the individual models starting from the C5.0 tree. The viewer tab displays the model in tree format. Its first three levels are shown in Figure 7.28.

The response rate was higher among customers of the Grocery, Childrenwear, and Accessories department, possibly indicating a target group of women with children who

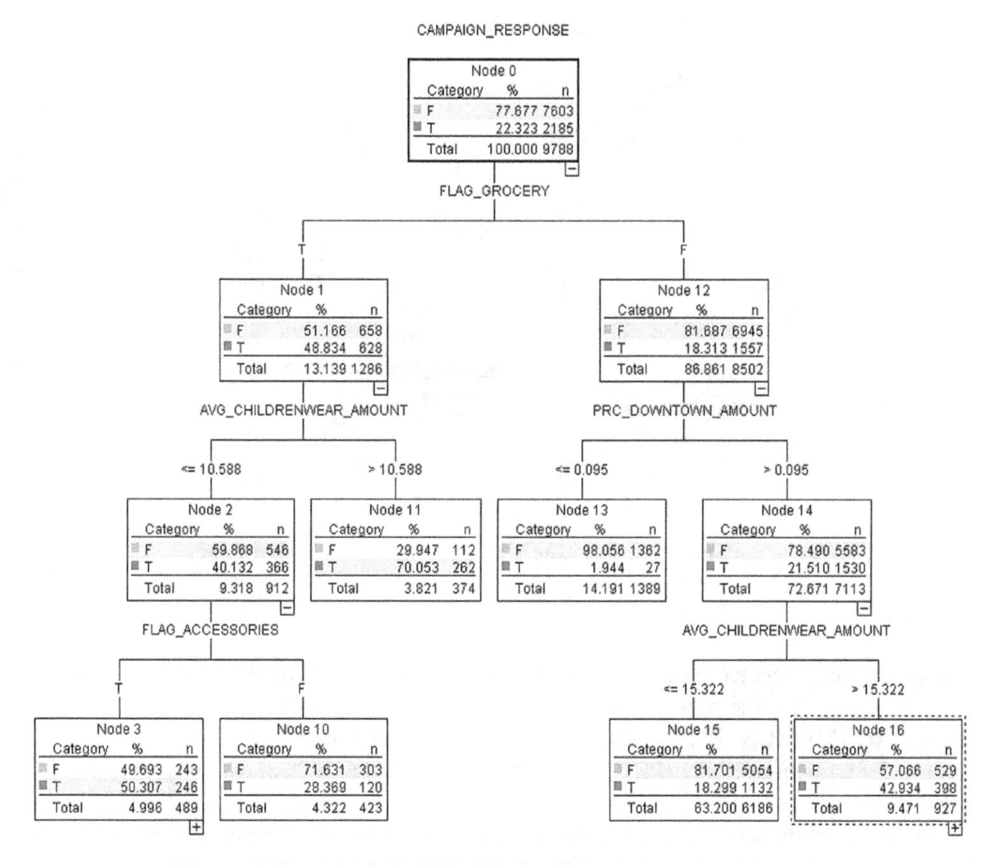

Figure 7.28 The first three levels of the generated C5.0 model

are also responsible for the house equipment. The overall response rate of 22% was tripled, reaching 70%, among customers who had already purchased grocery and childrenwear products in the past. Frequency and diversity of purchases (appearing in the next splits and in the subsequent levels of the tree) also seemed to have a positive relation with response, further increasing the percentage of those who accepted the offer. What is evident, even from the first splits of the model, is the gain from studying all aspects of purchase behavior instead of only using the RFM components as predictors. By introducing attributes which summarized the purchase preferences of the customers, the retailer had managed to reveal a richer profile of the target customers and boosted the predictive accuracy of the classification model.

Figure 7.29 presents the generated TAN (Tree Augmented Naïve Bayesian network) and its network's structure.

The same products which were identified by the C5.0 model to be associated with the response (Grocery, Accessories, Childrenwear) were also retained as significant predictors by the preprocessing feature selection procedure of the Bayesian network. Recency, frequency, and diversity of purchases (the NUM_PRODUCT_GROUPS field denoting the distinct number of product groups with purchases) were also included in the model. The significant predictors and the target field are represented as nodes in the Bayesian network.

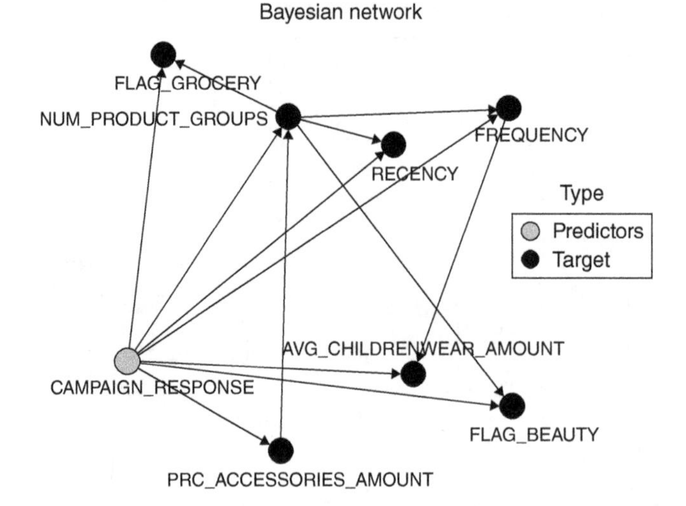

Figure 7.29 The TAN Bayesian network

The nodes are connected with arcs (directed edges). If an arc is drawn from attribute *X* to attribute *Y*, then *X* is the parent of *Y*. All predictors have two attributes as parents, the target attribute (CAMPAIGN_RESPONSE field) plus one more additional predictor, except the accessories amount attribute (PRC_ACCESSORIES_AMOUNT) whose only parent is the target attribute.

Our assumption here is that each attribute's probabilities only depend on the information provided by its parents. These probabilities are presented in conditional probability tables (CPTs). The CPT for each attribute is displayed when clicking on its node. Figure 7.30 displays the CPT for the grocery flag attribute (FLAG_GROCERY field).

Note that all numeric attributes had been discretized in five bands. Each row of the table presents the probability distribution of the attribute, given its parents. In other words, the table summarizes the probability of each value of the node attribute (grocery field) for each combination of values of its parents. Obviously, all row entries sum to 1. For instance, by examining the seventh row of Figure 7.30, we can see that the probability of a grocery buyer is 0.64 among responders of the test campaign with 4.6–5.8 product groups' purchases. This probability is lower (0.48) among nonresponders of the campaign with the same number of product groups' purchases (eighth row of the CPT).

Under the assumption stated above, the algorithm can use the CPTs to estimate the probability of response for each customer/instance by multiplying the CPT entries.

Although the CHAID model presented superior performance and seemed a perfect candidate for scoring customers, the final decision was to apply for deployment a model voting procedure based on all models except C&RT. After all, using the advices of more than one "experts" minimizes the risk of misguidance. The ensemble method applied was confidence-weighted voting. Furthermore, although the Gains chart provided valuable information, marketers also studied profitability and ROI charts (to be presented below) to optimize the campaign size according to the estimated financial gains. These charts were created by evaluating the ensemble model with an Evaluation node.

Conditional probabilities of FLAG_GROCERY

Parents		Probability	
NUM_PRODUCT_GROUPS	CAMPAIGN_RESPONSE	T	F
< 2.2	T	0.09	0.91
< 2.2	F	0.03	0.97
2.2 ~ 3.4	T	0.20	0.80
2.2 ~ 3.4	F	0.14	0.86
3.4 ~ 4.6	T	0.38	0.62
3.4 ~ 4.6	F	0.25	0.75
4.6 ~ 5.8	T	0.64	0.36
4.6 ~ 5.8	F	0.48	0.52
> 5.8	T	1.00	0.00
> 5.8	F	1.00	0.00

Figure 7.30 The conditional probabilities for the FLAG_GROCERY attribute

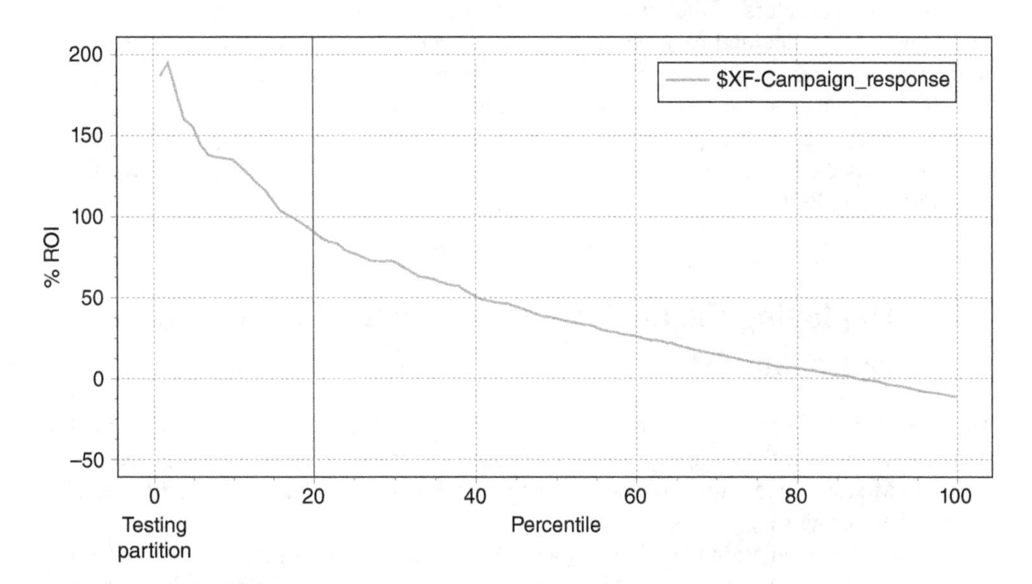

Figure 7.31 The ROI chart of the ensemble model

The ensemble model presented an increased AUC (Area under the ROC curve) measure of 0.753. Figure 7.31 presents its ROI chart. The estimated cumulative %ROI for each percentile is calculated as

$$\text{ROI\%} = (\text{estimated Profit per offer/Cost})\% = (((\text{Revenue} \times \text{Response }\%) - \text{Cost})/\text{Cost})\%$$

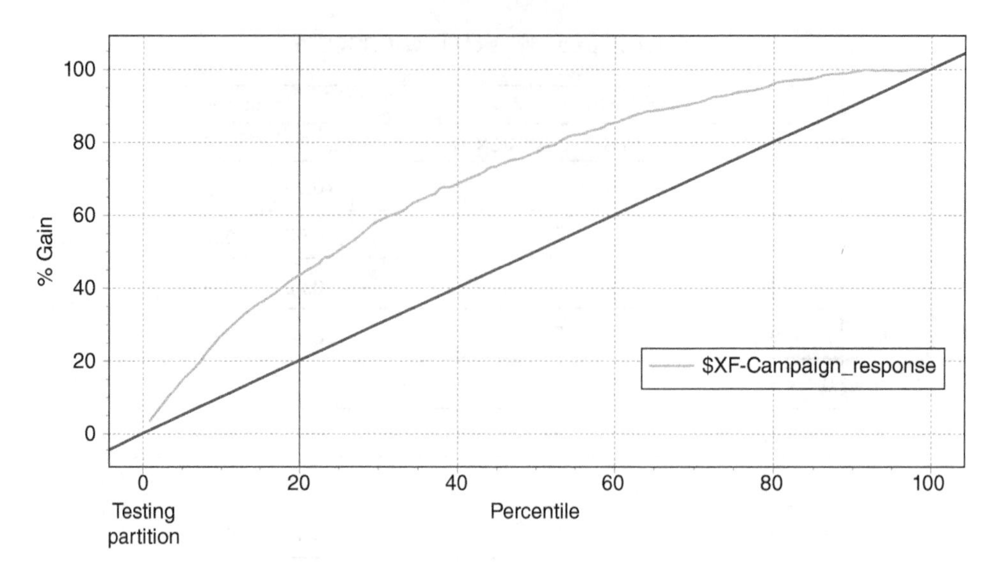

Figure 7.32 The Gains chart of the ensemble model

The ROI chart indicates the (probabilistic) return on investment per instance (per offer).

Since the marketers of the retailer wanted profits (revenue-cost) at least equal to their investments, they decided to target the top 20% percentile which presented an estimated ROI of about 100%. Thus, they expect an average profit of 50$ for each 50$ they spend.

The Gains chart of the ensemble model is presented in Figure 7.32.

The top 20% percentile presented a Gains value of 43.2%. Thus, by targeting these tiles, marketers expected to capture about 40% of the total responders and double their investment by achieving an ROI of about 100%.

7.13 Deploying the model and preparing the cross-selling campaign list

The final step of the procedure was the deployment phase. The trained model was applied to "unseen" cases, and the campaign list was constructed based on cross-selling propensities. The IBM Modeler stream for deployment is shown in Figure 7.33. The modeling file was also used as the scoring file.

The scoring population contained all active customers with no purchases at the House department who had not taken part in the test campaign. Through a Select node, customers with purchases at the House department had been filtered out. Then, only customers present in the scoring file but not present in the campaign responses file were retained.

Two new, estimated fields were derived for scored customers:

1. $XF-CAMPAIGN_RESPONSE: indicating the predicted class
2. $XFC-CAMPAIGN_RESPONSE: indicating the prediction confidence

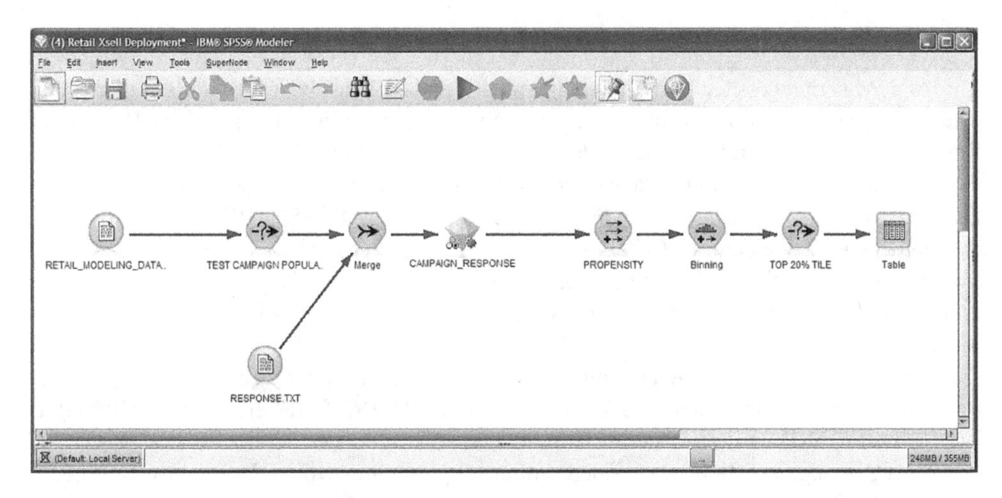

Figure 7.33 The Modeler deployment stream

Preview from Filter Node (5 fields, 10 records)

	CARD_ID	$XF-CAMPAIGN_RESPONSE	$XFC-CAMPAIGN_RESPONSE	PROPENSITY	PROPENSITY_TILE5
1	9000018	T	0.325	0.325	5
2	9000022	F	0.471	0.529	5
3	9000026	T	0.659	0.659	5
4	9000051	T	0.473	0.473	5
5	9000060	T	0.676	0.676	5
6	9000084	T	0.795	0.795	5
7	9000108	T	0.483	0.483	5
8	9000199	F	0.532	0.468	5
9	9000226	T	0.795	0.795	5
10	9000231	T	0.438	0.438	5

Figure 7.34 The model generated fields for the scored customers

The prediction confidence can be easily transformed into cross-selling propensity, that is, likelihood of accepting the offer, by subtracting the confidence from 1 in the case of customers predicted as "F" (nonresponders). The campaign list included the top 20% of the customers with the highest propensities, corresponding to a propensity cutoff value of about 0.30. Through binning, customers were grouped into quintiles (bins of 20%), and the members of bin 5 were selected for the campaign. The model estimated fields, namely, prediction, confidence, propensity, as well as the propensity tile for a sample of scored customers, are presented in Figure 7.34.

7.14 The retail case study using RapidMiner

In the next paragraphs, the development of the value segments, the RFM segments, and the cross-sell model is presented with the employment of RapidMiner.

7.14.1 Value segmentation and RFM cells analysis

The RapidMiner process for both value segmentation and RFM analysis is shown in Figure 7.35. The upper part of the process implements the value segmentation and starts with a Repository Retrieve operator which loads the modeling dataset presented in Table 7.7.

The value segmentation is a simple discretization task, and it was performed by a Discretize by Frequency operator which was applied to the MONETARY attribute. Initially, cardholders were binned into 20 tiles (vingitiles) of 5% each as shown in Figure 7.36. The top 5% of customers with the highest purchases were assigned to the "range20" bin, while on the other end of the monetary pyramid, the bottom 5% of customers with the lowest purchases were assigned to "range1" bin.

The discretization converted the numeric MONETARY field into a nominal one. In order to retain the amount information as well, the original MONETARY field was replicated as MONETARY_NUM with a Generate Copy operator.

The value segmentation was refined with a Map operator which further grouped the 20 bins into the final four value segments:

- 1_CORE (Bottom 50%)
- 2_BRONZE (Medium–Low 30%)
- 3_SILVER (High 15%)
- 4_GOLD (Top 5%)

The importance of the derived segments was disproportionate to their sizes as shown in the pie chart in Figure 7.37 which depicts the share of each segment in the total purchase amount.

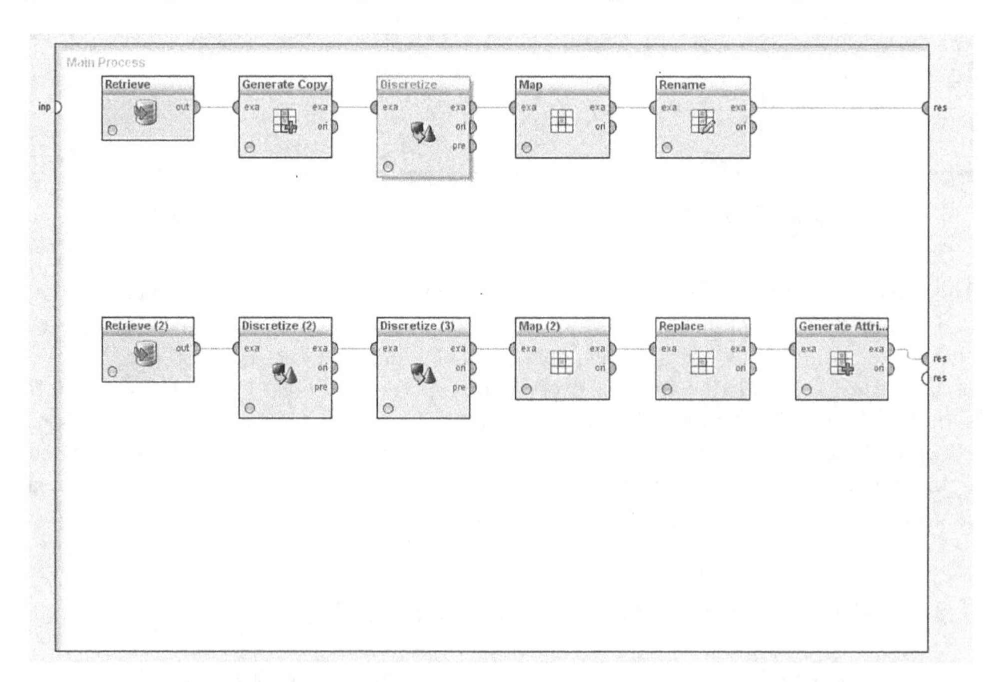

Figure 7.35 The RapidMiner process for value segmentation and RFM analysis

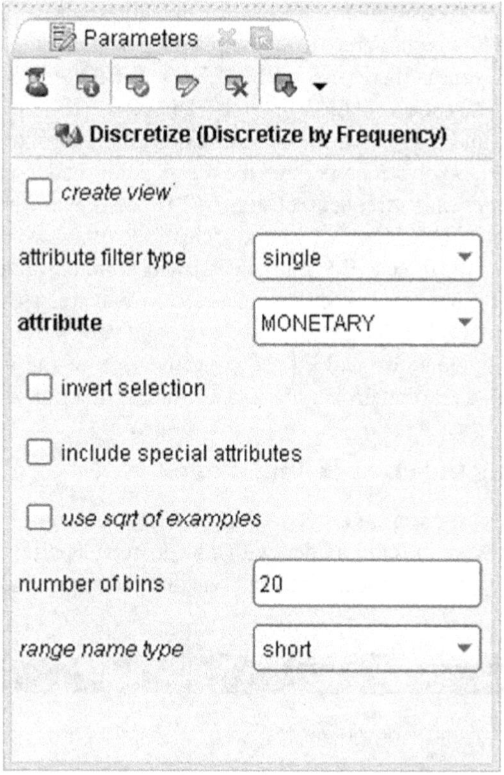

Parameters
Discretize (Discretize by Frequency)
create view
attribute filter type
single
attribute
MONETARY
invert selection
include special attributes
use sqrt of examples
number of bins
20
range name type
short

Figure 7.36 The Discretize by frequency operator applied for value segmentation

1_CORE
3_SILVER
4_GOLD
2_BRONZE
1_CORE 4_GOLD 2_BRONZE 3_SILVER

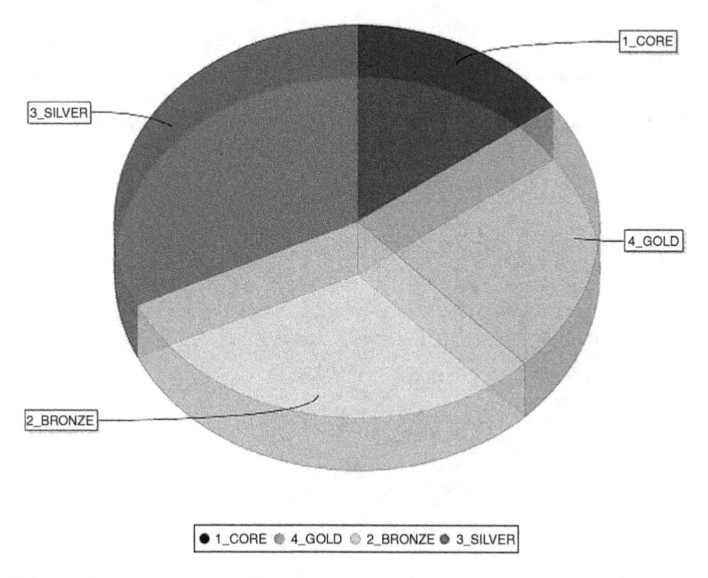

Figure 7.37 The percentage of the total purchase amount for each value segment

The lower part of the process builds the RFM segmentation scheme by using the same modeling dataset. The procedure is similar to the one applied for value segmentation and involved the discretization of the relevant numeric fields (RECENCY, FREQUENCY, MONETARY attributes) into five pools of 20% each (quintiles) through a Discretize by frequency operator (actually, due to lower dispersion and many value ties, the FREQUENCY attribute had to be binned into four bins of 25% each). Do you remember that the recency bins had to be inversed so that bin 5 would correspond to better (more frequent) customers? This task was carried out by a Map operator which aligned the RECENCY scale with the scale of the other two components.

Before concatenating the three RFM components, their nominal values of the form "range1" to "range5" were slightly modified. The "range" prefix which was assigned automatically by RapidMiner was trimmed with a Replace operator so that all values were of the form 1 to 5. The final step for RFM segmentation involved the concatenation of the individual *R*, *F*, and *M* values through a Generate Attributes operator and a concat() function as shown in Figure 7.38.

7.14.2 Developing the cross-selling model

The RapidMiner process file for the cross-selling model is shown in Figure 7.39.

The modeling data were initially loaded with a Repository Retrieve operator. Apparently, the model training requires instances with known outcome (purchase yes/no). Therefore,

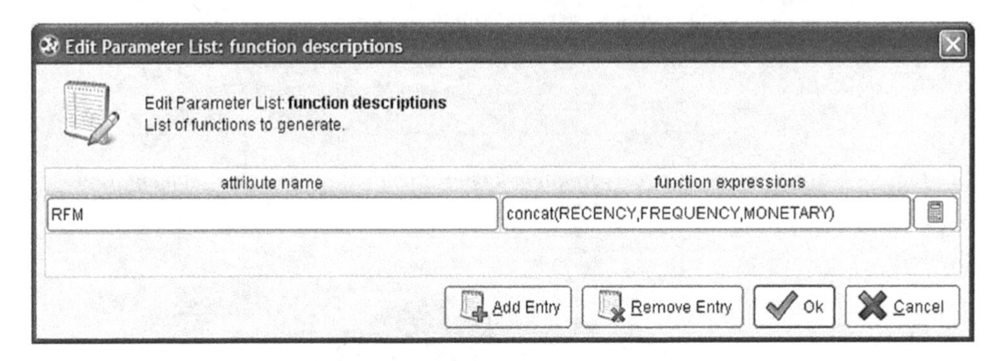

Figure 7.38 Constructing RFM cells by concatenating the relevant binned attributes

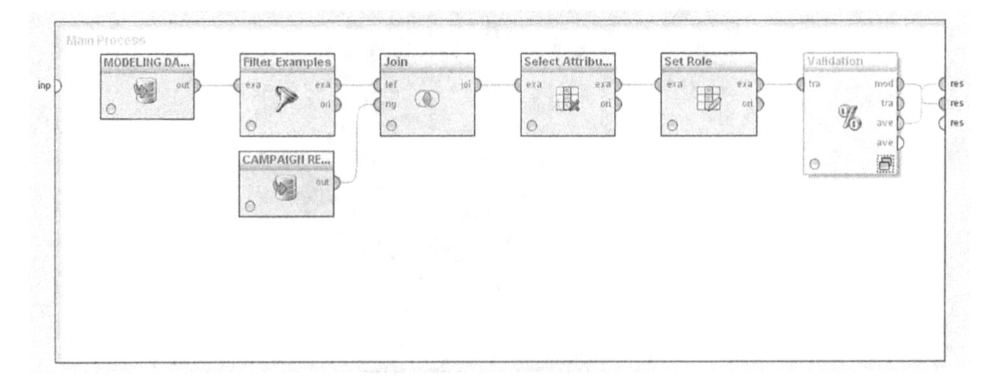

Figure 7.39 The RapidMiner process file for the cross-selling model

only customers who had participated in the test campaign were selected. Shoppers of the House department were filtered out with a Filter Examples operator. Then, in order to merge the modeling data with the campaign response data, a Join operator has been applied. The CARD_ID field was used as the key for merge and an inner join type was applied. Only customers present in both data sources, hence only customers who had participated in the test campaign, were retained.

A Select Attribute operator was then used to keep only the predictors and the target field (label field, in RapidMiner's terminology) as listed in Table 7.7. The rest of the attributes were omitted and did not contribute in model building.

The role of each field in model building has been defined using a Set Role operator. Specifically, the CAMPAIGN_RESPONSE field, recording the campaign responses, was assigned a label (target) role for the subsequent cross-selling model. The rest of the fields kept their default regular role to participate as predictors in the classification model.

7.14.3 Applying a Split (Holdout) validation

The evaluation of the classification model is optimistic when based on the training dataset. In order to evaluate the model on a sample different than the one used for model training, a Split (Holdout) validation method has been applied through a Split Validation operator. This operator partitioned the modeling dataset into a training and a testing part of 70% and 30%, respectively, as shown in Figure 7.40.

The dataset was split with a stratified random sampling and a 0.7 split ratio. Therefore, 30% of the training instances were reserved for evaluation and didn't participate in the model training. The sampling with stratification ensured that the distribution of the target field was maintained across both samples.

The Split Validation operator, contains two subprocesses as displayed in Figure 7.41.

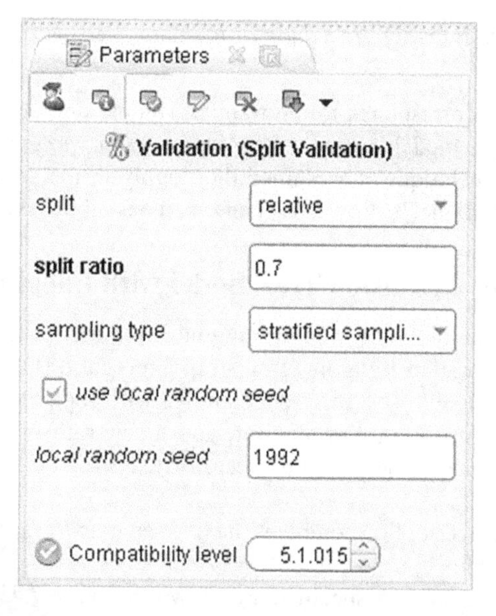

Figure 7.40 The Split validation settings

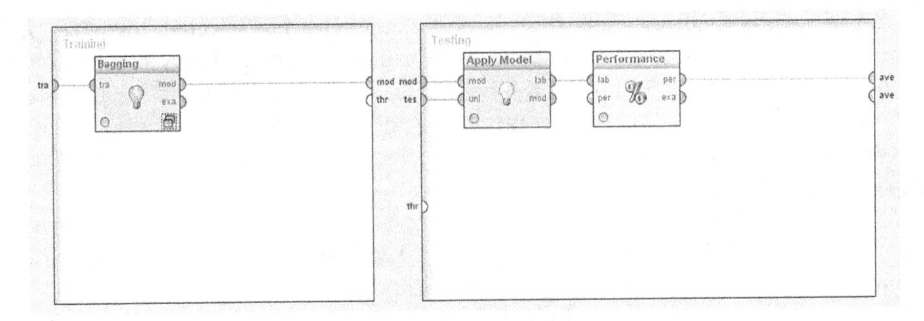

Figure 7.41 The Split Validation operator for partitioning the modeling dataset

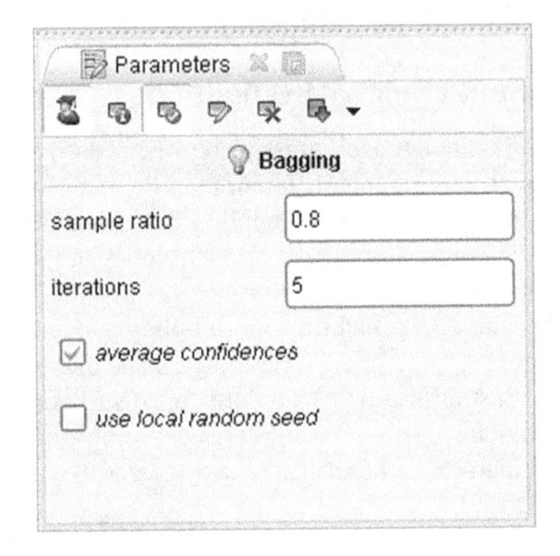

Figure 7.42 The Bagging procedure for building five separate Decision Trees

The left subprocess corresponds to the training sample and covers the model training phase. Instead of a single model learner, a Bagging operator was applied in order to build a set of classifiers. This technique is explained thoroughly in the next paragraph. The right subprocess corresponds to the Testing dataset and includes all the evaluation steps.

7.14.4 Developing a Decision Tree model with Bagging

A widely used ensemble method for combining multiple models in order to improve the classification accuracy, called Bagging (Bootstrap aggregation), has been applied in this case study. More specifically, five different samples with replacement have been chosen at random from the original training dataset, and a separate Decision Tree model has been built for each sample. The Bagging operator used for this procedure is shown in Figure 7.42.

The average confidences option combines the five separate trees, and their average confidences are used for scoring.

The Decision Tree parameters specified are shown in Figure 7.43, and they are summarized in Table 7.10.

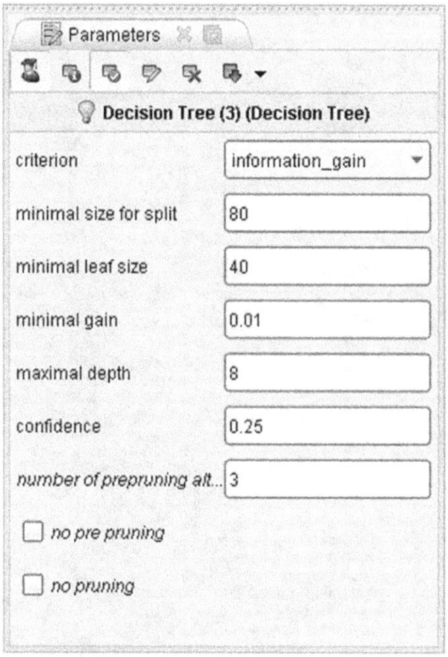

Figure 7.43 The Decision Tree model parameter settings

Table 7.10 Decision Tree model parameters

Decision Tree model parameter settings	
Parameter	Setting
Criterion	Information gain
Minimal size for split	80
Minimal leaf size	40
Minimal gain (to proceed to a split)	0.01
Maximal depth (levels of the tree)	8
Confidence	0.25

The execution of the RapidMiner process generated a bagged model, consisted of the five individual Decision Trees. By browsing the trees and their branches from the root node down to their leaf nodes, we can see the attributes and the patterns which were associated with increased response. In each node, the lighter shade in the bar represents the proportion of buyers. The width of the bar represents the number of customers at each node. From a first, visual inspection of the trees, we can see an increased number of responders among frequent customers with increased diversity of purchases (purchases of distinct product groups). Customers with purchases at the Grocery, Childrenwear, Accessories, and Beauty departments, especially in the Downtown Store, are also probable buyers. The first of the five Decision Trees is presented in Figure 7.44 in Graph view.

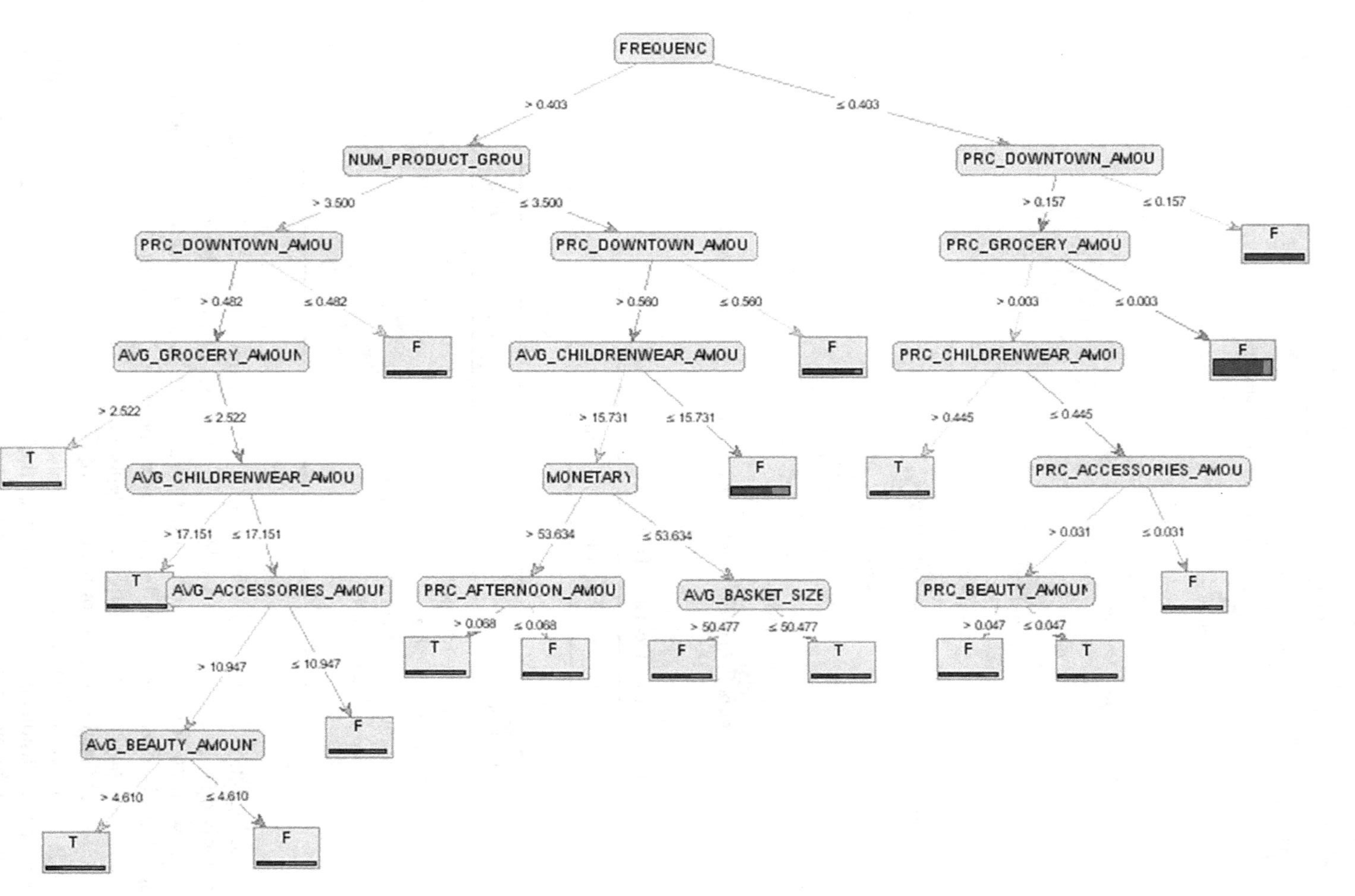

Figure 7.44 The Decision Tree model in tree format

7.14.5 Evaluating the performance of the model

The next step was to evaluate the performance of the bagged classifier. The Performance operator, placed in the testing subprocess of the validation, provided numerous performance metrics for assessing the classification accuracy of the model. Note that the Performance operator requires scored (labeled) instances; therefore, it was preceded by an Apply Model operator which classified each customer according to the model results. Also note that all evaluation measures are based on the 30% testing subsample.

The Confusion matrix of the ensemble classifier is presented in Figure 7.45. The overall accuracy (correct classification rate) was 79.19% since 3106 of the 3922 testing instances were classified correctly (2915 TN, True Negatives, and 191 TP, True Positives). The misclassification rate was 20.81% (134 FP, False Positives, and 682 FN, False Negatives).

The Precision of the model was 58.77%. It denotes the percentage of the predicted positives which were actual positives (TP/TP+FP). The Recall measure (true positive rate or sensitivity) was 21.88%. It denotes the percentage of actual positive instances predicted as such (TP/TP+FN). The *F* measure is a measure generated by combining Precision (a measure of exactness) and Recall (a measure of completeness) in a single metric. In our case, the *F* measure had a value of 31.89%.

The 325 customers predicted as responders are those with the highest response propensities, and they comprise the 8.3% top percentile. The percentage of total buyers appearing in the top 8 percentile rises to 21.88% (equal to the Recall measure). A random 8% sample would have captured 8% of the total responders. Hence, the Lift at this percentile was 2.64 (21.88/8.3). The Lift further increases at a value of 3.4 at the 2% percentile (although not presented here a detailed Lift curve can easily be plotted using the Create Lift Chart operator).

The ROC curve plots the model's true positive rate in the *Y*-axis against the false positive rate in the *X*-axis: in other words the proportion of positive instances in the *Y*-axis (percentage of actual responders) against the proportion of misclassified negative instances in the *X*-axis (percentage of nonresponders) at different propensity cutoffs and samples. The ROC curve for the bagged classifier is depicted in Figure 7.46.

It shows a steep incline at the left of the *X*-axis, hence at higher model propensities, indicating acceptable model performance. The AUC metric (Area Under the ROC Curve measure) was 0.755.

Since the bagged model presented satisfactory predictive performance according to all evaluation metrics examined, the next step was to deploy it on new instances to target the campaign.

7.14.6 Deploying the model and scoring customers

The final step of the procedure was the deployment of the ensemble model on customers who hadn't participated in the test campaign. These customers were filtered with a Set Minus operator which discarded the participants of the pilot campaign from the modeling dataset. Then, the model was applied, through an Apply Model operator, to customers who hadn't yet visited the House department. The respective RapidMiner process is displayed in Figure 7.47.

Table View   Plot View

accuracy: 79.19%

	true F	true T	class precision
pred. F	2915	682	81.04%
pred. T	134	191	58.77%
class recall	95.61%	21.88%	

Figure 7.45 The Confusion matrix

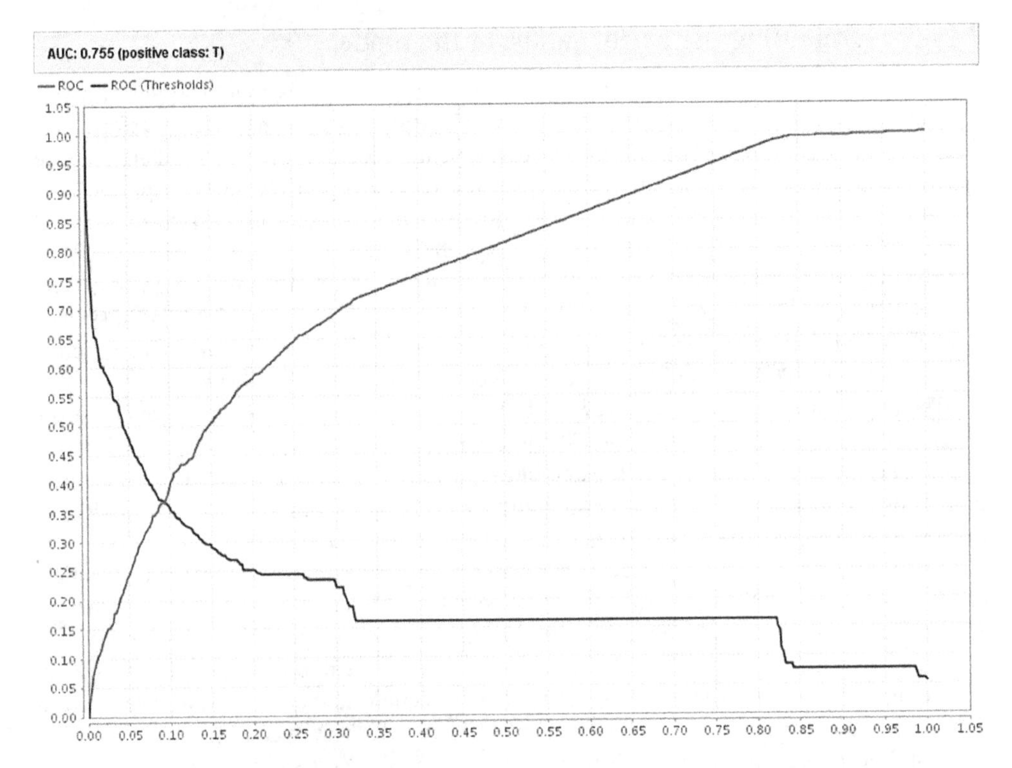

Figure 7.46 The model's ROC curve

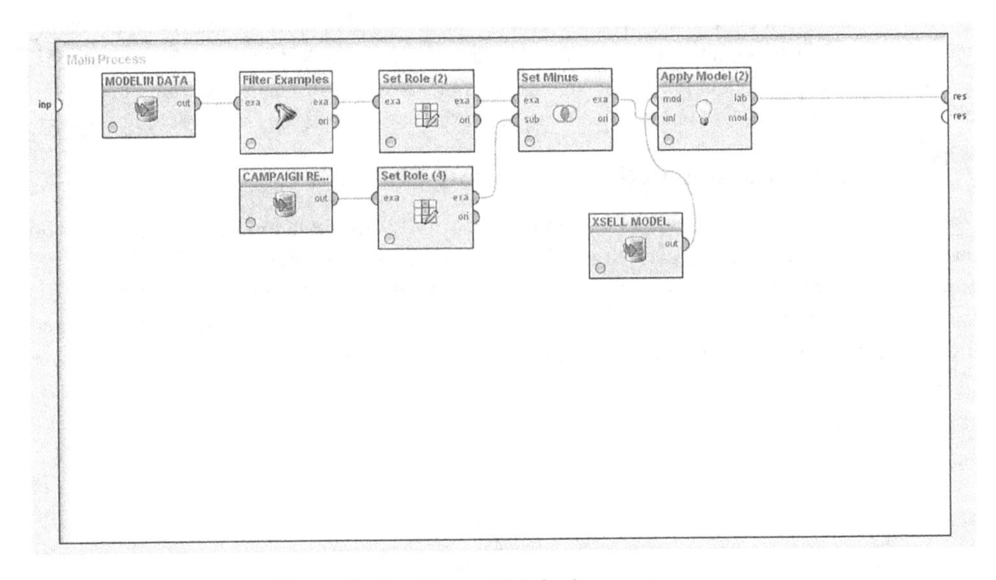

Figure 7.47 The model deployment process

ExampleSet (5666 examples, 4 special attributes, 83 regular attributes)iew Filter (5666 / 5666): all

Row No.	CARD_ID	confidence(F)	confidence(T)	prediction(CAMPAIGN_RESPONSE)	REGISTRATION_DATE
1	9000006	0.857	0.143	F	2008-03-23
2	9000010	0.842	0.158	F	2008-03-23
3	9000011	0.826	0.174	F	2008-03-24
4	9000018	0.396	0.604	T	2008-03-22
5	9000020	0.842	0.158	F	2008-03-25
6	9000022	0.502	0.498	F	2009-02-13
7	9000026	0.367	0.633	T	2008-03-29
8	9000031	0.744	0.256	F	2008-03-22
9	9000032	0.842	0.158	F	2008-05-29
10	9000049	0.857	0.143	F	2008-03-28
11	9000051	0.337	0.663	T	2008-03-23
12	9000060	0.325	0.675	T	2008-03-23
13	9000061	0.857	0.143	F	2008-03-28
14	9000064	0.811	0.189	F	2008-03-23
15	9000077	0.775	0.225	F	2009-02-13
16	9000079	0.857	0.143	F	2009-03-01
17	9000080	0.811	0.189	F	2008-03-22
18	9000084	0.266	0.734	T	2008-03-23
19	9000088	0.958	0.042	F	2009-02-22
20	9000094	0.777	0.223	F	2008-03-23
21	9000103	0.857	0.143	F	2008-03-23

Figure 7.48 The prediction fields derived by the RapidMiner Decision Tree model

Three new, model derived fields were created after the model deployment: the model prediction (field prediction(CAMPAIGN_RESPONSE) with values *T*/*F*), the estimated likelihood for nonresponse (confidence(*F*) field), and the response propensity, that is, the estimated likelihood for response (confidence(*T*) field). The values of the confidence fields range between 0 and 1. A screenshot of the prediction fields for a sample of customers is shown in Figure 7.48.

Customers were finally ranked according to their response propensities, and the cross-selling mailing list was constructed, including those with the highest likelihood to accept.

7.15 Building the cross-selling model with Data Mining for Excel

In the next paragraphs, the development of the cross-sell model is presented using this time the algorithms of Data Mining for Excel.

7.15.1 Using the Classify Wizard to develop the model

To build a classification model in Data Mining for Excel, we have to use the simple steps of the Classify Wizard. The modeling dataset includes those customers that had been involved in the test campaign. As outlined before, old shoppers of the House department had already been filtered out. The target field (CAMPAIGN_RESPONSE) records the campaign responses.

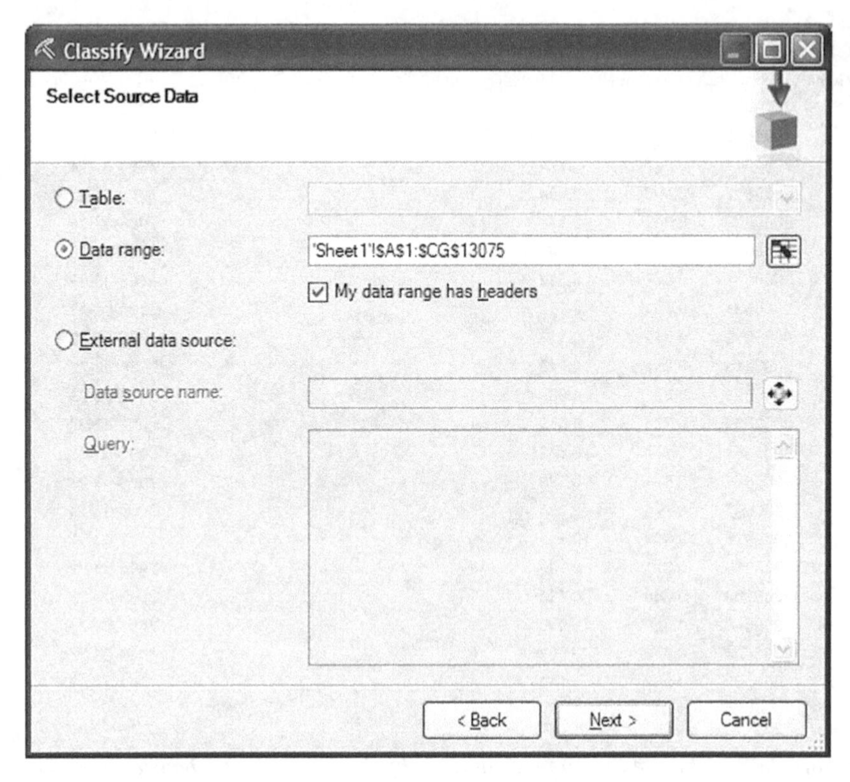

Figure 7.49 *Selecting the source data for model training in the Classify Wizard of Data Mining for Excel*

In the first step of the Classify Wizard, the source data for training the model has been selected as shown in Figure 7.49. In this example, a data range of the active datasheet was specified as the model dataset, although in general, data can also be loaded using an external data source and/or an SQL query.

The predictors ("Inputs") and the target field ("Column to analyze") were then selected as shown in Figure 7.50. The complete list of predictors is presented in the data dictionary shown in Table 7.7.

7.15.2 Selecting a classification algorithm and setting the parameters

Two different Decision Tree models were trained, using two different attribute selection methods ("Score methods"), the default BDE (Bayesian Dirichlet Equivalent with Uniform prior) and the Entropy. The algorithms and the respective parameters were set through the "Parameters..." menu as shown in Figure 7.51.

The default "Complexity" value, based on the number of 10+ used inputs, was 0.9. It was relaxed to 0.5 to yield a larger tree. The default "Split method" of "Both" was applied. This method produces optimal groupings of the inputs, combining multiway and binary splits. Finally, a "Minimum Support" value of 50 was specified, setting the minimum acceptable size of the leaf nodes to 50 records.

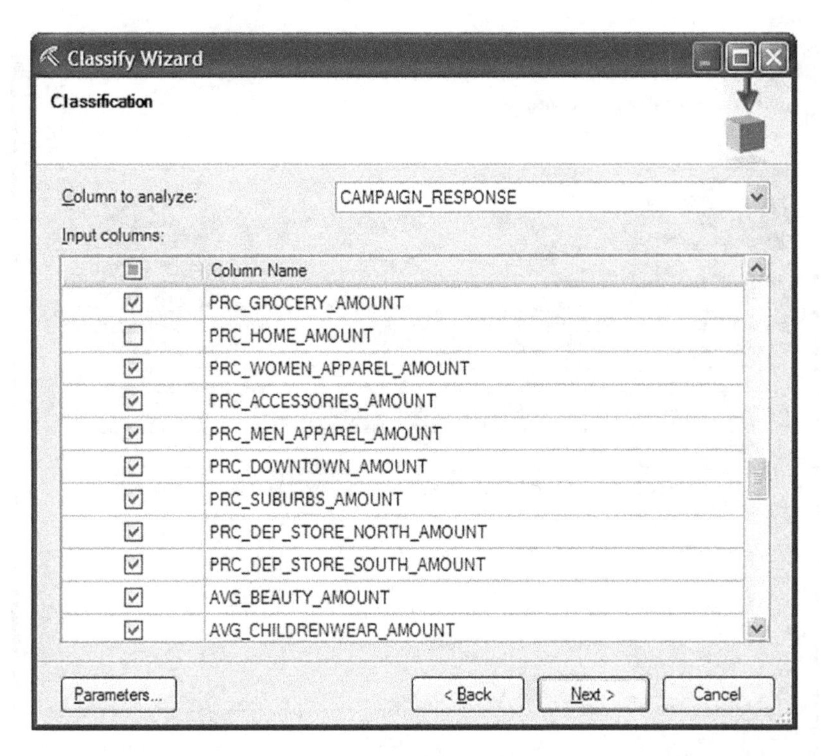

Figure 7.50 Assigning roles to the model fields in the Classify Wizard of Data Mining for Excel

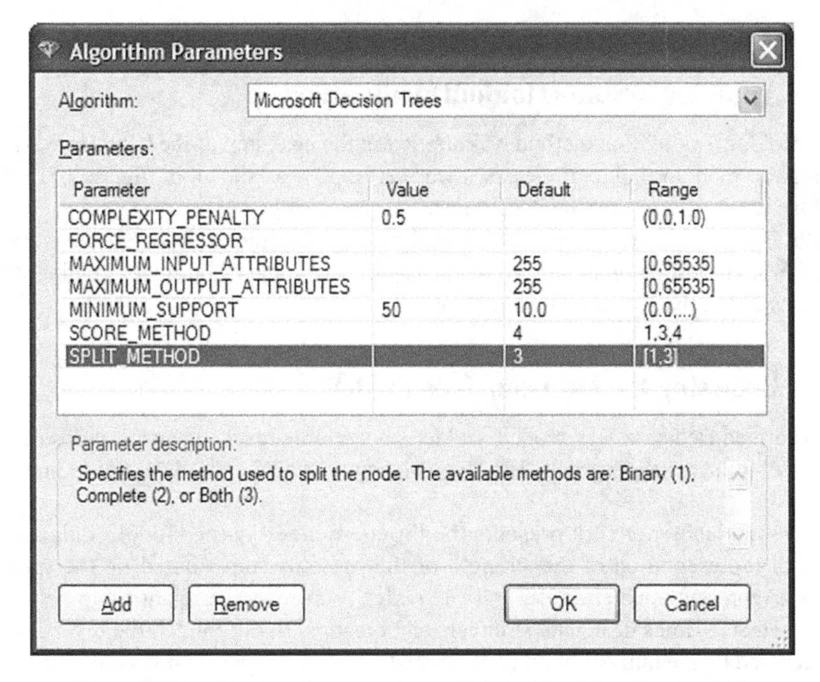

Figure 7.51 Setting the parameters for the Decision Tree models

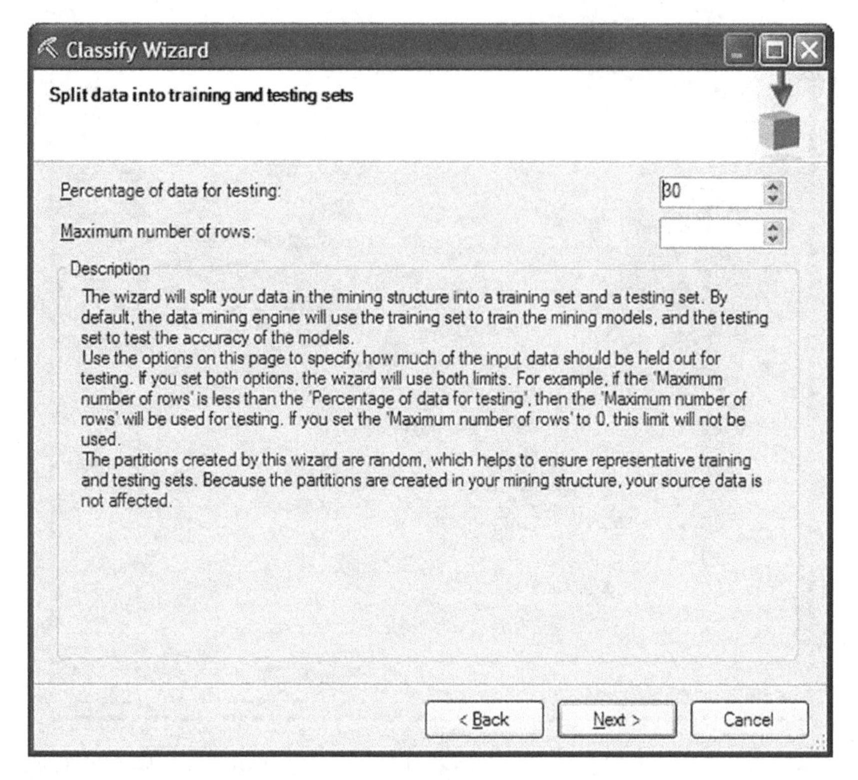

Figure 7.52 Applying a Split validation method

7.15.3 Applying a Split (Holdout) validation

A Split (Holdout) validation method was applied in the next step of the Classify Wizard. The percentage of random, holdout test cases was set to 30% as shown in Figure 7.52. The rest 70% of the data was used for the development of the model.

In the last step of the wizard, the created mining structure and the model were stored (Figure 7.53). The testing (holdout) part of the dataset was stored in the created mining structure, enabling its use in the subsequent validation.

7.15.4 Browsing the Decision Tree model

The two derived Decision Tree models yielded comparable results in terms of discrimination as we'll see in the next paragraph. The Dependency network of the BDE tree is presented in Figure 7.54.

The Dependency network presents the inputs retained in the model. It also provides a graphical representation of the strength of their relationships, based on the split score. Each node represents one attribute, and each edge represents the relationship between two attributes. Heavier lines designate stronger relationships. In our model, the diversity of purchases denoted by the number of distinct product groups with purchases and the frequency of purchases were proven significant predictors. Visits to a specific store (Downtown Store) as

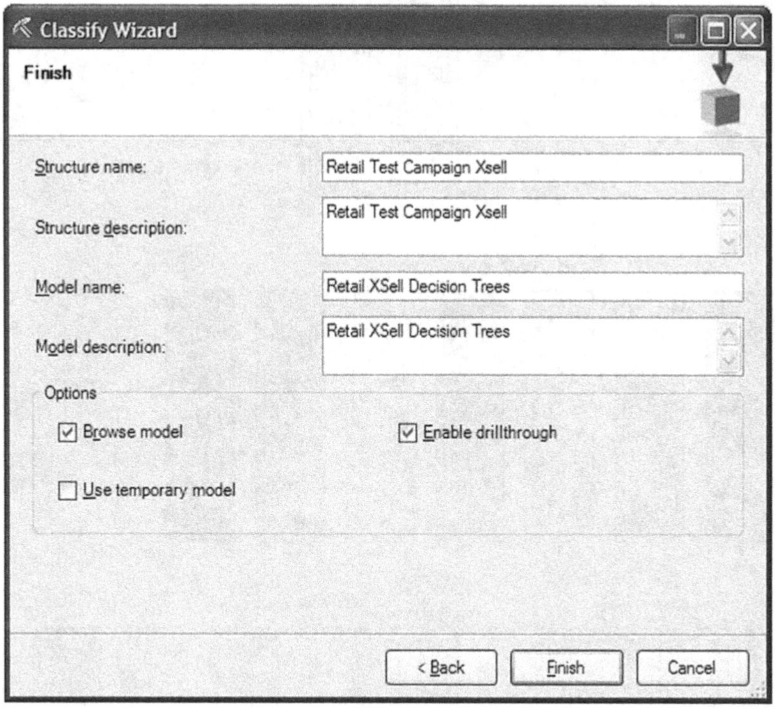

Figure 7.53 Storing the mining structure and model

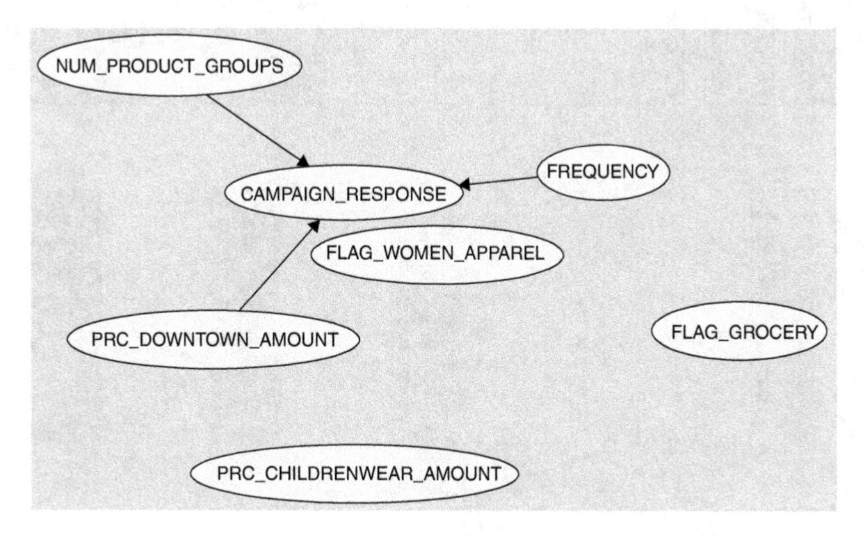

Figure 7.54 The Dependency network of the BDE Decision Tree model

well as previous purchases of women apparel, childrenwear, and groceries also appeared to be related with the campaign response.

The developed Decision Tree model is presented in Figure 7.55. By examining the tree branches, stemming from the root node down to the terminal, the leaf nodes, we can identify

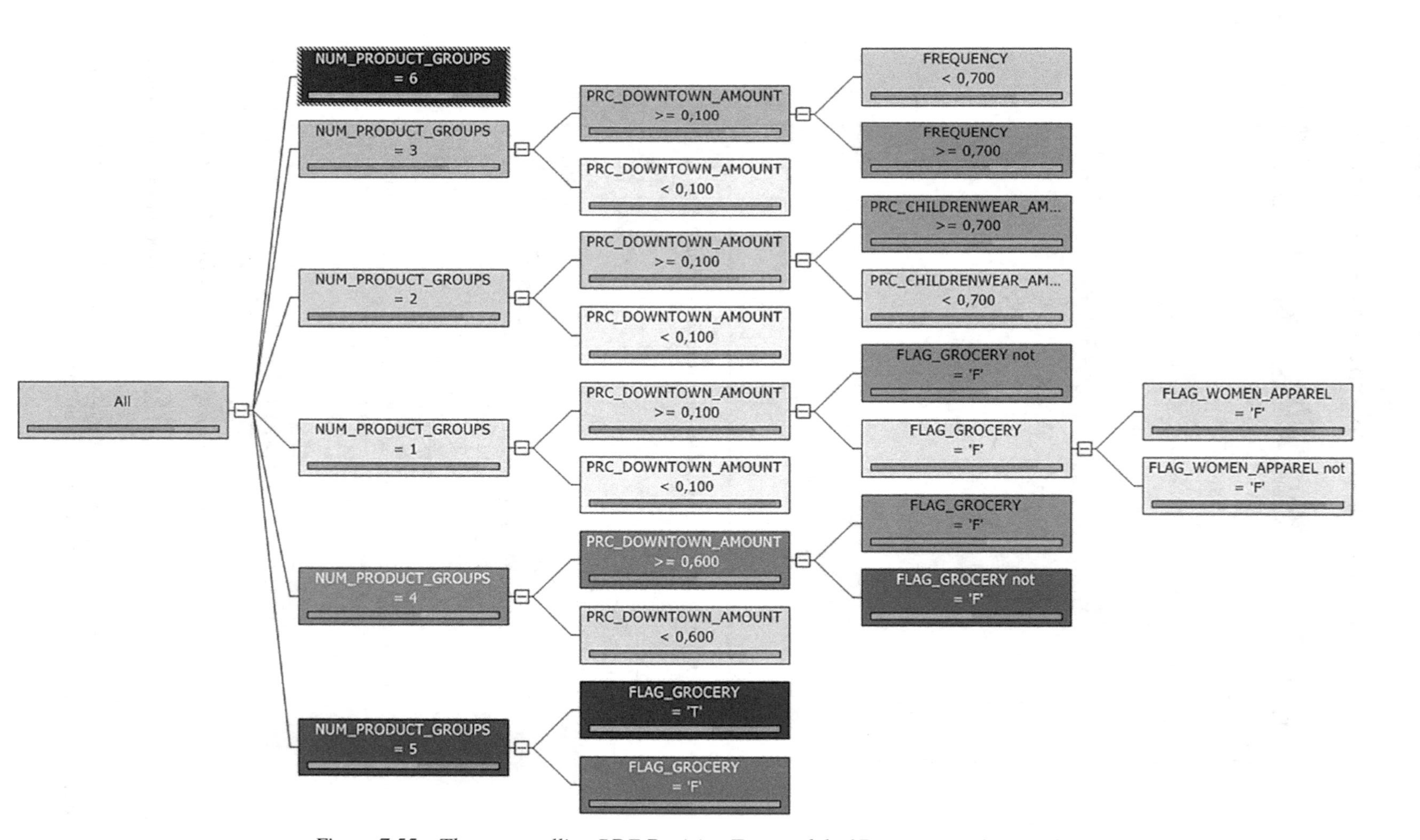

Figure 7.55 The cross-selling BDE Decision Tree model of Data Mining for Excel

the characteristics associated with increased response and the profile of responders. In each node, a darker background color denotes increased concentration of buyers.

Customers with diverse buying habits, especially those with purchases of at least four distinct product groups, presented an increased percentage of responders. Increased concentration of responders was also found among visitors of a specific store (Downtown Store) and among frequent customers and buyers of Groceries and Childrenwear.

7.15.5 Validation of the model performance

The model validation was performed using the Classification Matrix wizard and the Accuracy Chart wizard of the Data Mining for Excel.

The Classification Matrix wizard produces a Confusion (misclassification) matrix to summarize the accuracy and the error rate of the classification model. In the first step of the wizard, the entire structure was selected for validation, in order to evaluate and compare the two Decision Trees developed as shown in Figure 7.56.

The target field as well as the desired format of the results (in counts and/or percentages) was specified in the second step of the wizard as shown in Figure 7.57.

The model was selected to be tested on the testing (holdout) dataset which had been stored along with the entire mining structure (Figure 7.58). If the test data were from a different data source, then a mapping of the structure fields ("mining fields" in Excel terminology) to the external fields would be required. This mapping is done in the last step of the Classification Matrix wizard.

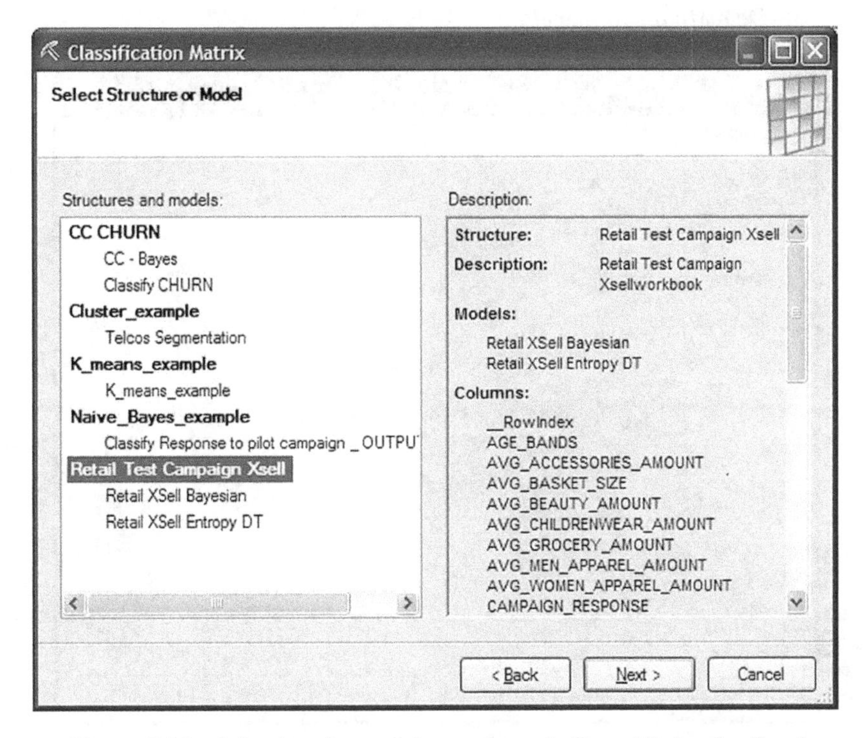

Figure 7.56 Selecting the model to evaluate in Data Mining for Excel

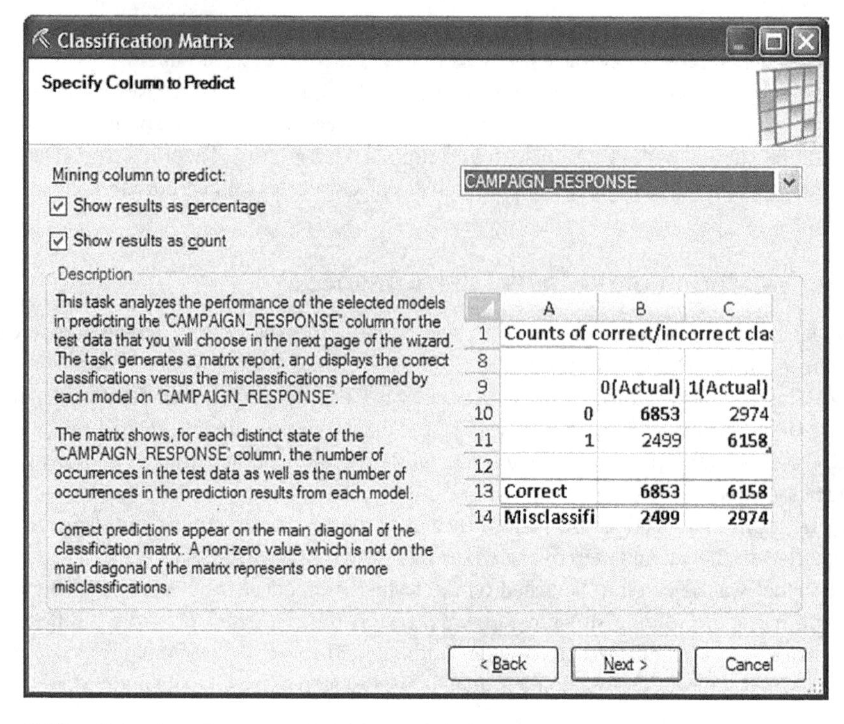

Figure 7.57 Selecting the target field for validation in the Classification Matrix wizard of Data Mining for Excel

Classification Matrix
Select Source Data
Test data from mining structure
Table:
Data range:
My data range has headers
External data source:
Data source name:
Query:
< Back
Finish
Cancel

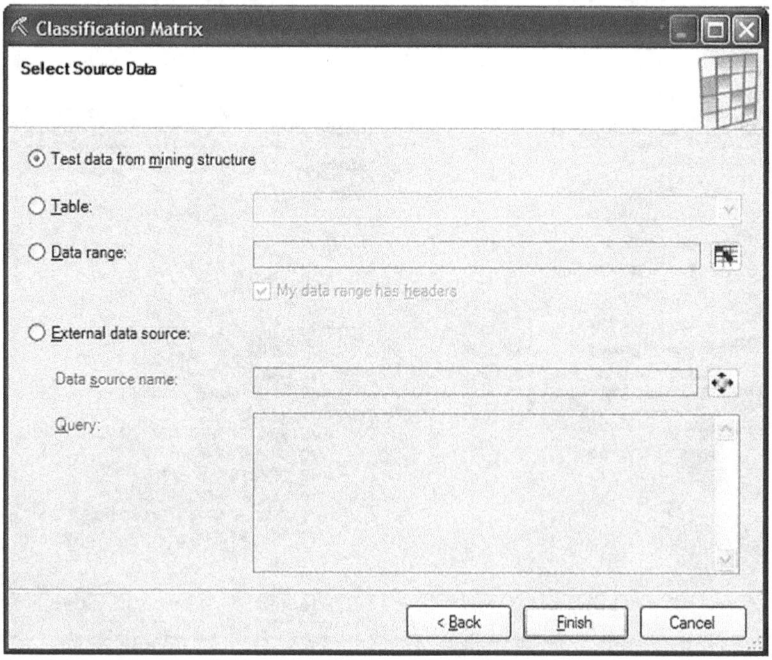

Figure 7.58 Selecting the validation dataset in the Classification Matrix wizard of Data Mining for Excel

Table 7.11 presents the accuracy and the error rate of the two Decision Tree models based on the testing dataset. The two models performed equally well. They both achieved an accuracy of 80% with an error rate of 20%.

The Confusion matrices for the BDE and the Entropy models are presented in Tables 7.12 and 7.13, respectively. The actual classes, in columns, are cross tabulated with the predicted classes. Both models presented a large false negative (FN) rate as they misclassified a large proportion of actual responders as nonresponders. Since the default propensity threshold for predicting an instance as a responder is 50%, the increased FN rate designates that the estimated response propensities were often below 50% among the actual responders. Does this mean that the model is not good? To answer that, we must ask a different question: does the model propensities rank well? Do they discriminate positives from negatives, responders from nonresponders? Are the actual responders assigned with higher estimated probabilities compared to nonresponders? The discrimination power of the model was examined with the Gains chart and the Accuracy Chart wizard.

The steps of the Accuracy Chart wizard are similar with those of the Classification Matrix wizard except from the definition of the target class which is required for producing Gains charts (Figure 7.59).

The Gains charts for the two models are presented in Figure 7.60.

Table 7.11 The accuracy and the error rate for the generated Decision Tree models

Model name	Retail XSell Bayesian	Retail XSell Bayesian	Retail XSell Entropy DT	Retail XSell Entropy DT
Total correct	80.06%	3140	80.32%	3150
Total misclassified	19.94%	782	19.68%	772

Table 7.12 The confusion matrix for the BDE Decision Tree model

Results as percentages for model “Retail XSell Bayesian”		
	F(Actual) (%)	T(Actual) (%)
F	**95.73**	80.32
T	4.27	**19.68**
Correct	**95.73**	**19.68**
Misclassified	**4.27**	**80.32**

Table 7.13 The Confusion matrix for the Entropy Decision Tree model

Results as percentages for model “Retail XSell Entropy DT”		
	F(Actual) (%)	T(Actual) (%)
F	**98.11**	88.24
T	1.89	**11.76**
Correct	**98.11**	**11.76**
Misclassified	**1.89**	**88.24**

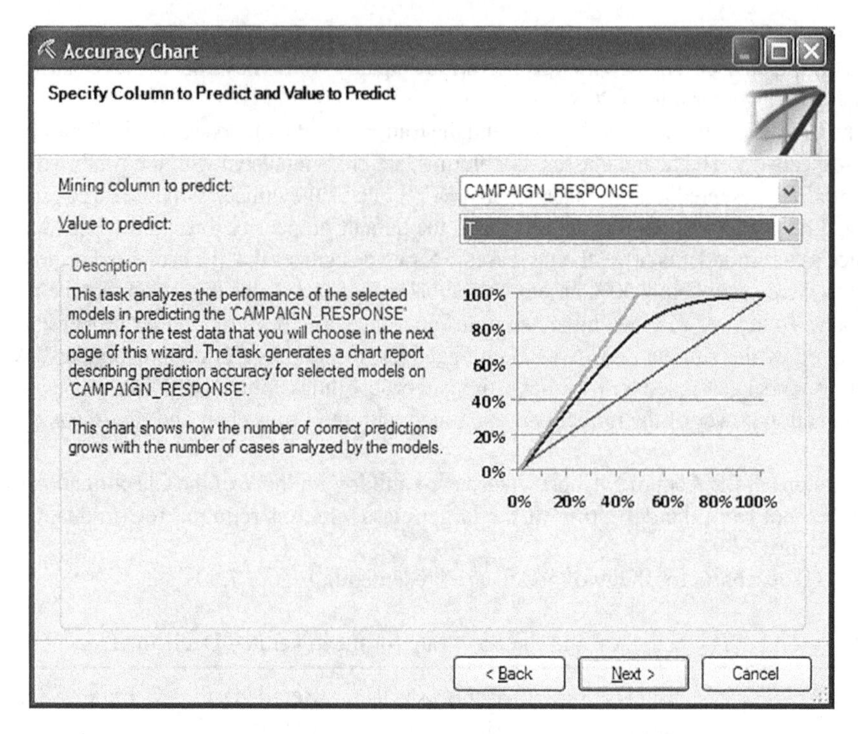

Figure 7.59 Selecting the target class in the Accuracy Chart wizard of Data Mining for Excel

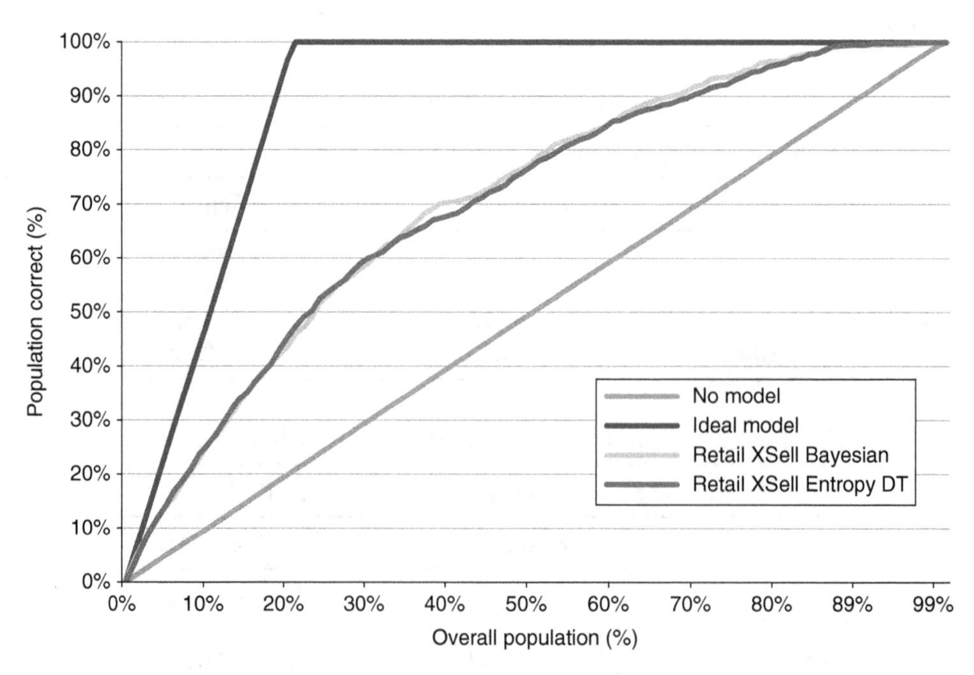

Figure 7.60 The Gains charts for the two Decision Tree models in Data Mining for Excel

Table 7.14 The Gains table and the top 20 percentiles of the two Decision Tree models

Percentile (%)	Ideal model (%)	Retail XSell Bayesian (%)	Retail XSell Entropy DT (%)
0	0.00	0.00	0.00
1	4.83	3.59	2.97
2	9.65	6.81	6.44
3	14.48	9.41	9.53
4	19.31	11.88	12.00
5	24.13	14.11	14.23
6	28.96	15.84	16.96
7	33.79	18.44	18.69
8	38.61	20.67	20.92
9	43.44	22.90	23.64
10	48.27	25.50	25.37
11	53.09	26.98	27.23
12	57.92	28.96	29.58
13	62.75	31.06	31.93
14	67.57	33.29	33.91
15	72.40	35.15	35.27
16	77.23	37.13	37.25
17	82.05	38.61	38.99
18	86.88	40.47	40.47
19	91.71	42.08	43.07
20	96.53	43.56	45.30

Once again, the two models presented comparable performance. Table 7.14 lists the cumulative Gain %, that is, the cumulative percentage of responders among the top 20 propensity percentiles.

A random 2% sample would have captured 2% of the total responders. The ranking of customers according to the BDE model propensities raised the Gain % of the top 2% percentile to 6.81%. Hence, the Lift at this percentile was 3.4 (6.81/2).

Figure 7.61 presents the cumulative distribution of responders and nonresponders across the propensity percentiles. The plot suggests acceptable discrimination. The point where the maximum separation is observed was the 37% percentile. The maximum separation was 39.7%, and it is equal to the KS statistic.

7.15.6 Model deployment

Finally, the BDE model was deployed on new customers, who hadn't participated in the test campaign. Using the "Query" wizard, the stored model was applied to the selected dataset. The model estimates that were selected to be derived included the predicted class (based on the 50% propensity threshold) and the prediction confidence, the estimated probability of the prediction (Figure 7.62).

The file of scored customers and the model derived estimates are shown in Figure 7.63.

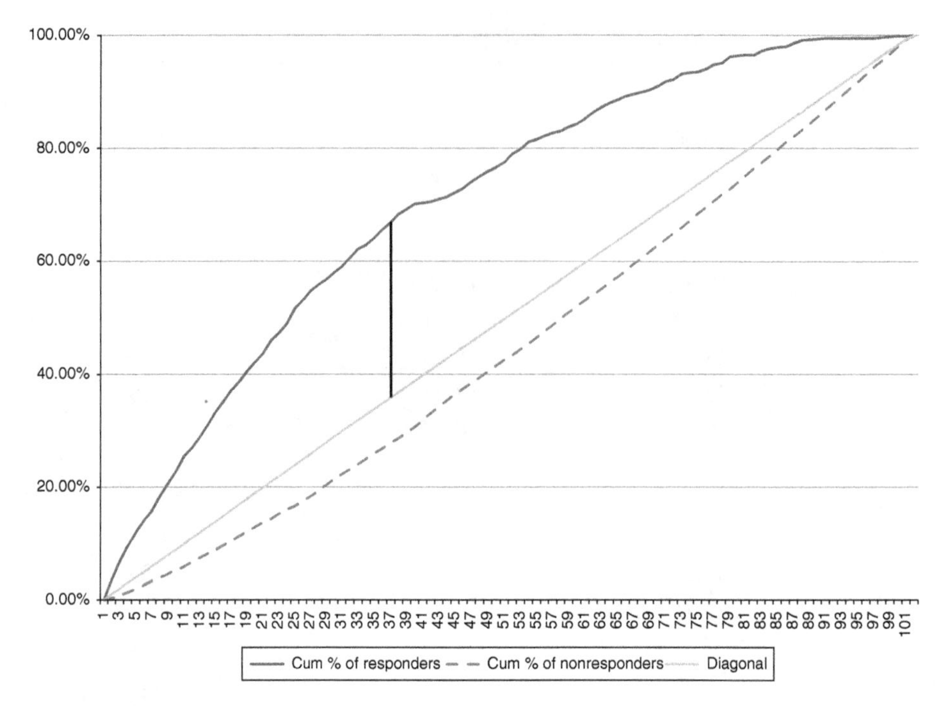

Figure 7.61 The cumulative distribution of responders and nonresponders across propensity percentiles

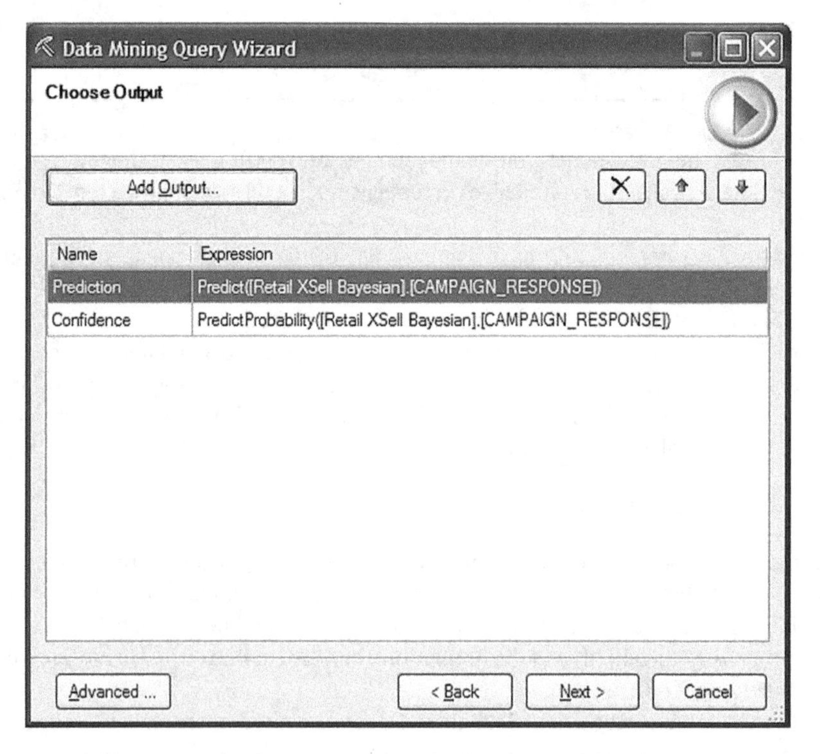

Figure 7.62 Using the Query wizard to deploy the model and score customers

CARD_ID	Prediction	Confidence
9000005	F	0.910121691
9000007	F	0.910121691
9000008	T	0.643835616
9000009	F	0.851131877
9000012	F	0.910121691
9000013	F	0.758218928
9000014	F	0.851131877
9000015	F	0.798381553
9000019	F	0.910121691
9000023	F	0.758218928
9000024	F	0.633705968
9000027	F	0.851131877
9000028	F	0.616189735
9000034	F	0.616189735
9000035	F	0.798381553
9000036	F	0.798381553
9000037	F	0.758218928
9000038	F	0.758218928
9000039	F	0.616189735

Figure 7.63 The model estimates and the scored file

7.16 Summary

The case studies presented in this chapter concerned a retailer who explored its wealth of available data to better understand its customers and gain insight on their behaviors. By starting from transactional data, we saw how we can transform raw information to valuable KPIs which summarize customer purchase patterns. We developed an RFM cell segmentation, which described the customer relationship in terms of spending amount, recency, and frequency of purchases. Finally, we used three different data mining tools to develop, build, and evaluate a classification model for targeting a cross-selling campaign for the promotion of a new department.

8

Segmentation application in telecommunications

8.1 Mobile telephony: the business background and objective

In an environment of hard competition, especially in the case of mature markets, offering of high-level quality services is essential for mobile phone network operators to become established in the market. In times of rapid changes and hard competition, focusing only on customer acquisition, which is nevertheless becoming more and more difficult, is not enough. Inevitably, organizations have to also work on customer retention and on gaining a larger "share of customer" instead of only trying to gain a bigger slice of the market. Growth from within is sometimes easier to achieve and equally important as winning customers from competitors.

Hence, keeping customers satisfied and profitable is a one way street for success. In order to achieve this, operators have to focus on customers and understand their needs, behaviors, and preferences. Behavioral segmentation can help in the identification of the different customer typologies and in the development of targeted marketing strategies.

Nowadays, customers may choose from a huge variety of offered services. The days of voice-only calls are long gone. Mobile phones are communication centers and it's up to the user to select the way of usage that suits his needs. People can communicate via SMS and MMS messages. They can use their phones for connecting to the Internet, for sending e-mails, for downloading games and ringtones, and for communicating with friends and family when they travel abroad. Mobile phones are perceived differently by various people. Some customers only use them in rare circumstances and mainly for receiving incoming calls. Others are addicted to their devices and cannot live without them. Some treat them as electronic gadgets, while for others, they are a tool for work.

Effective CRM using Predictive Analytics, First Edition. Antonios Chorianopoulos.

Companion website: www.wiley.com/go/chorianopoulos/effective_crm

As you can imagine, this multitude of potential choices results in different usage patterns and typologies. Once again, the good news is that usage is recorded in detail. *C*all *D*etail *R*ecords (CDRs) are stored, providing a detailed record of usage. They contain detailed information about all types of calls made. When aggregated and appropriately processed, they can provide valuable information for behavioral analysis.

All usage history should be stored in the organization's mining datamart. Information about frequency and intensity of usage for each call type (voice, SMS, MMS, Internet connection, etc.) should be taken into account when trying to identify the different behavioral patterns. In addition to this, information such as the day/time of the calls (workdays vs. nonworkdays, peak vs. off-peak hours, etc.), roaming usage, direction of calls (incoming vs. outgoing), and origination/destination network type (on-net, off-net, etc.) could also contribute in the formation of a rich segmentation solution.

In this section, we'll present a segmentation example from the mobile telephony market. Marketers of a mobile phone network operator decided to segment their customers according to their behavior. They used all the available usage data to reveal the natural groupings in their customer base. Their goal was to fully understand their customers in order to:

- Develop tailored sales and marketing strategies for each segment.
- Identify distinct customer needs and habits and proceed to the development of new products and services, targeting the diverse usage profiles of their customers. This could directly lead to increased usage on behalf of existing customers but might also attract new customers from the competition.

8.2 The segmentation procedure

The methodological approach followed was analogous to the general framework presented in detail in the relevant chapter. In this section, we'll just present some crucial points concerning the project's implementation plan, which obviously affected the whole application. The key step of the process was the application of a cluster model to segment the consumer postpaid customer base according to their behavioral similarities. The application of the clustering algorithm was preceded by a data reduction technique which identified the underlying data dimensions which were used as inputs for clustering. The entire procedure is described in detail in the following paragraphs.

8.2.1 Selecting the segmentation population: the mobile telephony core segments

Mobile telephony customers are typically categorized in core segments according to their rate plans and the type of the relationship with the operator. The first segmentation level differentiates residential (consumer) from business customers. Residential customers are further divided into postpaid and prepaid:

- Postpaid—contractual customers: Customers with mobile phone contracts. Usually, they comprise the majority of the customer base. These customers have a contract and a long-term billing arrangement with the network operator for the received services. They are billed on a monthly basis and according to the traffic of the past month; hence, they have unlimited credit.

- Prepaid customers: They do not have a contract-based relationship with the operator, and they are buying credit in advance. They do not have ongoing billing and they pay for the services before actually using them.

Additionally, business customers are further differentiated according to their size into corporate, SME, and SOHO customers.

The typical core segments in mobile telephony are depicted in Figure 8.1.

The objective of the marketers was to enrich the core segmentation scheme with refined subsegments. Therefore, they decided to focus their initial segmentation attempts exclusively on residential postpaid customers. Prepaid customers need a special approach in which attributes like the intensity and frequency of top-ups (recharging of their credits) should also be taken into account. Business customers also need a different handling since they comprise a completely different market. It is much safer to analyze those customers separately, with segmentation approaches such as value based, size based, industry based, etc.

Moreover, only MSIDNs (telephone numbers) with current status active or in suspension (due to payment delays) were included in the analysis. Churned (voluntary and involuntary churners) MSISDNs have been excluded from the start. Their "contribution" is crucial in the building of a churn model but trivial in a segmentation scheme mainly involving phone usage. In addition, segmentation population was narrowed down even further by excluding users with no incoming or outgoing usage within the past 6 months. Those users have been flagged as inactive, and they have been selected for further examination and profiling. They could also form a target list for an upcoming reactivation campaign, but they do not have much to contribute to a behavioral analysis since, unfortunately, inactivity is their only behavior at the moment.

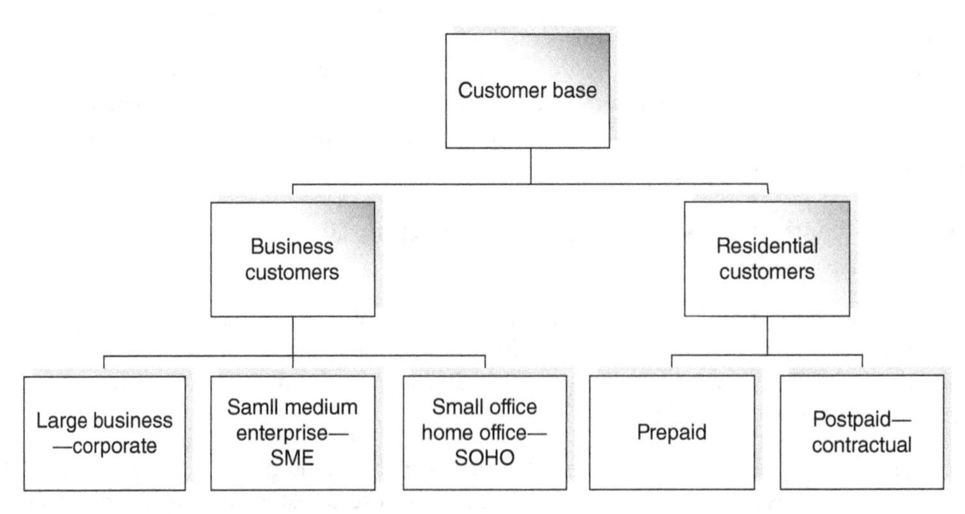

Figure 8.1 Core segments in mobile telephony. Source: Tsiptsis and Chorianopoulos (2009). Reproduced with permission from Wiley

Table 8.1 Mobile telephony usage aspects that were investigated in the behavioral segmentation

Number of calls	***Information by Core service type***	Voice SMS MMS Internet ...
Minutes/traffic **Community**	***Information by Call direction***	Incoming Outgoing International Roaming

8.2.2 Deciding the segmentation level

Customers may own more than one MSISDN which may use in a different manner to cover different needs. In order to capture all the potentially different usage behaviors of each customer, it has been decided to implement the behavioral segmentation at MSISDN level. Therefore, relevant input data have been preaggregated accordingly, and the derived cluster model assigned each MSISDN to a distinct behavioral segment.

8.2.3 Selecting the segmentation dimensions

Once again, the mining datamart tables, comprised the main sources of the input data. Table 8.1 outlines the main usage aspects that were selected as segmentation criteria.

The modeling phase was followed by extensive profiling of the revealed segments. In this phase, all available information, including demographic data and contract information details, were cross-examined with the identified customer groups.

8.2.4 Time frames and historical information analyzed

A 6-month observation period was analyzed, a time span that in general ensures the capturing of stable, nonvolatile behavioral patterns instead of random or outdated ones. Summary fields (sums, counts, percentages, averages, etc.) covering the 6-month observation period were used as inputs in the clustering model.

8.3 The data preparation procedure

Readers not interested in the data preparation procedure might skip the following paragraphs and proceed directly to Section 8.4. Note that the data preparation phase presented here is simplified and deals merely with the derivation of KPIs. The Modeler stream for the enrichment of the modeling file with derived KPIs is shown in Figure 8.2. A set of fields, with a PRC_ prefix, was constructed to summarize the relative usage (percentage of total calls) by service type, preferred day/time, as well as the international and the roaming usage. Additionally, the

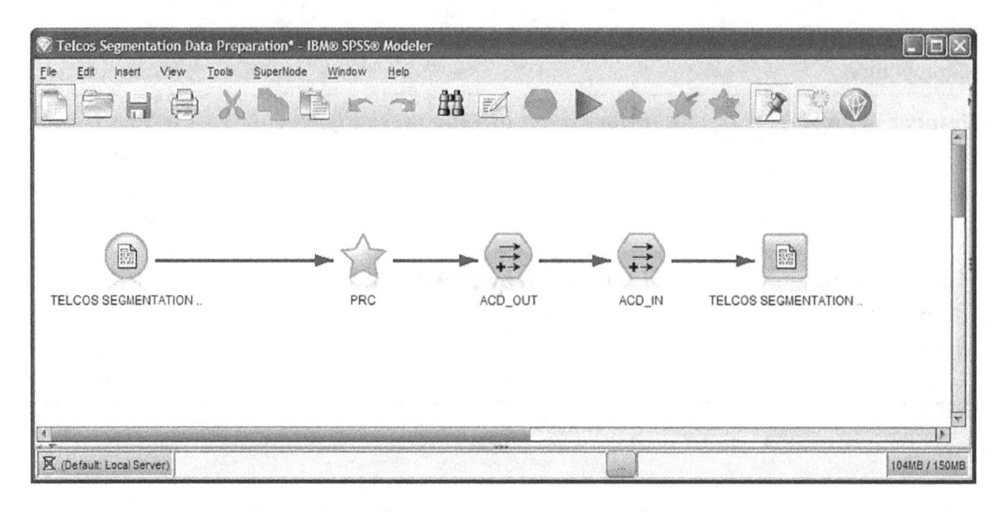

Figure 8.2 The data preparation Modeler stream

average call duration, both incoming and outgoing, was calculated as the ratio of minutes to number of calls (fields ACD_IN and ACD_OUT, respectively) for each MSISDN.

8.4 The data dictionary and the segmentation fields

The list of candidate inputs for behavioral segmentation initially included all usage fields contained in the mining datamart and the customer "signature" reference file. In a later stage and according to the organization's specific segmentation objectives, marketers selected a subset list of clustering inputs which are designated in the last column of Table 8.2.

Obviously, this list can't be considered a silver-bullet approach that can cover all the needs of every organization. It represents the approach adopted in the specific implementation but it also outlines a general framework of potential types of fields that could be proved useful in similar applications.

The selected fields indicate the marketers' orientation to a segmentation scheme that would reflect usage differences in terms of preferred *type of calls* (voice, SMS, MMS, Internet, etc.), *roaming* usage (calls made in a foreign country), and frequency of *international calls* (calls made in home country to international numbers).

The full attribute list of the modeling file is presented in Table 8.2. The last section of the table lists attributes derived through the data preparation phase. The role of each attribute is designated in the last column of the table. Attributes used as inputs in the subsequent models are flagged with an INPUT role.

8.5 The modeling procedure

The IBM SPSS Modeler stream (procedure) for segmentation through clustering is displayed in Figure 8.3. This stream uses the modeling file presented in Table 8.2.

Table 8.2 Mobile telephony segmentation fields

Data file: TELCOS SEGMENTATION MODELING DATA.txt		
Field name	Description	Role in the model
Community		
OUT_COMMUNITY_TOTAL	Total outgoing community: monthly average of distinct phone numbers that the holder called (includes all call types)	INPUT
OUT_COMMUNITY_VOICE	Outgoing voice community	INPUT
OUT_COMMUNITY_SMS	Outgoing SMS community	INPUT
IN_COMMUNITY_VOICE	Incoming voice community	
IN_COMMUNITY_SMS	Incoming SMS community	
IN_COMMUNITY_TOTAL	Total incoming community	
Number of calls by call type		
VOICE_OUT_CALLS	Monthly average of outgoing voice calls. Derived as the ratio of total voice calls during the 6-month period examined to 6 (months) or to customer tenure for new customers	INPUT
VOICE_IN_CALLS	Monthly average of incoming voice calls	INPUT
SMS_OUT_CALLS	Monthly average of outgoing SMS calls	INPUT
SMS_IN_CALLS	Monthly average of incoming SMS calls	INPUT
MMS_OUT_CALLS	Monthly average of outgoing MMS calls	INPUT
EVENTS_CALLS	Monthly average of event calls	INPUT
INTERNET_CALLS	Monthly average of Internet calls	INPUT
TOTAL_OUT_CALLS	Monthly average of outgoing calls (includes all call types)	INPUT
Minutes/traffic by call type		
VOICE_OUT_MINS	Monthly average number of minutes of outgoing voice calls	INPUT
VOICE_IN_MINS	Monthly average number of minutes of incoming voice calls	INPUT
EVENTS_TRAFFIC	Monthly average of events traffic	INPUT
GPRS_TRAFFIC	Monthly average of GPRS traffic	INPUT
International calls/roaming usage		
OUT_CALLS_ROAMING	Monthly average of outgoing roaming calls (calls made in a foreign country)	INPUT

(Continued)

Table 8.2 (**Continued**)

Field name	Description	Role in the model
OUT_MINS_ROAMING	Monthly average number of minutes of outgoing voice roaming calls	INPUT
OUT_CALLS_ INTERNATIONAL	Monthly average of outgoing calls to international numbers (calls made in home country to international numbers)	INPUT
OUT_MINS_ INTERNATIONAL	Monthly average number of minutes of outgoing voice calls to international numbers	INPUT
Usage by day/hour		
OUT_CALLS_PEAK	Monthly average of outgoing calls in peak hours	
OUT_CALLS_OFFPEAK	Monthly average of outgoing calls in off-peak hours	
OUT_CALLS_WORK	Monthly average of outgoing calls on workdays	
OUT_CALLS_NONWORK	Monthly average of outgoing calls on nonworkdays	
IN_CALLS_PEAK	Monthly average of incoming calls in peak hours	
IN_CALLS_OFFPEAK	Monthly average of incoming calls in off-peak hours	
IN_CALLS_WORK	Monthly average of incoming calls on workdays	
IN_CALLS_NONWORK	Monthly average of incoming calls on nonworkdays	
Days with usage		
DAYS_OUT	Monthly average number of days with any outgoing usage	INPUT
DAYS_IN	Monthly average number of days with any incoming usage	INPUT
Average call duration		
ACD_OUT	Average duration of outgoing voice calls (in minutes)	INPUT
ACD_IN	Average duration of incoming voice calls (in minutes)	INPUT
Demographics—profiling fields		
AGE	Age of customer	
Gender	Gender of customer	

Table 8.2 (**Continued**)

Field name	Description	Role in the model
Derived fields (IBM Modeler stream: "Telcos Segmentation Data Preparation.str")		
PRC_OUT_COMMUNITY_VOICE	Percentage of outgoing voice community: outgoing voice community as a percentage of total outgoing community	INPUT
PRC_OUT_COMMUNITY_SMS	Percentage of outgoing SMS community	INPUT
PRC_IN_COMMUNITY_VOICE	Percentage of incoming voice community	
PRC_IN_COMMUNITY_SMS	Percentage of incoming SMS community	
PRC_VOICE_OUT_CALLS	Percentage of outgoing voice calls: outgoing voice calls as a percentage of total outgoing calls	INPUT
PRC_SMS_OUT_CALLS	Percentage of SMS calls	INPUT
PRC_MMS_OUT_CALLS	Percentage of MMS calls	INPUT
PRC_EVENTS_CALLS	Percentage of Event calls	INPUT
PRC_INTERNET_CALLS	Percentage of Internet calls	INPUT
PRC_OUT_CALLS_ROAMING	Percentage of outgoing roaming calls: roaming calls as a percentage of total outgoing calls	INPUT
PRC_OUT_CALLS_INTERNATIONAL	Percentage of outgoing international calls: outgoing international calls as a percentage of total outgoing calls	INPUT
PRC_OUT_CALLS_PEAK	Percentage of outgoing calls in peak hours	
PRC_OUT_CALLS_OFFPEAK	Percentage of outgoing calls in nonpeak hours	
PRC_OUT_CALLS_WORK	Percentage of outgoing calls on workdays	
PRC_OUT_CALLS_NONWORK	Percentage of outgoing calls on nonworkdays	
PRC_IN_CALLS_PEAK	Percentage of incoming calls in peak hours	
PRC_IN_CALLS_OFFPEAK	Percentage of incoming calls in nonpeak hours	
PRC_IN_CALLS_WORK	Percentage of incoming calls on workdays	
PRC_IN_CALLS_NONWORK	Percentage of incoming calls on nonworkdays	

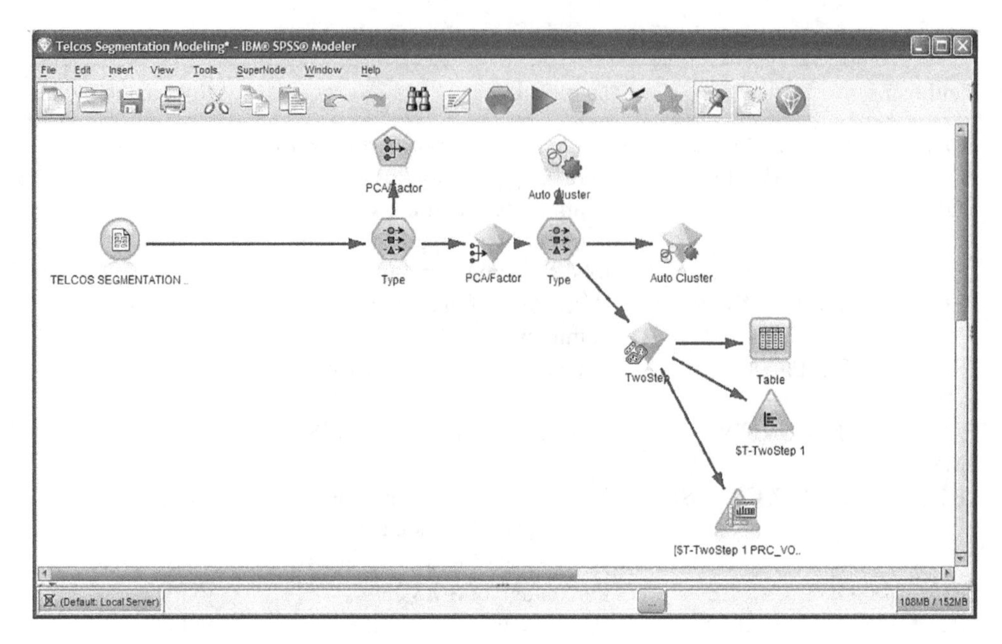

Figure 8.3 The IBM SPSS Modeler procedure for segmentation through clustering

The main modeling steps included:

- The application of a data reduction technique, PCA in particular, to reveal the distinct dimensions of information which were then used as clustering fields
- The use of a cluster model which analyzed the data components identified by PCA and grouped MSISDNs in clusters of similar behaviors
- The profiling of the revealed segments

These steps are explained in detail in the following text.

8.5.1 Preparing data for clustering: combining fields into data components

The number of the original segmentation inputs exceeded 30. Using all those fields as direct input in a clustering algorithm would have produced a complicated solution. Therefore, the approach followed was to incorporate a data reduction technique in order to reveal the underlying data dimensions prior to clustering.

This approach was adopted to eliminate the risk of deriving a biased solution due to correlated attributes. Moreover, it also ensures a balanced solution, in which all data dimensions contribute equally, and it simplifies the tedious procedure of segmentation understanding by providing conceptual clarity.

Specifically, a Principal Components Analysis (PCA) model with varimax rotation was applied to the original segmentation fields. A Type node was used to set as Inputs (Direction In) the designated attributes of Table 8.2. The rest of the attributes were assigned with a

Table 8.3 The PCA model parameter settings

PCA model parameter settings	
Parameter	**Setting**
Model	Principal Components analysis (PCA)
Rotation	Varimax
Criteria for the number of factors to extract	Initially eigenvalues over 1. Then set to eight components after examination of the results

None role and didn't contribute to the subsequent model. The modeling data were then fed into a PCA/Factor modeling node, and a PCA model was developed which grouped original inputs into components. The parameter settings for the PCA model are listed in Table 8.3.

Initially, based on the eigenvalues over 1 criterion, the algorithm suggested the extraction of nine components. However, the ninth component was mainly related with a single original attribute, the GPRS traffic. Moreover, it accounted for a small percentage of the information of the original inputs, about 3%. Having considered that, the analysts decided to go for a simpler solution of eight components, selecting to sacrifice a small part of the original information for simplicity. In the eight-component solution, the GPRS traffic was combined with Events usage to form a single component.

Before using the derived components and substituting more than 30 fields with a handful of new ones, the data miners of the organization wanted to be sure that:

- The PCA solution carried over most of the original information.
- The derived components, which would substitute the original attributes, were interpretable and had a business meaning.

Therefore, they started the examination of the model results by looking at the table of "Explained Variance." Table 8.4 presents these results.

The resulted eight components retained more than 73% of the variance of the original fields. This percentage was considered satisfactory and thus the only task left before accepting the components was their interpretation. In practice, only a solution comprised of meaningful components should be retained.

So what do those new composite fields represent? What business meaning do they convey? As these new fields are constructed in order to substitute the original fields in the next stages of the segmentation procedure, it is necessary to be thoroughly decoded before being used in upcoming models.

The component interpretation phase included the examination of the "rotated component matrix" (Table 8.5), a table that summarizes the correlations (loadings) between the components and the original fields.

The "interpretation" results are summarized in Table 8.6.

The explained and labeled components and the respective component scores were subsequently used as inputs in the clustering model. This brings us to the next phase of the application: the identification of useful groupings through clustering.

Table 8.4 Deciding the number of extracted components by examining the variance explained table

Total variance explained			
Components	Eigenvalue	% of variance	Cumulative %
1	6.689	21.576	21.576
2	4.000	12.903	34.479
3	2.550	8.227	42.705
4	2.443	7.881	50.587
5	1.962	6.329	56.916
6	1.791	5.779	62.695
7	1.742	5.618	68.313
8	1.662	5.362	73.675
9			
10			
11			
...	...	...	...
31	0.002	0.00	100.00

8.5.2 Identifying the segments with a cluster model

The generated components represented effectively all the usage dimensions of interest, in a concise and comprehensive way, leaving no room for misunderstandings about their business meaning. The next step of the segmentation project included the usage of the derived component scores as inputs in a cluster model. Through a new Type node, the generated components were set as Inputs for the training of the subsequent cluster model.

The clustering process involved the application of an Auto Cluster node for the training and evaluation of 2 TwoStep models with different parameter settings in terms of outlier handling. Figure 8.4 presents an initial comparison of the two cluster models provided by the generated Auto Cluster model.

The cluster viewer presents the number of identified clusters as well as the Silhouette coefficient and the size of smallest and largest clusters for each model. In our case, the first TwoStep model was selected for deployment due to its higher Silhouette value and, more importantly, due to its transparency and interpretability.

The parameter settings of the selected cluster model are presented in Table 8.7.

As shown in Figure 8.5, the model yielded a five-cluster solution with a fair Silhouette measure of 0.29.

The distribution of the revealed clusters is depicted in Figure 8.6.

As a reminder, we outline that these clusters were not known in advance, neither imposed by users, but uncovered after analyzing the actual behavioral patterns of the usage data. The largest cluster, Cluster 1, which as we'll see corresponds to "typical" usage, contained about 30% of customers. The smallest one included about 11.5% of total users.

Table 8.5 Understanding and labeling the components through rotated component matrix

Rotated component matrix								
	Components							
	1	2	3	4	5	6	7	8
OUT_COMMUNITY_TOTAL	0.918							
OUT_COMMUNITY_VOICE	0.916							
TOTAL_OUT_CALLS	0.912							
VOICE_OUT_CALLS	0.908							
VOICE_IN_CALLS	0.844							
VOICE_IN_MINS	0.769							
DAYS_OUT	0.740							
DAYS_IN	0.688							
VOICE_OUT_MINS	0.648							0.536
PRC_OUT_COMMUNITY_SMS		0.913						
PRC_SMS_OUT_CALLS		0.903						
PRC_VOICE_OUT_CALLS		−0.822						
OUT_COMMUNITY_SMS	0.415	0.758						
PRC_OUT_COMMUNITY_VOICE		−0.710						
SMS_OUT_CALLS	0.356	0.690						
OUT_CALLS_INTERNATIONAL			0.907					
OUT_MINS_INTERNATIONAL			0.894					
PRC_OUT_CALLS_INTERNATIONAL			0.857					
OUT_MINS_ROAMING				0.883				
OUT_CALLS_ROAMING				0.875				
PRC_OUT_CALLS_ROAMING				0.813				
EVENTS_CALLS					0.903			
PRC_EVENTS_CALLS					0.879			
EVENTS_TRAFFIC					0.447			
GPRS_TRAFFIC					0.364			
PRC_INTERNET_CALLS						0.921		
INTERNET_CALLS						0.903		
PRC_MMS_OUT_CALLS							0.923	
MMS_OUT_CALLS							0.918	
ACD_OUT								0.840
ACD_IN								0.683

The outlier cluster

Actually, the solution included an additional cluster, the "noise" cluster, labeled as −1 by Modeler. This small cluster was the outlier precluster, containing records identified in the first step of the TwoStep procedure as outliers. These cases, about 1.5% of total users, didn't contribute in the formation of the final clusters and were left out as a separate group. This group actually included superroamers, users with extremely high roaming and international usage and merits special investigation. Perhaps, it also represents a business opportunity. Those interested in a complete-coverage solution might consider merging it with Cluster 3, which as we'll see contains "normal" roamers.

8.5.3 Profiling and understanding the clusters

Each revealed cluster corresponds to a distinct behavioral typology. This typology had to be understood, named, and communicated to all the people of the organization in a simple and concise way before being used for tailored interactions and targeted marketing activities. Therefore, the next phase of the project included the profiling of the clusters through simple reporting techniques. The "recognize and label" procedure started with the examination of the clusters in respect to the component scores, providing a valuable first insight on their structure, before moving to profiling in terms of the original usage fields. The table of centroids is shown in Figure 8.7.

Since the component scores are standardized, their overall means are 0 and the mean for each cluster denotes the signed deviation from the overall mean. By studying these deviations, we can see the relatively large mean values of Factors 1 and 8 among cases of Cluster 5. Since Factors 1 and 8 represent voice usage and average call duration, it seems that Cluster 5 was comprised of high voice users. This conclusion is further supported by studying the distribution of the factors for Cluster 5 with the series of boxplots presented in Figure 8.8.

The background boxplot summarizes the entire population, while the overlaid boxplot refers to the selected cluster. In respect to Factors 1 and 8, the boxplots for Cluster 5 are at the right side of those for the entire population, indicating high values and consequently high voice usage.

After studying the distributions of the factors, the profile of each cluster had started to take shape. In the final profiling stage, analysts returned to the primary inputs, seeking for more straightforward differentiations in terms of the original attributes. Table 8.8 summarizes the five clusters in terms of some important usage attributes. It presents the mean of each attribute over the members of each cluster. The last column denotes the overall mean.

Large deviations from the marginal mean characterize the respective cluster and denote a behavior that differentiates the cluster from the typical behavior.

Although inferential statistics have not been applied to flag statistically significant differences from the overall population mean, we can see some large observed differences which seem to characterize the clusters. Based on the information summarized so far, the project team started to outline a first rough profile of each cluster:

- Cluster 1 contains average voice users.
- Cluster 2 seems to include users with basic usage. They also seem to be characterized by increased average call duration and very small communities.

Table 8.6 The interpretation and labeling of the derived components

Derived components	
Component	Label and description
1	**Voice calls**
	The high loadings in the first column of the rotated component matrix denote a strong positive correlation between Component 1 and the original fields which measure voice usage and traffic such as the number and minutes of outgoing and incoming voice calls (VOICE_OUT_CALLS, VOICE_IN_CALLS, VOICE_OUT_MINS, VOICE_IN_MINS) and the size of the voice community (OUT_COMMUNITY_VOICE). Thus, Component 1 seems to be associated with voice usage Because generally voice calls constitute the majority of calls for most users and they tend to dominate the total usage, a set of fields associated with total usage (total number of calls, TOTAL_OUT_CALLS) as well as days with usage (DAYS_IN, DAYS_OUT) are also loaded high on this component
2	**SMS calls**
	Component 2 seems to measure SMS usage and community since it is seems strongly correlated with the percentage of SMS out calls (PRC_SMS_OUT_CALLS) and SMS community (PRC_OUT_COMMUNITY_SMS) The rather interesting negative correlation between Component 2 and the percentage of voice out calls (PRC_VOICE_OUT_CALLS) denotes a contrast between voice and SMS usage. Thus, users with high positive values in this component are expected to have increased SMS usage and increased SMS to voice calls ratio. This does not necessarily mean low voice traffic, but it certainly implies relatively lower percentage of voice calls and increased percentage of SMS calls
3	**International calls**
	Component 3 is associated with calls to international networks
4	**Roaming usage**
	Component 4 seems to measure outgoing roaming usage (making calls when abroad)
5	**Events and GPRS**
	Component 5 measures Event calls and traffic. GPRS traffic (originally constituting a separate component) is also loaded moderately high on this component
6	**Internet usage**
	Component 6 is associated with Internet usage
7	**MMS usage**
	MMS usage seems to be measured by Component 7
8	**Average call duration (ACD)**
	Fields denoting average call duration of incoming and outgoing voice calls seem to be related, and they are combined to form Component 8. Unsurprisingly, this component is also associated with minutes of calls (VOICE_OUT_MINS)

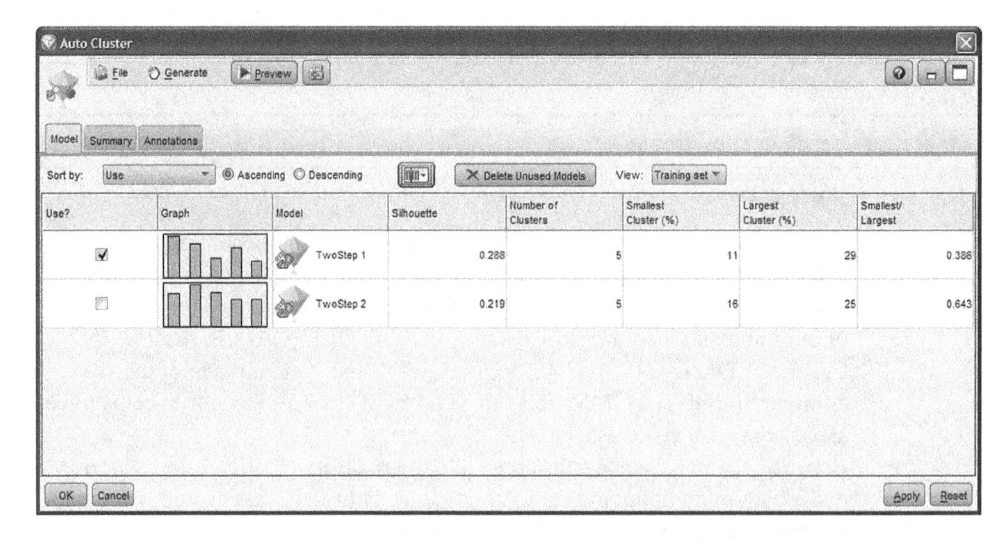

Figure 8.4 An initial comparison of the generated cluster models provided by the Auto Cluster model nugget

Table 8.7 The parameter settings of the cluster model

PCA model parameter settings	
Parameter	**Setting**
Model	TwoStep
Number of clusters	Automatically calculated
Exclude outliers option	On
Noise percentage	10%

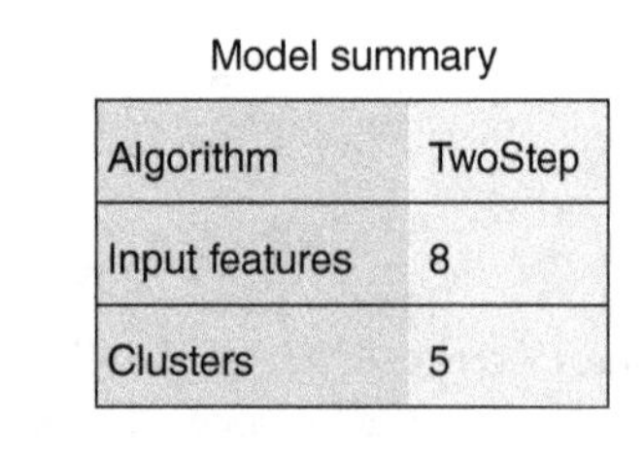

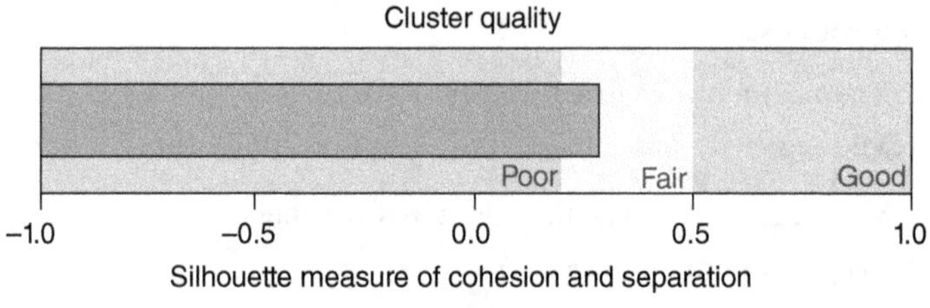

Figure 8.5 The Silhouette measure of the cluster model

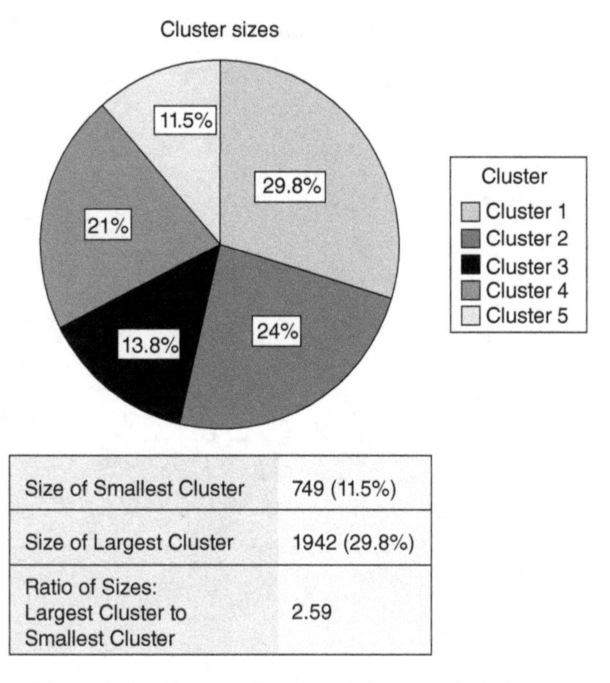

Size of Smallest Cluster	749 (11.5%)
Size of Largest Cluster	1942 (29.8%)
Ratio of Sizes: Largest Cluster to Smallest Cluster	2.59

Figure 8.6 The distribution of the revealed clusters

Clusters

Feature Importance

1.00000000 0.99999995 0.99999990 0.99999985

Cluster	cluster-1	cluster-2	cluster-3	cluster-4	cluster-5
Label					
Size	29.8% (1942)	24.0% (1565)	13.8% (897)	21.0% (1366)	11.5% (749)
Features	$F-Factor-1 0.16	$F-Factor-1 -0.85	$F-Factor-1 -0.05	$F-Factor-1 -0.10	$F-Factor-1 1.62
	$F-Factor-2 -0.55	$F-Factor-2 -0.38	$F-Factor-2 -0.13	$F-Factor-2 1.45	$F-Factor-2 -0.29
	$F-Factor-3 -0.18	$F-Factor-3 -0.02	$F-Factor-3 0.54	$F-Factor-3 -0.13	$F-Factor-3 -0.27
	$F-Factor-4 -0.20	$F-Factor-4 -0.07	$F-Factor-4 0.39	$F-Factor-4 -0.12	$F-Factor-4 0.07
	$F-Factor-5 -0.22	$F-Factor-5 -0.15	$F-Factor-5 0.74	$F-Factor-5 -0.16	$F-Factor-5 0.00
	$F-Factor-6 -0.09	$F-Factor-6 -0.20	$F-Factor-6 0.53	$F-Factor-6 -0.11	$F-Factor-6 -0.12
	$F-Factor-7 -0.06	$F-Factor-7 0.00	$F-Factor-7 -0.03	$F-Factor-7 0.01	$F-Factor-7 -0.13
	$F-Factor-8 -0.38	$F-Factor-8 0.31	$F-Factor-8 -0.32	$F-Factor-8 -0.09	$F-Factor-8 0.69

Figure 8.7 The table of centroids of the five clusters

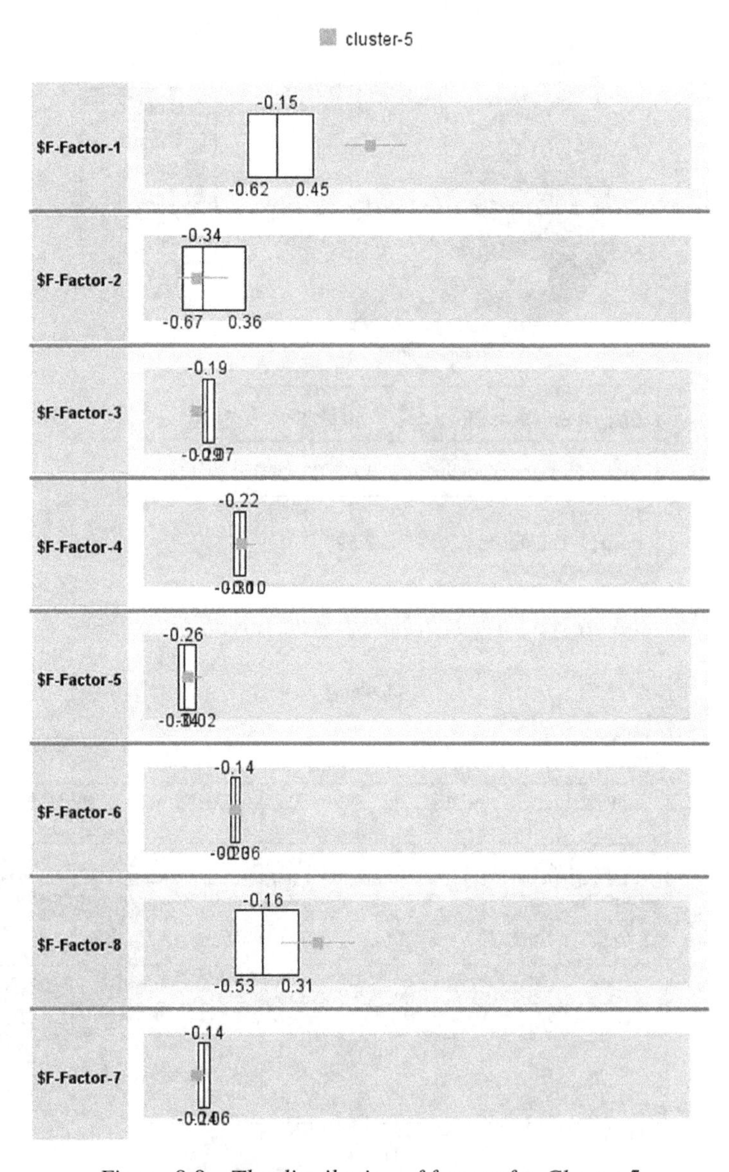

Figure 8.8 The distribution of factors for Cluster 5

- Cluster 3 includes roamers and users with increased communication with international destinations. They are also accustomed to Internet and event calls.
- Cluster 4 is mainly consisted of SMS users that seem to have an additional inclination to tech services like MMS, Internet, and event calls.

Table 8.8 The means of some original attributes of importance for each cluster

	Clusters					
	Cluster 1	Cluster 2	Cluster 3	Cluster 4	Cluster 5	Total
VOICE_OUT_CALLS	120	39	119	84	319	116
PRC_VOICE_OUT_CALLS	0.95	0.89	0.84	0.66	0.93	0.86
VOICE_OUT_MINS	92	37	108	77	357	108
VOICE_IN_CALLS	172	60	151	145	368	159
VOICE_IN_MINS	147	72	139	162	405	161
OUT_COMMUNITY_VOICE	31	12	30	22	67	29
SMS_OUT_CALLS	3	2	8	34	16	11
PRC_SMS_OUT_CALLS	0.02	0.04	0.05	0.25	0.05	0.08
OUT_COMMUNITY_SMS	1.57	1.06	3.44	8.89	5.36	3.67
OUT_CALLS_ROAMING	0.14	0.18	2.10	0.40	1.12	0.59
OUT_MINS_ROAMING	0.13	0.24	2.48	0.42	1.38	0.68
OUT_CALLS_INTERNATIONAL	0.29	0.21	2.81	0.53	1.06	0.76
OUT_MINS_INTERNATIONAL	0.36	0.32	4.42	0.74	1.86	1.16
MMS_OUT_CALLS	0.04	0.03	0.12	0.20	0.16	0.09
INTERNET_CALLS	0.07	0.03	0.93	0.15	0.17	0.21
EVENTS_CALLS	0.28	0.22	2.17	0.60	0.83	0.65
OUT_COMMUNITY_TOTAL	34	14	33	29	72	32
DAYS_OUT	25	16	23	23	28	23
DAYS_IN	26	18	24	24	28	24
ACD_OUT	0.75	0.93	0.89	0.86	1.17	0.88
ACD_IN	0.85	1.13	0.93	1.10	1.21	1.02

- Cluster 5 contains heavy voice users. They also have the largest outgoing communities. Perhaps, this segment included residential customers who use their phones for business purposes (self-professionals).

Although not presented here, the profiling of the clusters went on with demographical exploration and by assessing their revenue contribution. Additionally, the cohesion of the clusters was also examined mainly with boxplots and with dispersion measures, such as the standard deviations of the clustering fields for each cluster. Finally, a series of summarizing charts were built to graphically illustrate in an intuitive manner the cluster "structures." The bars in the graphs represent the means of selected, important attributes, after standardization (in z-scores) over the members of the cluster. The vertical line at 0 denotes the overall mean. Hence, a bar facing to the right indicates a cluster mean higher than the overall mean, while a bar facing to the left indicates a lower mean. The bar charts are followed by a more detailed profiling of each cluster which wrap ups all their defining characteristics. The clusters were labeled according to these profiles (Figures 8.9, 8.10, 8.11, 8.12, and 8.13).

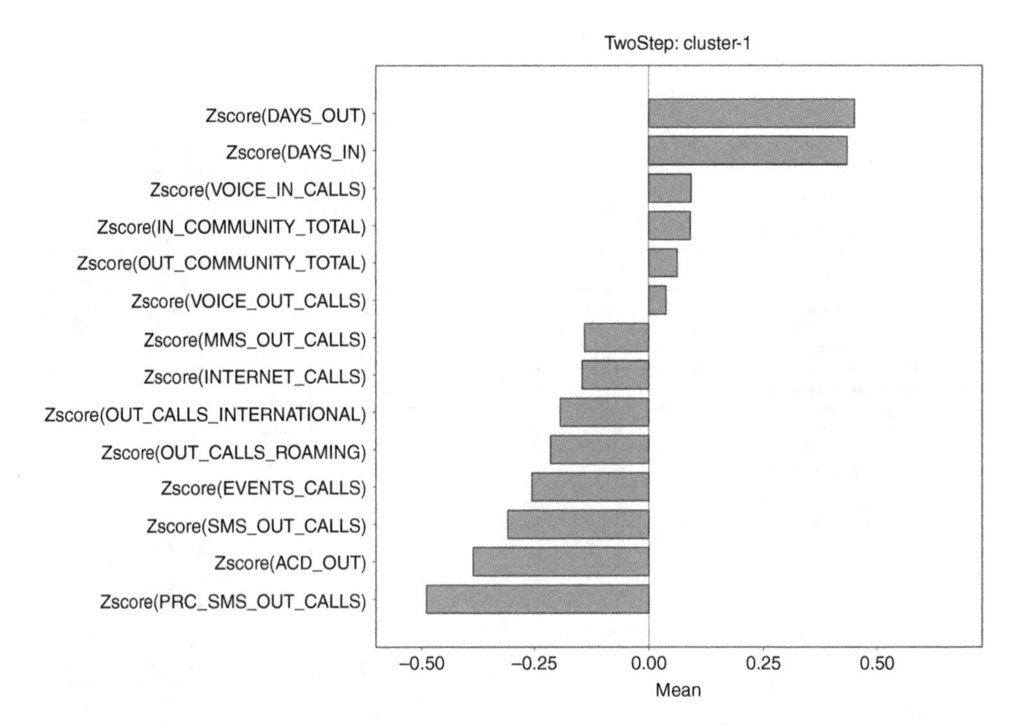

Figure 8.9 Cluster 1 profiling chart

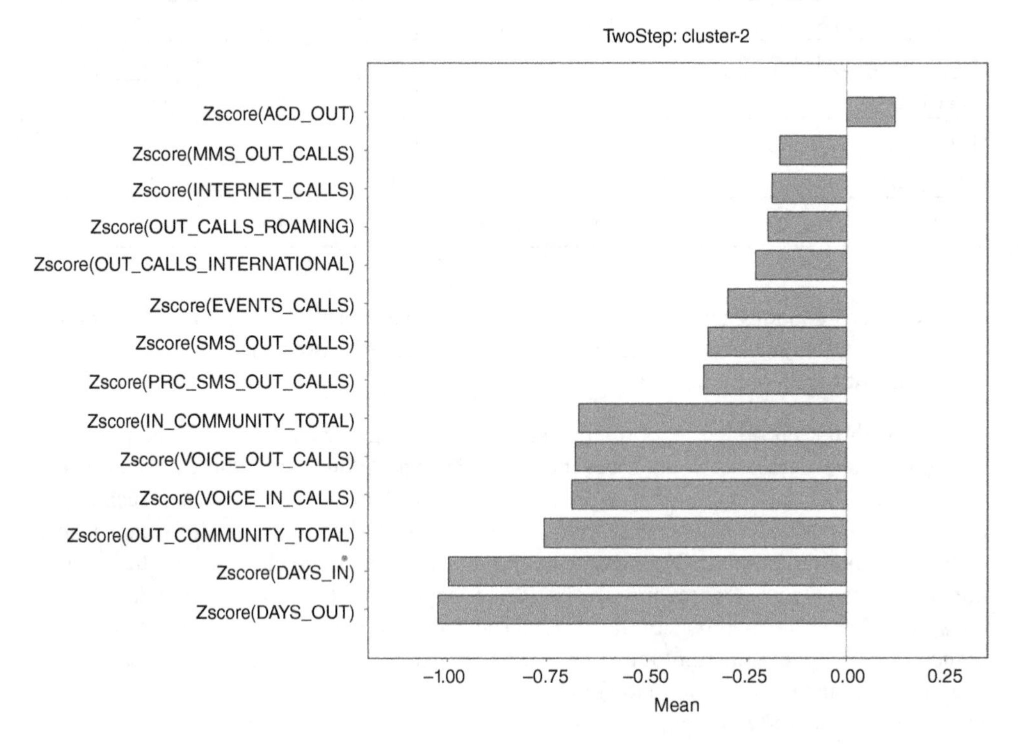

Figure 8.10 Cluster 2 profiling chart

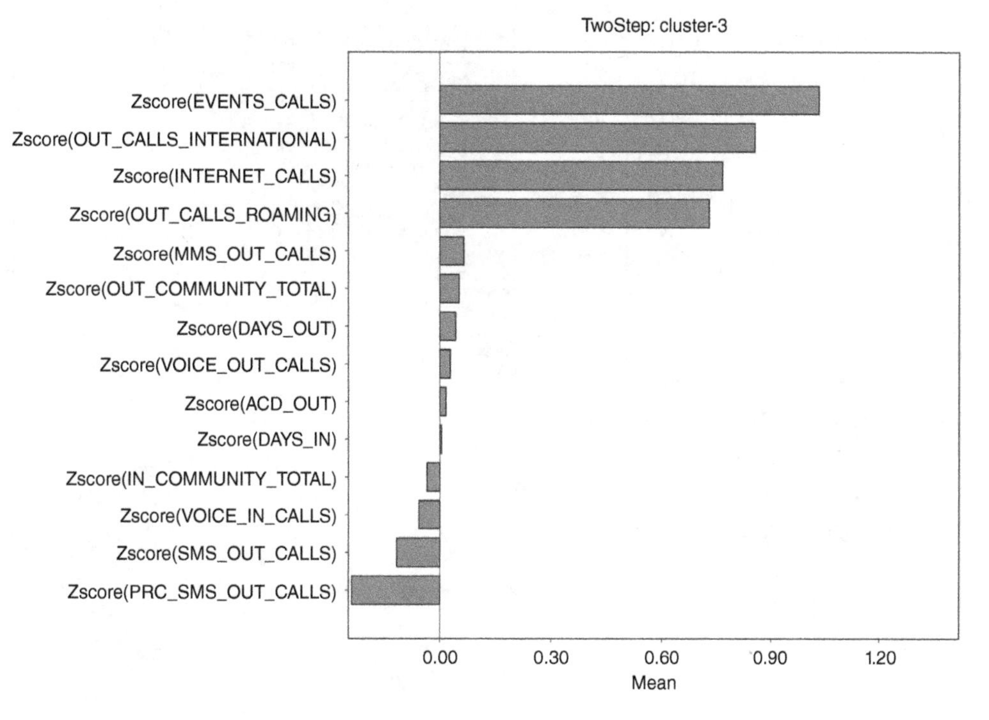

Figure 8.11 Cluster 3 profiling chart

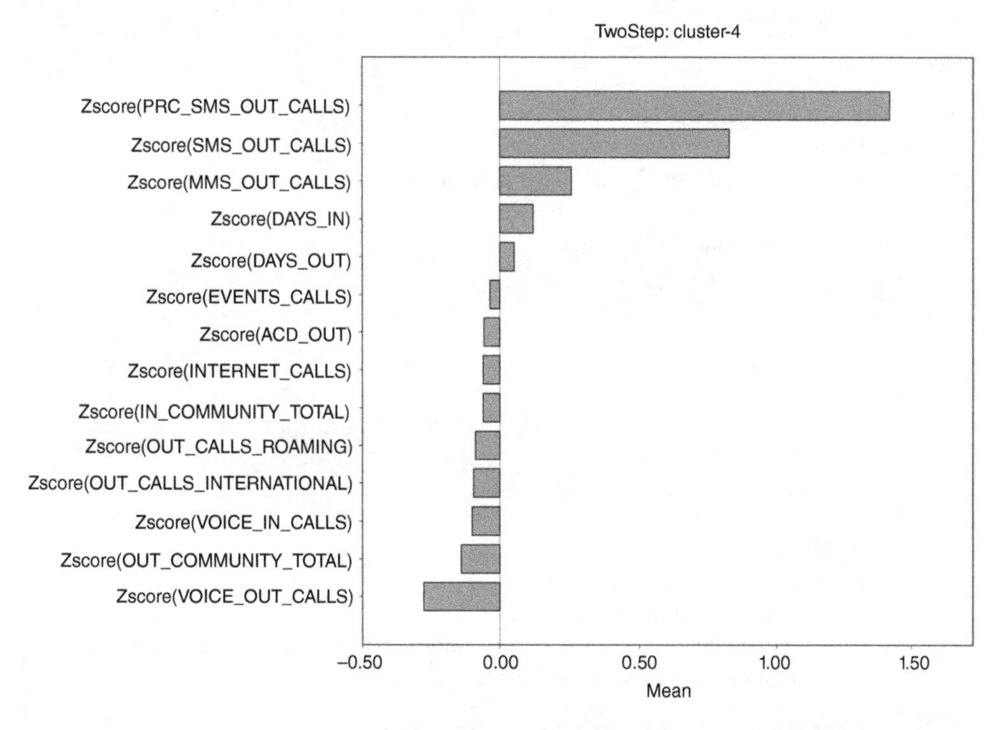

Figure 8.12 Cluster 4 profiling chart

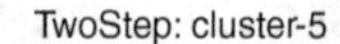

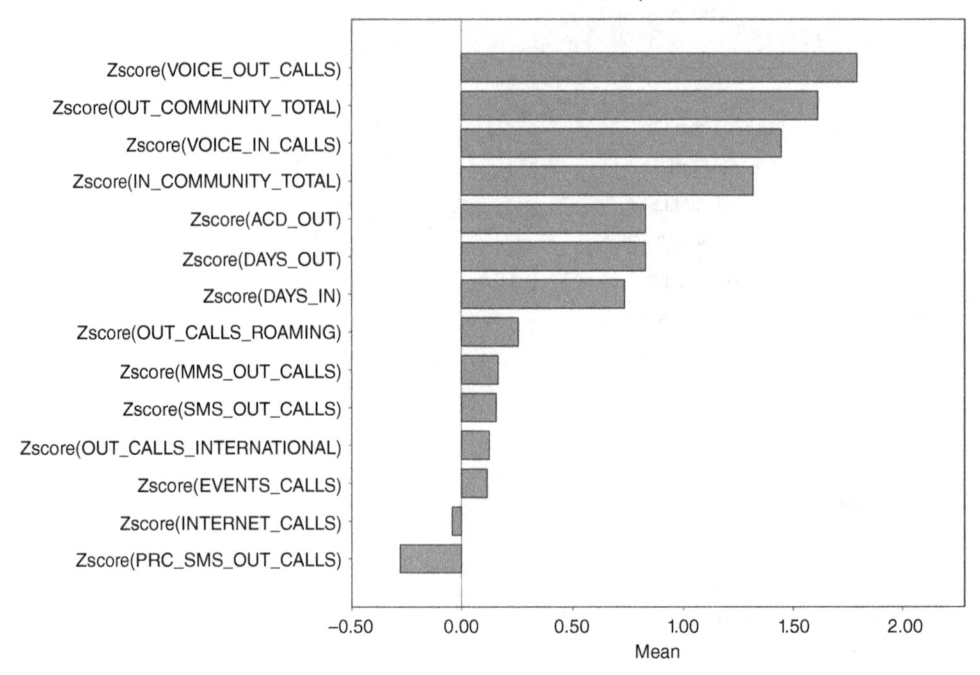

Figure 8.13 Cluster 5 profiling chart

Segment 1: typical voice users
Behavioral profile
• Average voice usage, in terms of number and minutes of calls • Low SMS usage • Low usage of tech services (MMS, Events, Internet) • Average communities • Relatively increased days with usage
Segment size
29.8%

Segment 2: basic users
Behavioral profile
• Lowest utilization of all services • Smallest outgoing and incoming communities • They use their phones a few days a month • However, when they make phone calls, the duration is quite long, as denoted by their average call duration, which is above the average
Segment size
24%

Segment 3: international users
Behavioral profile
• Frequent travelers with the highest roaming usage • Highest number of roaming calls and roaming traffic minutes • Frequent calling of international numbers. They have the highest number of international calls and minutes • They seem to be accustomed to the new types of communication: they have high Internet, Events, and MMS usage • Particularly in terms of Internet, they are the top users. Probably, they use it a lot when they travel abroad • Their voice usage is average
Segment size
13.8%

Segment 4: young—SMS users
Behavioral profile
• Heaviest SMS usage, much higher than the average • Low voice usage and community • Mostly use their handsets for sending SMS. They have by far the highest percentage of SMS calls and the lowest percentage of voice calls • Their voice usage is predominantly incoming • Their voice calls are brief, especially the outgoing ones • They also like to send MMS; they have the highest average number of MMS calls • Their Internet usage is average • Predominantly young users
Segment size
21%

Segment 5: professional users
Behavioral profile
• Heavy voice usage, incoming as well as outgoing • Highest number of voice calls and minutes. This segment includes many residential customers who seem to use their phone as a business tool as well. Their voice calls and minutes are almost three times above the average • They have by far the largest voice community • They use their phone on a daily basis and their voice calls quite long, as denoted by the relatively high average call duration • Relatively low SMS usage
Segment size
11.5%

8.5.4 Segmentation deployment

The final stage of the segmentation process involved the integration of the derived scheme in the company's daily business procedures. From that moment, all customers were characterized by the segment to which they were assigned. The deployment procedure involved a regular (on a monthly basis) segment update. More importantly, the marketers of the operator decided to further study the revealed segments and enrich their profiles with attitudinal data collected through market research surveys. Finally, the organization used all the gained insight to design and deploy customized marketing strategies which improved the overall customer experience.

8.6 Segmentation using RapidMiner and K-means cluster

An alternative segmentation approach using the K-means algorithm and RapidMiner is presented in the following paragraphs.

8.6.1 Clustering with the K-means algorithm

A two-step approach was followed for clustering with RapidMiner. Initially, a PCA algorithm was applied for reducing the data dimensionality and for replacing the original inputs with fewer combined measures. Subsequently, the generated component scores were used as the clustering fields in a K-means algorithm which identified the distinct customer groupings. The RapidMiner process is presented in Figure 8.14.

More specifically, a Retrieve operator was used to retrieve the modeling data from the RapidMiner repository. Then, through a Select Attributes operator, only the attributes designated as Inputs in Table 8.2 were retained for clustering. Since the PCA model in RapidMiner requires normalized inputs, the selected attributes were normalized with a Normalize operator. A z-normalization method was applied so eventually all attributes ended

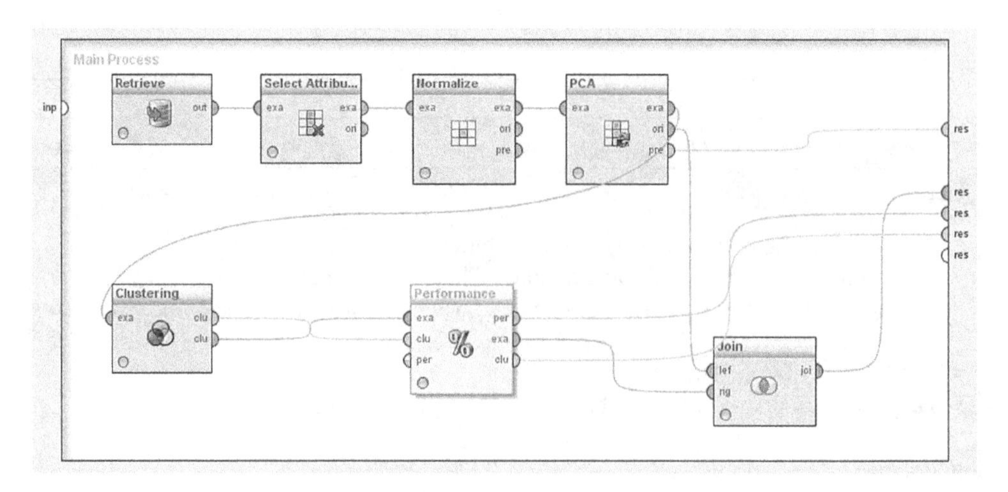

Figure 8.14 The RapidMiner process for clustering

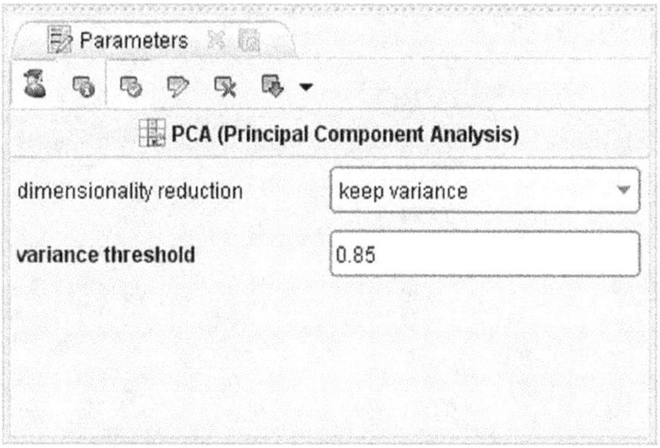

Figure 8.15 The PCA model settings

with a mean value of 0 and a standard deviation of 1. The z-scores were then fed into a PCA model operator with the parameter settings shown in Figure 8.15.

After trial and experimentation, a variance threshold of 85% was selected as the criterion for the extraction of components. This setting led to the extraction of 12 components which cumulatively accounted for about 85% of the total variance/information of the 31 original attributes, as shown in the rightmost column of the table in Figure 8.16. Initially, eight components were extracted which explained about 75% of the information of the original inputs. However, it turned out that the extraction of a larger number of components improved the K-means clustering, leading to a richer and more useful clustering solution. Therefore, the final choice was to proceed with the 12 components.

The 12 components were then used as clustering inputs in a K-means model. The parameter settings are shown in Figure 8.17.

The Euclidean distance was selected for measuring the similarities of the records. The add-cluster attribute option generated the cluster membership field and assigned each instance to a cluster. After many trials and evaluations of many different clustering solutions, a 5-cluster solution was finally adopted for deployment. Hence, a "k" of 5 was specified, guiding the algorithm to form five clusters. The cluster distribution is presented in Figure 8.18. Clusters 3 and 4 are dominant since they comprise almost 80% of the total customer base.

The generated K-means model was connected with a Cluster Distance Performance operator to evaluate the average distance between the instances and the centroid of their cluster. The respective results are shown in Figure 8.19.

The segmentation procedure was concluded with the profiling of the revealed clusters and with the identification of their defining characteristics. The first step was to study the centroids table, available in the results tab of the cluster model. To facilitate additional profiling in terms of the original attributes, the cluster membership field was cross-examined with the original inputs. Figure 8.20 presents a profiling chart which depicts the averages of some important (normalized) attributes over the clusters. The vertical line at 0 corresponds to the marginal means of the normalized attributes. Hence, bars and dots to the right of this reference line designate cluster means above the marginal mean, while bars and dots to the left designate lower cluster means.

Component	Standard Deviation	Proportion of Variance	Cumulative Variance
PC 1	2.649	0.226	0.226
PC 2	2.040	0.134	0.360
PC 3	1.791	0.103	0.464
PC 4	1.436	0.067	0.530
PC 5	1.341	0.058	0.588
PC 6	1.285	0.053	0.642
PC 7	1.247	0.050	0.692
PC 8	1.178	0.045	0.737
PC 9	1.004	0.033	0.769
PC 10	0.983	0.031	0.800
PC 11	0.931	0.028	0.828
PC 12	0.909	0.027	0.855
PC 13	0.833	0.022	0.877
PC 14	0.736	0.017	0.895
PC 15	0.709	0.016	0.911
PC 16	0.646	0.013	0.925
PC 17	0.607	0.012	0.936
PC 18	0.572	0.011	0.947
PC 19	0.535	0.009	0.956
PC 20	0.495	0.008	0.964
PC 21	0.465	0.007	0.971
PC 22	0.440	0.006	0.977
PC 23	0.425	0.006	0.983
PC 24	0.361	0.004	0.987
PC 25	0.336	0.004	0.991
PC 26	0.283	0.003	0.994
PC 27	0.272	0.002	0.996
PC 28	0.263	0.002	0.998
PC 29	0.221	0.002	1.000
PC 30	0.060	0.000	1.000
PC 31	0.042	0.000	1.000

Figure 8.16 The variance explained by the components

A brief profile of the clusters is presented below. The clusters were named according to their identified characteristics:

- **Cluster_0 SMS users**: Highest SMS usage. Highest SMS to voice ratio. Increased usage of MMS, Events, and Internet
- **Cluster_1 International users**: Highest roaming and international usage. Also increased usage of voice and MMS
- **Cluster_2 Professional users**: High voice usage. Increased community, frequency (days), and duration of voice calls

Parameters

Clustering (k-Means)

add cluster attribute

add as label

remove unlabeled

k	5
max runs	10

determine good start values

measure types	NumericalMeasures
numerical measure	EuclideanDistance
max optimization steps	100

use local random seed

local random seed	565656

Figure 8.17 The K-means parameter settings

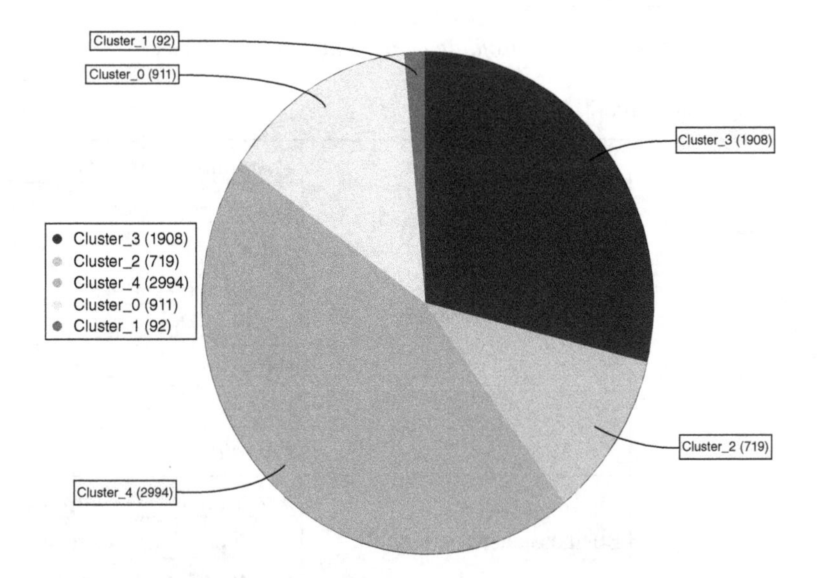

Figure 8.18 The distribution of the K-means clusters

PerformanceVector

```
PerformanceVector:
Avg. within centroid distance: -17.137
Avg. within centroid distance_cluster_0: -38.126
Avg. within centroid distance_cluster_1: -119.805
Avg. within centroid distance_cluster_2: -31.000
Avg. within centroid distance_cluster_3: -10.716
Avg. within centroid distance_cluster_4: -8.358
Davies Bouldin: -1.451
```

Figure 8.19 Evaluating the cluster solution using within centroid distances

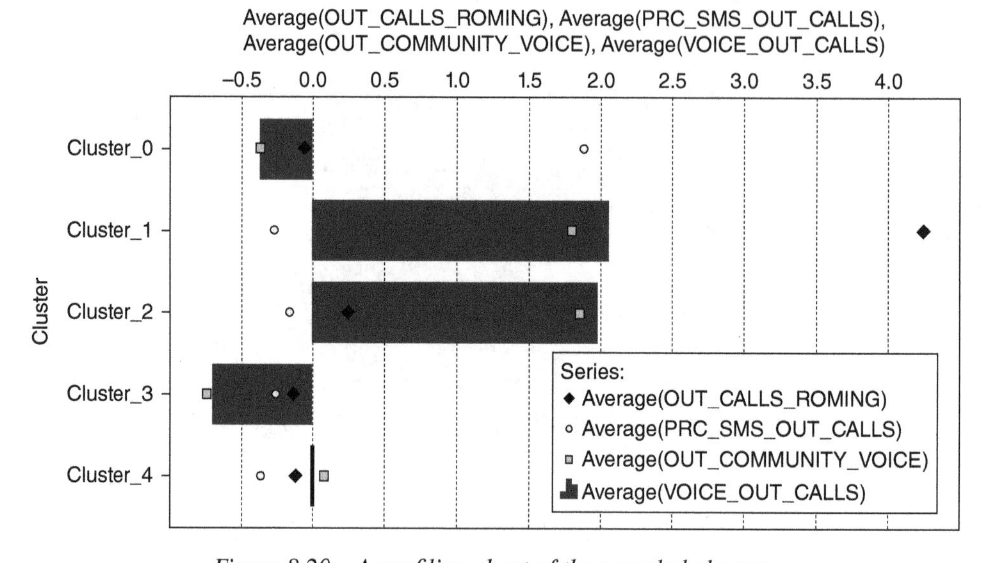

Figure 8.20 A profiling chart of the revealed clusters

Table 8.9 The mobile telephony segments

<table>
<tr><td rowspan="11">Core segments</td><td rowspan="8">Residential customers</td><td>Postpaid—contractual</td></tr>
<tr><td>Behavioral segments</td></tr>
<tr><td>Typical voice users</td></tr>
<tr><td>Basic users</td></tr>
<tr><td>International users</td></tr>
<tr><td>SMS users</td></tr>
<tr><td>Professional users</td></tr>
<tr><td>Prepaid</td></tr>
<tr><td rowspan="3">Business customers</td><td>Large business—corporate</td></tr>
<tr><td>Small medium enterprise—SME</td></tr>
<tr><td>Small office home office—SOHO</td></tr>
</table>

- **Cluster_3 Basic users**: Lowest usage of all services and lowest community
- **Cluster_4 Typical voice users**: Average usage, almost exclusively voice

The revealed clusters seem indeed similar to the ones identified by the TwoStep algorithm. Tempted to compare the two solutions? After joining and cross-tabulating the cluster membership fields, it appears that the two models present an "agreement" of about 70%. Specifically, after excluding the TwoStep "outlier" cluster, the two models seem to assign approximately 70% of the instances in analogous groups. Where do they not agree? Their main difference is in the "international" segment which in the K-means solution is smaller, including only those roamers with increased general usage. The similarity of the clusters nevertheless is a good sign for the validity of the solutions.

8.7 Summary

In this chapter, we've followed the efforts of a mobile phone network operator to segment its customers according to their usage patterns. The business objective was to group customers in terms of their behavioral characteristics and to use this insight to deliver personalized customer handling. The first segmentation effort took into account the established core customer segments and initially focused on the residential postpaid customers which were further segregated into five behavioral segments, as summarized in Table 8.9. The procedure followed for the behavioral segmentation included the application of a PCA model for data reduction and a cluster model for revealing the distinct user groups.

Bibliography

Anderson, Kristin. *Customer relationship management*. New York: McGraw-Hill, 2002.

Berry, Michael J. A., and Gordon Linof. *Mastering data mining: the art and science of customer relationship management*. New York: John Wiley & Sons, Inc., 1999.

Fernandez, George. *Data mining using SAS applications*. Boca Raton: Chapman & Hall/CRC, 2003.

Han, Jiawei, Micheline Kamber, and Jian Pei. *Data mining: concepts and techniques*. 3rd ed. Morgan Kaufmann series in data management systems. Amsterdam: Elsevier/Morgan Kaufmann, 2012.

Hughes, Arthur Middleton. *Strategic database marketing*. 4th ed. New York: McGraw-Hill, 2012.

IBM. IBM SPSS Modeler 16 algorithms guide. Armonk: IBM, 2014. http://www-01.ibm.com/support/docview.wss?uid=swg27038316.

IBM. IBM SPSS Modeler 16 applications guide. Armonk: IBM, 2014. http://www-01.ibm.com/support/docview.wss?uid=swg27038316.

IBM. IBM SPSS Modeler 16 modeling nodes. Armonk: IBM, 2014. http://www-01.ibm.com/support/docview.wss?uid=swg27038316.

IBM Redbooks. Mining your own business in banking using DB2 Intelligent Miner for Data (IBM Redbooks). Armonk: IBM, 2001. http://www.redbooks.ibm.com/.

IBM Redbooks. Mining your own business in retail using DB2 Intelligent Miner for Data. Armonk: IBM, 2001. http://www.redbooks.ibm.com/.

IBM Redbooks. Mining your own business in telecoms using DB2 Intelligent Miner for Data (IBM Redbooks). Armonk: IBM, 2001. http://www.redbooks.ibm.com/.

Larose, Daniel T. *Discovering knowledge in data: an introduction to data mining*. Hoboken: Wiley-Interscience, 2005.

Linoff, Gordon S. *Data analysis using SQL and excel*. New York: John Wiley & Sons, Inc., 2007.

Linoff, Gordon, and Michael J. A. Berry. *Data mining techniques: for marketing, sales, and customer relationship management*. 3rd ed. Indianapolis: Wiley, 2011.

MacLennan, Jamie, Tang, Zhaohui, and Bogdan Crivat. *Data mining with Microsoft SQL server 2008*. Indianapolis: Wiley Publishing, Inc., 2009.

Matignon, Randall. *Data mining using SAS enterprise miner*. Hoboken: John Wiley & Sons, Inc., 2007.

Microsoft. SQL Server 2012 Data Mining. Washington, DC: Microsoft, 2012. https://msdn.microsoft.com/en-us/library/bb510516(v=sql.110).aspx.

Nisbet, Robert, John F. Elder, and Gary Miner. *Handbook of statistical analysis and data mining applications*. Amsterdam: Academic Press/Elsevier, 2009.

North, Matthew. *Data mining for the masses*. US: Global Text Project, 2012.

Olson, David, and Dursun Delen. *Advanced data mining techniques*. Berlin: Springer-Verlag, 2008.

Peelen, E. *Customer relationship management*. Upper Saddle River: Financial Times/Prentice Hall, 2005.

Effective CRM using Predictive Analytics, First Edition. Antonios Chorianopoulos.

Companion website: www.wiley.com/go/chorianopoulos/effective_crm

RapidMiner. RapidMiner operator reference guide. Cambridge, MA: RapidMiner, 2014. http://docs.rapidminer.com/.

RapidMiner. RapidMiner studio manual. Cambridge, MA: RapidMiner, 2014. http://docs.rapidminer.com/.

Rud, Olivia Parr. *Data mining cookbook: modeling data for marketing, risk and customer relationship management*. New York: John Wiley & Sons, Inc., 2001.

Shmueli, Galit, Nitin R. Patel, and Peter C. Bruce. *Data mining for business intelligence: concepts, techniques, and applications in Microsoft Office Excel with XLMiner*. 2nd ed. Hoboken, NJ: John Wiley & Sons, Inc., 2010.

Tsiptsis, Konstantinos, and Antonios Chorianopoulos. *Data mining techniques in CRM: inside customer segmentation*. Chichester: John Wiley & Sons, Ltd, 2009.

Witten, Ian H., Eibe Frank, and Mark A. Hall. *Data mining: practical machine learning tools and techniques*. 3rd ed. [Morgan Kaufmann series in data management systems]. Burlington: Morgan Kaufmann, 2011.

Index

Note: Page numbers in *italics* refer to Figures; those in **bold** to Tables.

Effective CRM using Predictive Analytics, First Edition. Antonios Chorianopoulos.

Companion website: www.wiley.com/go/chorianopoulos/effective_crm

CPSIA information can be obtained
at www.ICGtesting.com
Printed in the USA
BVHW050929120121
597594BV00007B/89

9 781119 011552